Regional Knowledge Economies

NEW HORIZONS IN REGIONAL SCIENCE

Series Editor: Philip McCann, *Professor of Economics, University of Waikato, New Zealand and Professor of Urban and Regional Economics, University of Reading, UK*

Regional science analyses important issues surrounding the growth and development of urban and regional systems and is emerging as a major social science discipline. This series provides an invaluable forum for the publication of high-quality scholarly work on urban and regional studies, industrial location economics, transport systems, economic geography and networks.

New Horizons in Regional Science aims to publish the best work by economists, geographers, urban and regional planners and other researchers from throughout the world. It is intended to serve a wide readership including academics, students and policymakers.

Titles in the series include:

Spatial Dynamics, Networks and Modelling
Edited by Aura Reggiani and Peter Nijkamp

Entrepreneurship, Investment and Spatial Dynamics
Lessons and Implications for an Enlarged EU
Edited by Peter Nijkamp, Ronald L. Moomaw and Iulia Traistaru-Siedschlag

Regional Climate Change and Variability
Impacts and Responses
Edited by Matthias Ruth, Kieran Donaghy and Paul Kirshen

Industrial Agglomeration and New Technologies
A Global Perspective
Edited by Masatsugu Tsuji, Emanuele Giovannetti and Mitsuhiro Kagami

Incentives, Regulations and Plans
The Role of States and Nation-states in Smart Growth Planning
Edited by Gerrit J. Knaap, Huibert A. Haccoû, Kelly J. Clifton and John W. Frece

New Directions in Economic Geography
Edited by Bernard Fingleton

The Management and Measurement of Infrastructure
Performance, Efficiency and Innovation
Edited by Charlie Karlsson, William P. Anderson, Börje Johansson and Kiyoshi Kobayashi

Knowledge Externalities, Innovation Clusters and Regional Development
Edited by Jordi Suriñach, Rosina Moreno and Esther Vayá

Regional Knowledge Economies
Markets, Clusters and Innovation
Philip Cooke, Carla De Laurentis, Franz Tödtling and Michaela Trippl

Regional Knowledge Economies

Markets, Clusters and Innovation

Philip Cooke

University Research Professor in Regional Development and Director, Centre for Advanced Studies, University of Wales, Cardiff, UK

Carla De Laurentis

Researcher, Centre for Advanced Studies, University of Wales, Cardiff, UK

Franz Tödtling

Professor, Department of City and Regional Development, Vienna University of Economics and Business Administration, Austria

Michaela Trippl

Researcher, Department of City and Regional Development, Vienna University of Economics and Business Administration, Austria

NEW HORIZONS IN REGIONAL SCIENCE

Edward Elgar

Cheltenham, UK • Northampton, MA, USA

Published by
Edward Elgar Publishing Limited
Glensanda House
Montpellier Parade
Cheltenham
Glos GL50 1UA
UK

Edward Elgar Publishing, Inc.
William Pratt House
9 Dewey Court
Northampton
Massachusetts 01060
USA

A catalogue record for this book
is available from the British Library

Library of Congress Cataloguing in Publication Data

Regional knowledge economies: markets, clusters and innovation/
    by Philip Cooke ... [et al.]
        p. cm. — (New horizons in regional science)
    Includes bibliographical references and index.
    1. High-technology industries—Location—Great Britain.
    2. High-technology industries—Location—Australia.
    3. Technological innovations. 4. Knowledge management.
    5. Industrial clusters.
    HC260.H53R44   2007
    338.60941—dc22
                                            2006034553

ISBN 978 1 84542 529 6 (cased)

Printed and bound in Great Britain by MPG Books Ltd, Bodmin, Cornwall

Contents

Preface

The origins of this book lay in a somewhat contorted attempt, that was eventually successful, to access research funding to conduct a comparative research project entitled Collective Learning in the Knowledge Economy: Milieu or Market? The aim was to examine advanced technology industry, including that located in clusters, in different countries with liberal and coordinated regulatory regimes, to see if firms in a proximate milieu such as a cluster performed better or worse than similar firms not located in clusters. A first effort, including German and Norwegian partners as well as the Austrian and UK ones, was in 2001 sent to the European Science Foundation collaborative research scheme, in which at least three needed to succeed with their own national research councils for the research to proceed. Only the Austrian proposal succeeded as the other three were not supported, hence the successful proposal also 'failed'. But in coordinated market Austria the applicants were encouraged to submit a further proposal, whereas in liberal market UK that was barred. This was discovered when an improved version, helped by taking account of assessor comments, was submitted only to be refused entry because revised proposals were disallowed. Clearly no opportunities for making learning gains in the UK research funding arena then.

Determined to keep trying with what peer-reviewers had thought a worthy project and spurred on by the success of the Austrian proposal to acquire national research foundation funding, the further improved proposal was sent to a UK private research foundation where it was turned down moderately brutally, and a little depressingly. Meanwhile the Austrian work was about to commence – utilizing almost exactly the same methodology. At Franz Tödtling's final urging, a completely revised version was written, which given its multiple polishings was probably the best the UK applicant had ever put together. It was submitted, accepted for review and, finally, in 2003 was successful. So ESRC Award Number RES-000-23-0192 is duly acknowledged, as is the Austrian Science Foundation for enabling the research on which this book is based to be conducted. Work started about six months after the Austrian project, we met for catch-up and planning meetings, and got on with the research. The Austrian partners then had to seek further funding from a different source (the Foundation of the City of Vienna for the Support of Sciences) to conduct interviews with biotechnology firms and intermediaries, and the research finished in early 2006. Hence the results readers will see were the product of hard-fought struggle and not a little tenacity of the kind entrepreneurs are often familiar with but which academics too often file away and forget. Entrepreneurs in our experience will press on against waves of failure because they simply believe in an idea and that eventually someone somewhere will buy it. This is because they usually do not have many ideas to play with, whereas for academics ideas are the coin of the realm. This argument about the importance of tenacity in academic research is one we are happy to share with and urge upon readers of this book. We also wish to express thanks to Edward Elgar and especially Matthew Pitman for their help and understanding as we brought this publishing project to fruition.

Philip Cooke
Carla De Laurentis
Franz Tödtling
Michaela Trippl

Cardiff and Vienna
September 2006

PART I

Conceptual Issues

1. Introduction: aims of the book

The research on which this book was based aimed to establish the extent to which firms performed better when co-located within clusters than outside them. The research projects in Austria and the UK also sought to throw comparative light on this issue in economies often thought to represent versions of the two classical European models of industry organization; the first, corporatist or 'coordinated market', the second 'liberal market' after Hall and Soskice's widely cited classification of 'varieties of capitalism' (Hall and Soskice, 2001) and, in particular, their associated business systems.[1] The precise way in which these issues were tackled involved the use in both countries of an identical set of research instruments and questions inquiring about relative firm performance in two modern, advanced technology industries, information and communication technology (ICT) and biotechnology. The research questions therefore sought to tease out, for both settings, the extent to which innovative milieux (after Maillat, 1998) or markets were the most performance-enhancing settings for such high-technology businesses. This question had particular salience because of the following. Many continental European economies, from Austria to Germany, Switzerland and the Nordic countries operate a 'social partnership' model of economic organization in which interaction occurs by negotiation across boundaries between otherwise divergent interests associated with government, industry and labour. At a lower scale than the national economy, this resonates with the 'networking propensity' associated with regional milieux of innovative small firms as analysed in Maillat's pioneering work in the Swiss Jura region of watchmakers.

Contrariwise, a liberal market regime of economic governance might be hypothesized to have few of these milieu effects, being more associated with economic individualism and a competitive ethos than the 'networking propensity'. Indeed, books such as that of Best (1990) showed in detail that this characterized the distinctive furniture and other comparable industries of Italy, a variant on the coordinated market type of economy, and the UK. The Italian furniture industry prospered from its industrial district, small firm milieux while the UK one fared less well in pursuit of scale economies amid cutthroat competition.[2] Despite this, and returning to the Austria–UK comparison, for the industries under investigation, the overall UK performance is significantly better. For example, keeping in mind the fact that Austria is about one-sixth the demographic size of the UK, its ICT and biotechnology shares in economic activity are weaker.

As elaborated in more detail elsewhere Austria's share of knowledge-intensive services (including software, research, media, and so on) was 27 per cent of total employment in 2001, while in the UK it was 49 per cent. This echoes Peneder (1999) who showed Austrian high-technology industry to contribute 14 per cent of Austrian value-added compared to 24 per cent in the UK. In biotechnology, the UK in 2004 had nearly 200 'pipeline' products in various trialling stages, whereas Austria had none. Having said that, Austria had in 2003 some 30 dedicated biotechnology firms (DBFs), 64 per cent of them located in Vienna (Tödtling and Trippl, 2005), whereas the UK had in 2003 some 455, down from 494 in 2002 due to firm closures and merger and acquisition activity in the sector (DTI, 2005). A comparable proportion of the UK's DBFs are located within a 50-mile radius of London but, unlike Vienna, most are concentrated in university towns like Cambridge and Oxford, and in the Guildford area. The university towns host what are officially denoted 'biotechnology industry clusters'; Guildford is known as a 'biotechnology agglomeration'. Whereas leading edge bioscientific research and talent support 'academic entrepreneurship' in the university towns, Guildford is a case of 'localization economies' spatial concentration due mainly to experienced technical labour and excellent transportation links to London and internationally, something the research reported here sought further to test (DTI, 1999; Cooke, 2001).

Thus the aims of the book are fivefold, involving first, conducting comparative research in the UK and Austria as instances of, respectively, liberal market and coordinated market economies. Second, comparing the two countries in respect of high-technology industry performance and whether market or milieu emphases in firm practices predominated in explaining firm performance. Performance is measured in terms of growth measures of revenues, profitability, innovation and employment. Third, the book explores the extent to which the regulatory context affects the ways economic spillovers, especially those related to knowledge diffusion, operate. Particular attention is devoted to testing claims in the literature of the advantages accruing to firms located in proximity to sectoral neighbours and whether sector or spatial location is the salient factor in firm performance. Fourth, continuing the theoretical interest of the authors in knowledge transfer, especially considering the role of propinquity in facilitating this, the book aims to investigate the processes by which tacit knowledge becomes codified. It is hypothesized that the binary relation between these 'senses of knowing' (Orlikowski, 2002) are more complex than suggested in the literature begun by Polanyi (1966). As hinted by Zitt et al. (2003: 296):

in many cases both tacit and codified information are linked in a number of ways. For example tacit information may act as a facilitator for the transmission of codified knowledge. If direct information flows matter then the availability of skilled manpower (universities as recruitment base, mobility of researchers) also plays an important role in localised dissemination of knowledge.[3]

A broad conceptual model of the intermediary cognitive and institutional steps from implicit to explicit knowledge in bioscience, referred to as distinct kinds of *complicit* knowledge, identified at least 15 of these (Cooke, 2005). More detailed empirical analysis would undoubtedly yield more. Harmaakorpi (2005) further identifies the role of pre-tacit and self-transcending knowledge as a crucial element in the knowledge creation process.[4] Some light will be shed also on variations among ICT and biotechnology knowledge in exploring these complex developmental issues. Finally, the book aims to explore the influence of different sub-national economic governance regimes upon sector and firm performance, including economic development policies, academic entrepreneurship and outsourcing client–supplier interactions. Accordingly, we hope to illuminate numerous currently hidden corners of the knowledge exploration to innovation exploitation process in distinctive advanced technology industries and varying governance contexts: a robust test of current conventional wisdom about markets, proximity and the leveraging of value from knowledge, we believe.

EUROPE AND THE INNOVATION CHALLENGE IN ICT AND BIOTECHNOLOGY

While as recently as 1995 the European Commission's *Green Paper on Innovation* (EC, 1995) bemoaned the European Union's industrial innovation performance compared with those of Japan and the USA, nowadays the exemplar role of Japan has receded due to its modest performance in biotechnology and failure to innovate around the Internet, two spheres in which the USA predominates globally. The technology stocks boom of the 1990s ended with the crash of 2000–01, spurred, it has been argued, by the injunction to rein back on aspects of 'irrational exuberance'. United States (US) Federal Reserve Bank Chairman, Alan Greenspan, observed this questionable spirit in the momentum built by his and other US commentators' postulation that a 'new economy' had been born. This was fantasized as a self-refuelling 'ideas market' that meant conventional dogma about the importance of profitability in stock valuation could be jettisoned in favour of reliance largely upon investor gambling instincts, sometimes interlarded with fraudulent advice from self-interested market advisers and other 'experts'. Despite the legal fallout and wrecked hopes of easy riches that ensued from the bursting of the tech-stocks bubble, ICT and to a lesser extent biotechnology have by no means gone away. The main change in the ten years since 1995 is that Japan has, in relative terms, disappeared from the innovation radar to be replaced by the new and different development challenges to both Europe and the USA, let alone the rest of the world's economies, by the rise of the sleeping economic giants of India and China.

Europe has slipped back in some ways against its main surviving innovation benchmark, the USA, since 1995. This is despite the European Union (EU) Commission's injunctions, formulated at the Lisbon and Barcelona summits, to make Europe the world's most competitive knowledge-based economy by 2010. Consider the data for recent biotechnologically derived innovations in Table 1.1. These were researched for the UK Department of Trade and Industry's biotechnology monitoring assessments and represent most of the significant commercial innovations for 2003. A number of points arise from these entries. First, some 76 per cent had a US involvement in the origination of the drug, usually but not always by a dedicated biotechnology firm. Second, where the originator or partner was from outside the USA, it was 30 per cent from Switzerland, a non-EU member, though of course a European country. Third, the UK was the leading EU company source, at 30 per cent, with one entrant, Celltech, entering Belgian ownership shortly after. Finally, the 15 per cent Food and Drug Administration (FDA) approvals from wholly non-US partnerships involved one UK-German and one UK-French alliance, while Serono from Switzerland was the only non-US singleton originator. In some respects this is a less forbidding performance than might be imagined from inspection of other data (BIGT, 2004) that shows the USA far ahead of Europe in the number of drugs in the pipeline. So much so that a new model of commercial exploitation from globally leading UK bioscientific research called the 'conveyor belt' is emerging, the basic idea being that commercialization of UK discoveries will increasingly be conducted in the USA through cooperative licensing deals with US DBFs.[5]

Thus it may generally be concluded that for biotechnology, US DBFs are far more innovative and commercially successful than those of the rest of the world, including Europe and that the technological lead, while showing some slight signs of being eroded since Bioscience Innovation and Growth Team (BIGT, 2004) shows Europe has marginally more drugs in late stage trials than does the USA, the acid test is how many, comparatively speaking, receive FDA approval to enter the market. Thus far, US firms have proved superior in that respect.

With regard to innovation in ICT, Table 1.2 summarizes key indicators that show the scale of the US lead. Thus US ICT patents registered at the

Table 1.1   Selected US Food and Drug Administration drug approvals 2003

Company	Drug	Approval body	Indication
Actelion (Switzerland)/ Celltech (UK now USA)	Zavesca	FDA	Gaucher's disease
Cubist Pharma (USA)	Cubicin	FDA	Psoriasis
Genentech (US)/ Novartis (Switzerland)/ Tanox (USA)	Xolair	FDA	Asthma
Genzyme (USA)/ BioMarin (USA)	Aldurazyme	FDA/EMEA	MPS1 deformity
Genzyme (USA)	Fabrazyme	FDA	Fabry disease
Gilead (USA)	Emtriva	FDA/EMEA	HIV
GSK (UK)/ Bayer (Germany)	Levitra	FDA/EMEA	Erectile dysfunction
GSK (UK)/ Corixa (USA)	Bexxar	FDA	Non-Hodgkin's lymphoma
Medimmune (USA)/ Wyeth (USA)	Filumist	FDA	Influenza vaccine
Millennium (USA)	Velcade	FDA	Bone cancer (multiple myeloma)
Roche (Switzerland)/ Trimeris (USA)	Fuzeon	FDA/EMEA/ Swissmedic	HIV
Serono (Switzerland)	Serosotin	FDA	HIV
SkyePharma (UK)/ Sanofi-Synthélabo (France)	Uroxatral	FDA	Prostate

Note: EMEA – European Medicines Evaluation Authority; Swissmedic – Swiss equivalent.

Source: DTI (2005).

European Patent Office rose during the 1990s by 43 per cent to 20 per cent of total US EPO patenting. At 62 per cent the increase was greater in the EU albeit from a lower base, rising to 13 per cent of total European Patent Office (EPO) patents registered. The UK increase was higher at 157 per cent but from a yet lower base than the EU and half the 1990 share of the US.

Table 1.2   Shares of ICT in total patent and value-added shares

Country	ICT patent share (EPO) %		Value-added share %		Business sector value-added share %	
	1990	1998	1995	2001	1995	2000
USA	14.0	20.0	2.1	2.3	10.0	12.0
EU	8.0	13.0	1.3	1.2	7.0	8.0
UK	7.0	18.0	1.7	1.4	10.0	11.0
Austria	5.0	7.0	NA	NA	NA	8.0

Sources: European Patent Office (EPO); Organisation for Economic Co-operation and Development (OECD) STAN (Structural Analysis) Indicators, 2004.

The UK catch-up in ICT patenting in the 1990s was accordingly somewhat spectacular. This could not be said for Austria, whose 40 per cent rise rivalled that of the USA but from one-third the US base.

The value-added shares in the broader economies and specifically their business services sectors also betray a better US than EU performance, though the US gap with the UK in business services share is much less. Here it may be said that the UK shows some similarity with its 'liberal market' counterpart in ICT business service value-added, while Austria, at least in 2000, mimicked a more continental profile. Either way, utilizing different indicators from the biotechnology data, where the US lead is more pronounced,[6] these confirm also that it has a substantially higher ICT patent share and a marginally growing economy-wide value-added share over the EU. The latter contains divergent trends, well represented by the UK and Austria, with the former displaying strong catch-up performance towards the USA in ICT patenting and Austria improving marginally from a low base. The implications of this for economic growth and particularly the role of total factor productivity in explaining growth are the subject of significant policy and academic interest.

Two views on the importance of ICT production and, crucially, *utilization* in economic activity are representative of attempts to place ICT's contribution in perspective. The results, from somewhat divergent starting points tend towards convergence regarding the contribution of ICT to productivity growth and the recently widening gap between the USA and EU. An approach that seeks to place the contribution of ICT to growth in perspective is offered by Kaloudis et al. (2005). In line with the narrower ICT share in value-added shown for both EU and the USA in Table 1.2, the broader group of high-technology (high-tech) manufacturing industries utilizing OECD categorizations[7] reveal high-tech manufacturing for leading OECD countries to contribute less than 10 per cent of total value-added in

1980 and almost exactly 10 per cent in 2002, a slight narrowing of the gap, affected notably by the ICT downturn in 2000–01. It rose to 13 per cent of *manufacturing* value-added by 2002 from less than 10 per cent in 1980. However, in correlating gross domestic product (GDP) growth rates with high-tech shares in manufacturing value-added, the USA is at the leading edge of the upwardly sloping regression line, with Sweden, Finland, Japan and the UK following. Co-ordinated market stalwarts like Germany, Austria and Italy follow some way behind.

So the data tend to support the high ICT and high GDP growth argument. However, the authors also show high-tech to have contributed only slightly more to overall *manufacturing* growth 1980–98 than OECD-defined medium-tech sectors.[8] Moreover, when annual compound growth rates for the 1991–2001 period are examined the correlations lose significance and lower-tech economies like Spain and medium-higher smaller economies like Norway show better growth. Meanwhile, the large coordinated group of Germany, Italy and France rival Japan in the low-growth performance category with those EU countries also displaying only middle-ranking high-tech shares in manufacturing value-added. The authors conclude that different economic strokes work for different manufacturing folks, namely an economy like Norway, rich in oil and gas resources and the upstream marine engineering associated with it can show a better yearly compound growth than higher-tech economies like the UK, the USA and, massively, Japan. In this analysis Austria performs more like Denmark, Finland and Canada – showing a better high-tech share of manufacturing value-added and yearly growth rate than Germany, Italy or France.

A different take on the important question of the extent to which high-tech intensity in manufacturing value-added is positively associated with productivity and GDP growth is provided by Van Ark and Inklaar (2005). The authors first show comparative GDP and labour productivity change between the USA and EU 15 between 1987 and 2004 (Table 1.3). This shows that the USA displayed a consistently higher GDP growth rate than the EU over the 1987–2004 period, and a growing if fluctuating labour productivity improvement over the same period. Meanwhile labour productivity in the EU declined significantly and seems to be structural. In contrast, US labour productivity grew after the slump in technology stock markets around 2000, improving by +0.8 per cent from 2000 to 2004, while that of the EU declined by the same magnitude as the USA grew in that period. Unit labour cost reductions and more efficient use of labour through investment in equipment and skills account for most labour productivity gains. In the US large-scale job-shedding in high value-adding sectors and greater use of, for example, e-commerce, largely account for this. Thus, for instance, in the ICT heartland of Silicon Valley some 400 000 jobs disappeared

Table 1.3   *US and EU comparative real GDP and labour productivity*
            change, 1987–2004

	Real GDP (%)		Labour productivity (%)	
	USA	EU	USA	EU
1987–1995	2.7	2.2	1.1	2.3
1995–2004	3.4	2.2	2.5	1.5
Of which:				
1995–2000	4.1	2.7	2.1	1.8
2000–2004	2.5	1.4	2.9	1.0
Of which:				
2004	4.4	2.5	3.0	1.3
Change 2000–04				
Over 1995–2000	−1.6	−1.3	+0.8	−0.8

Source: Based on Van Ark and Inkelaan (2005).

between 2000 and 2003 (Bazdarich and Hurd, 2003). This was the largest peacetime metropolitan job-loss aggregate in US history. In the EU, such labour market carnage was unknown, although recessions were registered, but labour market rigidities probably accounted for most of the decline in labour productivity after 2000.

Moving on, the key question of relevance to this analysis is the extent to which ICT contributed to productivity growth in the US and moderated its decline in the EU. These effects may occur in three ways: ICT investment deepening ICT capital;[9] significant ICT equipment innovation; and knowledge spillovers for intensive ICT user industries. In the EU 1987–2004 non-ICT capital deepening far outweighed that in ICT, though the two came into greater balance between 1995 and 2004. However in the USA there was a greater ICT capital deepening and a miniscule non-ICT capital deepening process, especially in the 1987–95 period. Concomitantly, *total* factor productivity in ICT production grew in the USA at more than double the EU rate in both periods. Also non-ICT total factor productivity was vestigial after 1995 in the EU, while it more than doubled in the USA. After 2000 ICT innovation fell back markedly in the EU and to a lesser extent in the USA, while only the USA, UK and Sweden showed improvement in non-ICT total factor productivity. Austria showed none, in contrast to the pre-2000 period. Finally, since 2000, knowledge spillovers have begun to show an effect upon total factor productivity from ICT use. In other words, *service* industry is the key to faster productivity growth in the USA compared with the EU. This is so in both the pre- and post-2000 periods, especially the latter when, in the EU, banking productivity declined absolutely and it halved in retail. By contrast, in the USA it grew across the key economic sectors, but especially in *business services*. Europe's much vaunted productivity advantage in telecommunications had largely disappeared by the post-2000 period, the USA having caught up, primarily through the contribution of ICT capital deepening. In *business services*, productivity growth is strongly negative in many EU countries. Where it is not, as in the UK, ICT explains a large part of growth, as it does in the USA. Business services productivity growth largely explains total services productivity growth in the USA post-2000 whereas in the EU it peaked in 1996.

To summarize, the advantages accruing from biotechnology and ICT excellence in the USA compared to the EU work in distinctive ways. In biotechnology, the DBFs and their 'big pharma' partners produce far more innovative biotechnology products and services on the health-care market. In the USA more drugs are in trials, though in the EU (plus Switzerland) in 2004 there were more in late-stage trials than in the USA, with what likely success is currently unknown. The US lead in this sector is nevertheless decisive. In ICT the EU problem appears to be complex, involving absolute declines in ICT investment in services, especially the value-adding producer services. These have witnessed a great surge of ICT capital deepening and spread of knowledge spillovers but less actual innovation in the USA since 2000. In particular, the important and burgeoning *business services* sector in the USA has been the key beneficiary of non-ICT industry investment in ICT hardware and software. Business services are many and varied, ranging from varieties of consultancy to e-commerce. This may resonate with the remarkable switch by large US enterprises towards research and development (R&D) outsourcing to such 'knowledge entrepreneurs' to be discussed later in this chapter. Whether it is happening belatedly in the EU seems anecdotally to be the case but if so it may be largely a post-2004 phenomenon and one that increasingly involves Asian host companies.

LIBERAL AND COORDINATED MARKETS: IDEAS, DEFINITIONS AND EXEMPLARS

At this point of this introduction to the book, we wish to offer a little more justification and explain why the 'varieties of capitalism' thesis attracted us more than any other institutionalist approach to analysing the influence of governance regimes upon the evolution of space economies.[10] Recall we are studying generic phenomena – innovative high-technology firms and their specific economic geographies – seeking explanation for performance variance that are both internal to the firm and sector, and external to both. As a key focus of this kind of research is *innovation* and the premier theoretical and empirical work in the field is neo-Schumpeterian innovation systems theory, set within an evolutionary economics macro-framework. Accordingly, we are naturally seeking theoretical compatibilities rather than choosing two or three theoretical frameworks from a portfolio to see which offers the most persuasive explanation of the empirical patterns observed. In this field of research there is no such portfolio. Business systems research is a compatible candidate but all its emphases lean towards the study of large firms, especially multinational companies and their intersections with the regulatory regimes at the macro-level. A striking feature of *innovation* systems theory and empirical research is that it has never been especially exercised about large firms per se but rather with institutions that originate and support the origination of innovation. Accordingly, while the R&D laboratory practices of large firms might be of interest, it is increasingly the case that interest equally falls upon non-profit public research institute laboratories, or university research laboratories, and increasingly the research activities of small and medium-size enterprises (SMEs) as they *interact* with each other and with the *innovation* environment of specialist consultants, incubator facilities, patent lawyers, venture capitalists, innovation *users*, and so on, since it is the system as much as, say, the firm that motivates the quest for understanding.

A different, though again compatible perspective, on regulatory regime comparisons is the now rather dated work that arose more from political sociology than evolutionary economics associated with the study of 'corporatism'[11] or in the USA 'trilateralism'. That terminology instantly takes us back to the pre-Thatcher, pre-Reagan era when corporatism, with its institutional consensus mechanisms involving 'social partnership' and 'social contracts' was often associated with economic success and market economies like the UK and even, to some extent, the USA, were seen to be flagging. But again, as well as concentrating heavily upon political institutions, more than even large corporations which were seldom objects of study, it too concerned itself with the macro-ecology of institutional accommodation and stabilization rather than the wellsprings of techno-economic networks and the sometimes disruptive innovations they brought forth as forms of Schumpeterian 'creative destruction'. Uniquely, therefore, 'varieties of capitalism' overcomes these drawbacks, and indeed research from its perspective even accommodates study of the kind with which this book is concerned, as testified to in, for example, the work of Casper et al. (1999) published in one of the house journals of the neo-Schumpeterian innovation studies community.[12]

Hence, for these reasons we feel the macro-framework we have chosen to help organize the comparative dimension of our work is the most appropriate. It now remains to expostulate its main elements and show how it is made to work on empirical material of the kind we will be presenting in Chapters 7 and 8 in Part II. We may begin by paraphrasing Casper et al. (1999) contrasting economic performance on grounds of innovative performance in high-technology industry between Germany as a coordinated market economy and the USA as a liberal market economy, something we attempt to do similarly for Austria and the UK. Thus Germany is characterized as a coordinated market economy by virtue of the following characteristics. First, the economy is underpinned by a private law system that regulates business and labour contracts. Business itself is organized in ways that embed large firms within networks of trade and industry associations, industrial relations and stakeholder or interest group rules and conventions. These associations are lobbyists for business with government and they often supply useful business services such as research, overseas representation and advice. Government policy relies on the legal system to maintain collective industry agreements between firms and regarding, for example, wage-bargaining, delegating some specific rights to self-managing unions or stakeholder groups. Workforce training is apprenticeship based, focused on specific skill-formation and organized through chambers of industry and commerce. The financial system is primarily bank based with stakeholder corporate governance and banks as major shareholders in larger firms. In Germany market regulation and non-market firm-level modes of business coordination predominate. Hence Germany's market is coordinated.

The USA is a liberal market economy with business organization managed through market interactions supported by a flexible and facilitative legal system. However, in itself it does not rule on the content of legal contracts, this being the responsibility of the contract partners. It has a shareholder not a stakeholder business culture, with minimal legal constraints on firm organization. Different historic and sectoral patterns of market regulation mean business coordination varies across different sectors, historically ranging from legal monopoly to relatively unregulated arrangements. Wage-bargaining is unionized in some industries, and negotiated individual compensation packages in others, with effectively no government intervention. Workforce training is not systematized and not apprenticeship based. The financial system is primarily one of capital markets and historically comparatively weak and regionalized banking. Hence most aspects of economic interaction are largely deregulated with markets coordinating business activity. Accordingly, the USA has a liberal market form of economic regulation.

We will see shortly how accurate a description of regulatory norms these are for, respectively, Austria and the UK. Before that, however, we wish (Table 1.4) to relate the positions of Germany and the USA to innovation

Table 1.4   Coordinated and liberal market structures for innovation in Germany and the USA

	Germany	USA
Science and technology (S&T) labour market	Long-term employment, often with a single employer. Formal rights under co-determination to training, work organization and unlimited employment contracts	Labour markets are deregulated and mobile. Limited, often individual employment contracts. Widespread labour poaching and 'head-hunting'
Firm organization	Consensus decisions, few individual employee incentives. Stock options uncommon, (illegal to 1998). Group reward schemes. Careers well defined and not short-term performance based	Financial 'property rights' ownership structures. No co-determination rules. Strong incentive structures and performance-based compensation. Rapidly shifting firm competencies
Financial ownership	Strong corporate governance rules. Bank not equity firm financing. Bank representatives on boards. Long-term, low-risk investment	Large capital markets (e.g. NASDAQ) fund investment. Equity-based financing, short term. Venture capital
Innovation	Banks do not fund R&D. Incremental innovation the norm. Niche competition. Few radical innovations. These rigidities hamper high-tech start-ups	Investment profile favours novelty and radical innovation. Disruptive technologies create broad, new global markets. Start-up friendly

Source: Based on Casper et al. (2003).

according to their distinctive regulatory regimes, stylizing the findings of Casper et al. (1999) on this. The stylization guides the analysis of comparative innovative performance of Austria and the UK in equivalent industries that are analysed and reported in the rest of this book. Commenting upon these contrasts, it can be said with confidence that these regulatory regimes differ markedly and that this appears to have a determinate effect upon variations in their respective software and biotechnology innovation profiles.

Thus, biotechnology marked a revolution in industrial organization whereby the R&D epicentre shifted from the internal R&D laboratories of large pharmaceuticals firms to university research centres of excellence, foundation-funded research institutes and DBFs. This echoed the shift in scientific primacy from chemistry to biology and the methodology of research from 'chance discovery', classically as with penicillin, to 'rational drug design'. Germany was the home of in-house chance discovery R&D, the paradigm case being Bayer's nineteenth-century breakthrough in synthesizing aspirin from coal tar, the first effective, affordable modern drug. In more recent times German pharmaceutical firms proved particularly incapable of coping with this disruptive triple break in the innovation trajectory of their industry. They responded by seeking aid from government programmes that simultaneously denied such funding to DBFs. Failing to master biotechnology, they later acquired or made alliances with US DBFs but gradually these proved a disappointment as 'star' scientists pocketed the investments and left to form new businesses as serial entrepreneurs. Hence German pharmaceuticals dominance that was pronounced in 1970s had by the 2000s disappeared, with Bayer suffering numerous technological setbacks and Hoechst ultimately being absorbed, along with its French merger partner (Rhone-Poulenc) that created Aventis, into the ownership in 2004 of successful French DBF Sanofi-Synthélabo. Table 1.5 shows some effects of the belated attempt by the federal government to stimulate DBF start-ups through the BioRegio innovative cluster support programme.

This more than doubled the population of German DBFs, but as the substance, content and performance of these admittedly youthful firms shows, this still leaves Germany orders of magnitude behind other, sometimes smaller, European economies let alone the USA. This comparison (Table 1.5) is notable in a number of ways. First, it shows just how effective, at least on the input side of the process, a specialized policy can be in creating firms, indeed in the German case BioRegio aimed expressly to create clusters of biotechnology firms in Munich, Heidelberg and the Rhineland. This is by no means an easy task and the 1995–2000 achievement was a significant boost to those who recognize the important role of policy faced with a

Table 1.5   Main international bioscience competitors, 2003

Country	Companies	Public cos	Market cap.*	Revenues*	Employees*	Pipeline*
USA	1457	307	€205.0 bn	€27.0 bn	191 000	872
UK	331	46	€9.4 bn	€3.0 bn	22 000	194
Switzerland	129	5	€7.3 bn	€2.0 bn	8 000	79
France	239	6	€0.5 bn	€0.3 bn	9 655	31
Germany	369	13	€0.5 bn	€0.5 bn	13 386	15

Note: * Public company data only.

Source: BIGT (2004).

situation of major market failure. Second, moreover, this occurred in a
strong coordinated market economy where such achievements are supposed
to be extra difficult. However, third, in biotechnology it is unusual even
under a liberal market regulatory regime for growth to be rapid. So
Germany's weak performance regarding number and value of public companies, the latter an acid test of achievement of, at least potential, revenues
and products in the trialling 'pipeline' bears witness to the possible
artificiality of the achievement. After the stock market downturn and the
ending of BioRegio many German firms disappeared – 14 alone in Munich
in 2004.[13] In conclusion the analysis confirms the association between innovative performance in biotechnology and regulatory regime since the USA
and the UK are way ahead of the coordinated markets in Table 1.5.

For ICT, specifically software, the analysis produces comparable though
larger-scale discrepancies in performance by high-tech businesses as between the liberal and coordinated regimes. In Germany software is strong in
applications, notably customer-specific solutions in engineering processes,
particularly automotive. Backed again by large federal government programmes, this niche was to be strengthened further still by knowledge transfer into industry from advanced university and institution research in
software. So, as in biotechnology there are many very small firms (some
5000 according to Casper et al., 1999) with fragmented capabilities.
Meanwhile one firm, SAP, bucked that market profile by becoming a global
leader in enterprise resource planning (ERP), meaning generic programs for
efficient reorganization of large companies around standardized integrated
software solutions. Software AG is another such standardized German firm,
indeed all others in the global top 20 are US firms. One reason why few
service-providing German firms are prominent is because most large
German firms meet their own software service requirements in-house. It is
in standardized markets that innovation is more disruptive and, though
Germany challenges the USA on a narrow corporate front, it is massively

outweighed by the likes of Microsoft, Oracle, JD Edwards and Peoplesoft, the leading ERP providers. The rest of German software is an adjunct to the German engineering tradition to a large extent, and locked-in to the fortunes of that globally strong but not unchallenged sector. So to conclude, in this analysis of the interaction between ICT and biotechnology in distinctive regulatory regimes, the postulated outcomes are fairly well predicted from the two stylizations. That is, German institutional arrangements tend to sustain incremental, not breakthrough, innovation. These also tend to favour the emergence of a few large, globally successful corporations with flotillas of state-dependent and fragmentary smaller firms and start-ups. Policy can be effective in augmenting numbers of the latter but their small scale and immaturity in Germany's regulatory context mean they do not grow swiftly as many US 'breakthrough' innovators often do.

KNOWLEDGE FLOWS AND THEIR ECONOMIC GEOGRAPHY: INNOVATIVE CLUSTERS

We come now to a brief overview of a key dimension of the book that will feature strongly in later chapters. This concerns a new and by no means totally clear reconfiguration of key global knowledge flows and their new economic geography. This arises from some of the issues already raised regarding the context for this book, notably global outsourcing and specifically *R&D outsourcing*. Of particular interest here is the extent this may have *metamorphosed* global knowledge interactions.[14] We are interested in how its realization evolves institutionally under the two master regulatory narratives with which we are here exercised, and to what extent, hypothetically, they are influenced by this. The policy dimensions of this analysis are briefly highlighted for Austria and the UK in the final section of this introductory chapter. The implications of what has been discussed already for knowledge flows of the kind in which we are interested, are the following. Foremost is the changing status of multinational corporations with respect to the origination of knowledge arising from *exploration* activity in their in-house R&D laboratories. It has recently become clear that the degree to which this happens nowadays has declined, at least for US industrial R&D, from around 71 per cent of all R&D being in-house in 1981 to only 39 per cent being so performed by 2001. This massive shift in R&D outsourcing has benefited overwhelmingly knowledge entrepreneurs, notably DBFs, but also their clustered equivalents in ICT, and university research centres of excellence that are often co-located with the aforementioned.

In these two industries, in particular, R&D outsourcing is very high and increasingly international. Thus from 1994 to 2000 there was a fivefold

Table 1.6   *Growth in R&D expenditures by small US companies,*
              1997–2001

Size of firm	1997	1998	1999	2000*	2001*	% change
<25	2 536	3 804	5 579	6 176	4 346	71
25 to 49	2 455	2 525	3 824	4 507	3 375	37
50 to 99	3 415	5 155	5 779	6 533	7 382	116
100 to 249	5 907	6 622	5 707	8 118	11 634	114
250 to 499	5 229	5 522	6 463	6 731	7 832	50
Total	19 542	23 628	27 352	32 065	34 569	77

Note: Millions of constant $1992/ *$1996.

Source: NSF (2005).

increase in R&D outsourcing to the rest of the world from the USA and a tenfold increase to China and Singapore (up from roughly $50–$500 million in both). The data in Table 1.6 reveal the massive increase in R&D outsourcing as a key knowledge flow under redirection towards SMEs in the USA alone. Moreover, other data from the same source show such knowledge-intensive SMEs to be far larger recipients in share of the total for ICT and biotechnology than, for example, universities.[15] This means that for an increasingly large share of exploration[16] knowledge in two sectors that are high R&D spenders, possibly accounting for half the $198.5 billion US total listed for 2001, SMEs and universities along with other independent research institutes are nowadays major providers. In fact NSF data on this show the relationship between firm and university recipients of R&D outsourcing to be in the ratio of 4:1. Clearly, from Table 1.6, firms in the USA employing fewer than 500 people account for some 18 per cent of US industrial R&D across all industrial sectors, a virtual doubling in five years. Such high-grade data are not available for European countries, but close reading of anecdotally based research such as that provided by the European Association of Contract Research Organisations (EACRO, 2005) shows a consistent trend. The hypothesis that this activity locates in the leading clusters for ICT and biotechnology is intimated in a study by Chesbrough (2003) and this is consistent with work by the likes of Norton (2000) and Florida (2002) who identify such R&D megacentres[17] as Boston, San Francisco, San Jose and San Diego as America's most entrepreneurial, most knowledge-intensive and most culturally talented and innovative of locations. For the present, space does not permit much more to be said although we shall return with further evidence about this *metamorphosis* in the control of exploration knowledge through R&D

outsourcing. The most that can be said about differences between our two stylized kinds of economy and economic governance is that large firms seem equally prone to R&D outsourcing in both regimes but that in, for example, Germany where Fraunhofers, Max Plancks and varieties of research centres from Helmholz to 'Blue List' have for long conducted industrial as well as publicly funded R&D, the most notable 'branching' is that which sees corporations like Siemens outsourcing to affiliates or simply opening internal R&D laboratories in Asia, notably China and India.[18]

POLICY SHIFTS IN INNOVATION GOVERNANCE: AUSTRIA AND UK

We come now to the final segment of the introduction to this book. Here we are interested in alerting ourselves and the reader to signs that the changes alluded to in the foregoing have been or are in the process of influencing the institutional and regulatory arrangements of economic governance in the two types of regime. We switch attention now from the larger stereotypes, Germany and the USA, to their smaller brethren, Austria and the UK. Clearly the last two are by no means carbon copies of the others although, as we shall see, the overlaps are striking. Let us perform a stylization (Table 1.7) around innovation system environments comparable to that performed for Germany and the USA in Table 1.4. Since this research involves a *regional* dimension in the analysis, a summary of key regional innovation support characteristics is also inserted.

Three things are immediately apparent regarding the contrasting institutional and regulatory regimes in question. The first is that Austria and the UK are by no means carbon copies of the stylized portrayals of German and US varieties of capitalism. However, the similarities outweigh the differences. The main difference is the predominance of bank-based funding and 'steering' of companies in Austria and of equity-based funding in the UK. Of course there are differences, depending on scale and nature of firms in question under each regime, but probably in no way do UK firms have bank-based control as in coordinated market economies, and a minority of Austrian firms would traditionally have been mainly equity based, though this is changing. Political change, dismantling of the social partnership, consensus-based public-private 'Austrian model' has occurred, removing from power the once hegemonic conservative and socialist coalitions of the past. Privatization of many public utilities and other holdings has occurred and a new impulse given to innovation, against the kind of economic stability, some would say stagnation, typical of the old model and its associated conventions.

Table 1.7   Coordinated and liberal market structures for innovation in Austria and the UK

	Austria	UK
S&T labour market	University-led; some large firms public-private R&D organizations. Until recently: long-term employment, collective contracts rights to training. Now: more workforce flexibilization and segmentation.	S&T labour markets are university-led, some government R&D, some corporate. Contracts national and local. Increasing 'headhunting'.
Firm organization	Large firms: consensus decisions, union bargaining. Small firms; paternalistic. Careers well defined and not short term and performance based. Few individual employee incentives, stock-options not common. Recently: more flexible incentive schemes.	Equity-based ownership structures. Weak union bargaining. Incentive schemes, performance-related pay. Flexible, global supply-chain configurations.
Financial ownership	Bank not equity firm financing. Bank representatives on boards. Strong corporate governance rules. Long-term, low-risk investment. Some role of public funds.	Large capital markets (e.g. AIM) fund investment. Equity-based financing, short term. Venture capital.
Innovation	Large firms: R&D investment internal. Small firms: little R, some D. Niche competition and incremental innovation. Few radical innovations. Public funds innovation support R&D, innovation collaboration and high-tech start-ups.	Investment profile favours incremental and radical innovation. Research strong, commercialization weak. Moderately start-up friendly.
Regions	Federal system, innovation part of *Land* brief. Corporatist management of services. Unions run labour	Asymmetric devolution. Economic development agencies common. They manage much innovation

Table 1.7 (continued)

	Austria	UK
	market service, industry runs Chambers. Regional economic development agencies.	support, regional venture funds, cluster strategies.

The second point of note is that Austria has a distinctive science and technology infrastructure and associated career path for scientists. It differs even from the German system in the following ways. Key here is the large Austrian Research Centre (ARC) network of national technological institutes such as Seibersdorf, where research, knowledge and technology transfer to firms is conducted. The ARC covers *inter alia* systems, nuclear, weaponry, rail, metallic, electrochemical, and computer vision research. Other research is naturally conducted in firms and universities especially, for industry-related research, in Austria's technical universities, which also frequently have technology parks and start-up programmes. In the UK most leading research is conducted in universities. There are few independent foundations and no research institute network comparable to the ARC. Government research and establishment research is not generally advanced or innovative, while business-funded research budgets have, in recent years, been in decline and/or increasingly outsourced to university laboratories or 'knowledge entrepreneur' research, design and engineering businesses.

Finally, regarding innovation and regions: innovation is prized in both countries but neither is particularly outstanding at commercialization of new knowledge. We have seen how UK biotechnological research of world class is likely to be exploited more in future by US DBFs. In this respect the Austrian and the UK innovation profiles may be closer than those between the USA and the UK. Having said that, the UK has many elements that are also found in the USA, namely, entrepreneurially minded university policies, abundant venture capital, patent lawyers, management accountants and specialist consultancies. However Austria's more publicly funded innovation system may in some respects be superior, given scale differences, to that of the UK. If we move to the regional level of governance, a region like Styria in Austria has excellent technical universities such as the one at Graz, producing numerous start-ups from its graduating PhD candidates, housing them on its technology park and linking them through collaborative activities connected to development agency cluster programmes. However, these innovators are often focused in medium-technology rather than high-technology sectors. Nevertheless a similar 'region' like Scotland or Wales finds many barriers to forming such a seam-less innovation system effect. Moreover, in those cases, most of the inno-vation support activity is public in origin owing to market weaknesses and failures. So, in conclusion, the UK is a markedly 'European' variety of liberal market institutional capitalism with its weak commercialization performance, while Austria is, as we shall see, in Chapter 8 in greater detail, moving away from its stylization as a coordinated market model. Both have solid regional institutional support for innovation – a point at which their divergences are possibly less marked than in other spheres. However, with the establishment of regional development agencies for the regions of England – but not Scotland or Wales – this was itself an economic govern-ance innovation dating only from 1999.

This concludes the introductory chapter. All the issues raised are addressed in some depth in the chapters that follow. Immediately following are two that comprise conceptual and methodological reviews of key con-cepts from economics and contemporary regional science such as the changing nature of the 'knowledge economy', the role of asymmetric knowledge in regional uneven development, clustering theory and practice, and how we researched comparatively the same industries in two such different regulatory regimes. The various models for commercialization of new knowledge and the mediating influences among markets, knowledge and other spillovers, networks and milieux are then assessed prior to an in-depth analysis of some of the global–local/regional knowledge flows issues informing current debate, such as that touched upon in relation to R&D outsourcing in this introduction. Part I of the book concludes with a chapter reviewing the contribution of an *innovation systems* perspective to understanding the complexities involved in the kind of regional research dealt with in this book, considering varieties of these in line with the general tenor of the book. Part II then moves into empirical analyses of the key research results for the two sectors and countries emphasizing sub-national innovation practices and economic governance interactions, again highlighting the varieties of regional innovation system effects observable from the data. The experiences of the two countries are compared in some analytical and empirical detail before a final chapters draws out the book's key lessons and conclusions for both theory and policy in respect of the issues raised by the research that has been reported.

NOTES

1. On 'national business systems' there is a copious literature from Europe, see for example, Whitley (2002), and from the USA, Hollingsworth and Boyer (1997). These are

'institutionalist' socio-economic analyses and compatible in conceptual terms with the approach taken in this book.

2. It is worth noting, in passing, the significant contribution made to the Italian economy in the post-war years by such small and medium enterprise industrial districts. In a historical analysis by Becattini and Dei Ottati (2006), it is shown that from 1951 to 2001 manufacturing districts grew in employment from 72000 to over 2 million while provinces dominated by larger enterprises remained at 1.8 million, having peaked at 2.2 million in 1961 and again in 1981. On most other performance indicators, such as value-added, export value, unemployment rates and, even, demographics, small firm districts outperformed large enterprises by a factor of at least two to one.

3. Despite the important insight regarding the practical and conceptual content between tacit (implicit) and codified (explicit) knowledge, these authors badly mix up 'information' and 'knowledge' as if they were identical, a problem by no means confined to them. In this book we will generally refer to 'knowledge' in relation to 'information' according to the following: *information* is a passive resource given *meaning* by the application of *knowledge*, which facilitates *action*. Thus if the actor has *knowledge* of her destination, *meaning* is given to the *information* in the train timetable, enabling appropriate travel *action* to be taken.

4. See also Scharmer (2001). This can take many forms, some no doubt deeply embedded and culturally as well as psychologically shaped. But an instance, discussed by Metcalfe (2005), might be the sense of unease, dissatisfaction or disagreement with the status quo that must be the first step in the knowledge-creation process that may result in a commercial innovation.

5. Foley (2005). A quote from Sir Chris Evans, head of Merlin Biosciences, envisaging a way out of the UK's entrepreneurship deficit in biotechnology. This is a good example of the rise of a specialist 'research industry' in an economy that is modest in its biotechnology commercialization achievements. Of course, time will tell if this happens and whether it is confined only to biotechnology.

6. Further data supporting this contention are provided in Chapter 6.

7. Aerospace, computers and office machinery, electronics-communications and pharmaceuticals.

8. Automotives, chemicals, machinery and scientific instruments.

9. Capital deepening means that in a growing economy, it raises capital per worker for all workers.

10. The approach is not immune from criticism. There are at least three dimensions of this. The first is that it is manufacturing-centric; second that, as we have noted, there is a great deal of variety within the two types; and third, that it postulates no sense of how coordinated market economies can compete in the long run against the 'monetarist assault' of the liberal market economies. Cogent answers, mostly showing how these points are dealt with in the perspective already, are supplied in Hall and Soskice (2003).

11. See, for example, Crouch and Streeck (1996).

12. Casper et al. (1999).

13. This statistic given verbally by BioM, the Munich BioRegio policy vehicle, in a meeting of the EU-BioLink bioincubation project at Genopole, Evry, Paris in January 2005. See www.cordis.lu/paxis/src/val_projec.htm.

14. This term is used in the preface to Penrose (1995) in reference to the manner in which innovation demands the mobilization of growth firms in global knowledge and innovation networks, something she suggests is metamorphosing what we might think of in terms of the structures of globalization. Prominent amongst these restructurings are geographical hierarchies of research accomplishment (see Cantwell and Iammarino, 2003).

15. The data source is the report by the US National Science Foundation (NSF, 2005).

16. This terminology of *exploration* and *exploitation* knowledge originated with March (1991).

17. For a definition of the French concept of *megacentres* as more than mere market-driven clusters because they embody also localized exploration research and talent formation institutions that are mostly publicly funded, see Cooke (2004).

18. See for example, Merchant (2005) on Siemens' investment of $500 million in an Indian export hub. The article notes this is 'to expand research and development, raising its force of Indian software engineers to 4,000'. The move followed similar announcements by Swedish/Swiss rival ABB, Finland's Nokia and pharmaceuticals companies Novartis and AstraZeneca to India in 2005.

REFERENCES

Bazdarich, M. and Hurd, J. (2003), *Anderson Forecast: Inland Empire & Bay Area*, Los Angeles, CA: Anderson Business School.

Becattini, G. and Dei Ottati, G. (2006), 'The performance of Italian industrial district and large enterprise areas in the nineties', *European Planning Studies*, **14**, 1139–62.

Best, M. (1990), *The New Competition: Institutions of Industrial Restructuring*, Cambridge: Polity.

Bioscience Innovation and Growth Team (BIGT) (2004), *Bioscience 2015*, London: Bioscience Innovation and Growth Team.

Cantwell, J. and Iammarino, S. (2003), *Multinational Corporations & European Regional Systems of Innovation*, London: Routledge.

Casper, S., Lehrer, M. and Soskice, D. (1999), 'Can high-technology industries prosper in Germany? Institutional frameworks and the evolution of the German software and biotechnology industries', *Industry & Innovation*, **6**, 5–24.

Chesbrough, H. (2003), *Open Innovation*, Boston, MA: Harvard Business School Books.

Cooke, P. (2001), 'Biotechnology clusters in the UK', *Small Business Economics*, **17**, 43–59.

Cooke, P. (2004), 'The accelerating evolution of biotechnology clusters', *European Planning Studies*, **12**, 915–20.

Cooke, P. (2005), 'Rational drug design, the knowledge value chain and bioscience megacentres', *Cambridge Journal of Economics*, **29**, 325–42.

Crouch, C. and Streeck, W. (eds) (1996), *Modern Capitalism or Modern Capitalisms?*, London: Pinter.

Department of Trade and Industry (DTI) (1999), *Biotechnology Clusters*, London: DTI.

Department of Trade and Industry (DTI) (2005), *Comparative Statistics for the UK, European & US Biotechnology Sectors*, London: DTI.

European Association of Contract Research Organisations (EACRO) (2005), *The Changing World of Industrial R&D: The Challenges for Industrial Research and Technology Organisations and Governments*, Brussels: European Association of Contract Research Organisations.

European Commission (EC) (1995), *The Green Paper on Innovation*, Brussels: European Commission.

Florida, R. (2002), *The Rise of the Creative Class*, New York: Basic Books.

Foley, S. (2005), 'Can Merlin's chief cast his spell on investors again?', *Independent*, 16 July, p. 40.

Hall, P. and Soskice, D. (eds) (2001), *Varieties of Capitalism: The Institutional Foundations of Comparative Advantage*, Oxford: Oxford University Press.

Hall, P. and Soskice, D. (2003), 'Varieties of capitalism and institutional change; a response to three critics', *Comparative European Politics*, **1**, 241–50.

Harmaakorpi, V. (2006), 'Regional Development Platform Method (RDPM) as a tool for regional innovation policy', *European Planning Studies*, **13**, 1085–114.

Hollingsworth, R. and Boyer, R. (1997), *Contemporary Capitalism: The Embeddedness of Institutions*, Cambridge: Cambridge University Press.

Kaloudis, A., Sandven, T. and Smith, K. (2005), 'Structural change, growth and innovation: the roles of medium and low-tech industries, 1980–2002', presentation to conference on 'Low-tech as a Misnomer: the role of non-research intensive industries in the knowledge economy', Brussels, 29–30 June.

Maillat, D. (1998), 'Interactions between urban systems and localised productive systems; an approach to endogenous regional development in terms of innovative milieu', *European Planning Studies*, **6**, 117–30.

March, J. (1991), 'Exploration and exploitation in organizational learning', *Organization Science*, **21**, 71–87.

Merchant, K. (2005), 'Siemens to invest $500m. in Indian export hub', *Financial Times*, 10 February, p. 30.

Metcalfe, S. (2005), 'Knowledge and enterprise: what makes capitalism work?', plenary presentation to 5th Triple Helix Conference 'The Capitalisation of Knowledge', Turin, 18–21 May.

National Science Foundation (NSF) (2005), *Research & Development in Industry 2001*, Washington, DC: NSF.

Norton, R. (2000), *Creating the New Economy: the Entrepreneur & the US Resurgence*, Cheltenham, UK and Northampton, MA, USA: Edward Elgar.

Orlikowski, W. (2002), 'Knowing in practice: enacting a collective capability in distributed organizing', *Organization Science*, **13**, 249–73.

Peneder, M. (1999), 'The Austrian paradox: "old" structures but high performance?', *Austrian Economic Quarterly*, **4**, 239–47.

Penrose, E. (1995), *The Theory of the Growth of the Firm*, 3rd edn, Oxford: Oxford University Press.

Polanyi, M. (1966), *The Tacit Dimension*, London: Routledge.

Scharmer, C. (2001), 'Self-transcending knowledge: organising around emerging realities', in I. Nonaka and D. Teece (eds), *Managing Industrial Knowledge: Creation, Transfer and Utilisation*, London: Sage.

Tödtling, F. and Trippl, M. (2005), 'Knowledge links in high technology industries: markets, networks or milieu? The case of the Vienna biotechnology cluster', paper presented to DRUID Tenth Anniversary Summer Conference 'Dynamics of Industry and Innovation: Organisations, Networks and Systems', Copenhagen, 27–29 June.

Van Ark, B. and Inklaar, R. (2005), 'Catching up or getting stuck? Europe's troubles to exploit ICT's productivity potential', EU KLEMS project presentation on 'The Determinants of Europe's Productivity Revival', Helsinki, 13 May.

Whitley, R. (ed.) (2002), *Competing Capitalisms: Institutions & Economies*, Cheltenham, UK and Northampton, MA, USA: Edward Elgar.

Zitt, M., Ramanana-Mahary, S., Bassecoulard, E. and Laville, F. (2003), 'Potential science-technology spillovers in regions: an insight on geographical co-location activities in the EU', *Scientometrics*, **57**, 295–320.

2. The emergent knowledge economy: concepts and evidence

This chapter first deals with the questions of what the knowledge economy is, how we can understand it, and how the process of knowledge generation and application works. Also the role of codified and tacit knowledge, their interplay and the spatial implications will be discussed. Second, we will bring forward empirical evidence from EU and OECD sources regarding the knowledge economy. We will try to find out, how far we have moved towards a knowledge economy and which countries have a lead in this respect. A particular focus will be on differences between EU countries and the USA, as well as on Austria and the UK, the two countries further analysed in greater detail in later chapters.

In the past few years we have seen an ever-growing literature on the rise of a knowledge-based economy (OECD, 1996; 2001; Drucker, 1998; Smith, 2002; David and Foray, 2003) or a learning economy (Lundvall and Johnson, 1994). A major driving force of this development have been the globalization process and tendencies of deregulation and liberalization, putting pressure on companies to innovate in order to stay competitive (Lundvall and Borrás, 1999; Dunning, 2000; Archibugi and Lundvall, 2002). Also the progress in ICT has allowed new forms of exchange and storage of information and has stimulated the codification of knowledge (Soete, 2002). The latter has been reflected in the boom and subsequent bust of the New Economy at the end of the 1990s. Another stream of literature (Keeble and Wilkinson, 2000; Storper, 1997; 2002; Dunning, 2000; Cooke, 2002) has been dealing with spatial aspects of the knowledge economy, in particular with high-technology clusters, local knowledge-spillovers and with local–global relationships.

Knowledge in one way or the other, mainly 'practical', has always been at the heart of economic growth, and innovations, that is, new ideas embodied in products, processes and organizations have traditionally fuelled economic development. The knowledge economy in its more recent understanding has to do with a continuing transformation towards more knowledge-intensive activities rather than a radical change and rupture of economies and societies. It refers to an emerging economy where productivity and growth are less based on the abundance of natural resources than on the capacity to improve the quality of human capital and factors of production, and to create new knowledge and ideas and incorporate them into equipment and people (David and Foray, 2003). Increasingly the growth and performance of national and regional economies seem to depend on the generation, dissemination and application of new knowledge and on innovations, particularly involving science and technology.

There are different views in the understanding of a knowledge economy, as has been pointed out by Smith (2002). The first is that *knowledge as an input* is becoming quantitatively and qualitatively more important than before. This is reflected in an increase of knowledge-related investment, such as R&D, education, software and information technologies, as observed for example, by the OECD (2001). Another perspective reflects the idea that *knowledge as a product* is getting more important than in the past. Along this line we can see, for example, the growth of knowledge-intensive business services and of high-tech industries. Often such companies are set up based on a new idea, incorporating and applying new knowledge into products. A third view argues that, in particular, *codified knowledge* as opposed to tacit personal skills has become more significant (Cowan et al., 2000). In particular, for science-based industries such as biotechnology this has been put forward by Cooke (2004; 2005a) and others. And then there is the argument that the knowledge economy rests on *technological developments in ICT*, allowing new forms of knowledge management and exchange. Electronic knowledge management systems, electronic knowledge exchange and 'platforms' and the use of the Internet are an expression of this. Castells (1996: p. 17) has characterized the knowledge economy in another view as 'the *action of knowledge upon knowledge itself* [being] the main source of productivity' (italics added). Knowledge using, such as in the modern food industry, is thus not enough; knowledge *creation* is necessary for a knowledge economy. As Cooke (2002: 4f) puts it

> Knowledge economies are not defined in terms of their use of scientific and technological knowledge, including their willingness to update knowledge and creatively forget old knowledge through learning. Rather, they are characterised by *exploitation of new knowledge in order to create more new knowledge*. This need not be scientific and technological knowledge alone, it can be creative knowledge in the artistic, design or musical senses of knowledge (italics added).

Each of the above views has weaknesses if they are applied alone as tools to understand the knowledge economy (Smith, 2002). Combined, however, they may throw light from different angles onto the complex phenomenon of the knowledge economy, and they are also strongly related to each other.

Rather prominent is the view of identifying the knowledge economy with the growth of *high-technology sectors* such as microelectronics and computers, ICT, pharmaceuticals, instruments and aircraft, or with knowledge-based services such as software, R&D or consultancy (knowledge as output). But a closer look reveals that sectors such as these are also relying more on science and R&D, thus knowledge is their key input. They are characterized by strong growth and firm formation, high innovation activity regarding products and processes, and they use both internal and external knowledge sources to a large extent (Keeble and Wilkinson, 2000). These industries often have strong links to universities and research, for example, through spin-offs, cooperations or joint use of facilities, and because of these characteristics they can be regarded as knowledge based.

It is important to bear in mind, however, that the knowledge economy is not confined to those sectors. *Also in many medium- and low-technology sectors* such as food, materials or textiles, *learning and innovation* may be significant (Maskell et al., 1998; Lundvall and Borrás, 1999). An increasing importance of knowledge and innovation in these industries can be seen from rising levels of qualifications or from innovation expenses, more broadly defined (OECD, 2001; EC, 2003a). Relevant knowledge sources in the latter case are different from the high-tech sectors, however. They are more often along the value chain, that is, customers, suppliers or service providers, in that they are more related to the operative activities of companies (Tödtling et al., 2006). Nutraceuticals or 'functional foods' are examples here (Cooke, 2007).

Despite the recent problems of the 'New Economy' and the collapse of thousands of Internet firms, the knowledge economy is, in fact, also characterized by an increasing role of ICT as an instrument of knowledge generation and transmission. This is related to a pressure towards the codification of knowledge and the use of computerized knowledge management systems. In particular, the Internet and 'search engines' such as Google have become a widely used source of relevant information also in science and research. The 'death of distance' has been proclaimed by Cairncross (1997) and others. However, there is broad agreement in the literature that there are limits to codification and to electronic knowledge transmission (Lundvall and Borrás, 1999; Johnson et al., 2002; Gertler, 2003). A large and important share of knowledge cannot be codified, or is too expensive to codify, and remains tacit. This also limits the use of ICT and the Internet in the knowledge process (Kaufmann et al., 2003; Morgan, 2004).

TYPES OF KNOWLEDGE AND THE KNOWLEDGE PROCESS

Knowledge has to be distinguished from *information*. The first can be regarded as a cognitive capacity which empowers its possessors with the capacity for intellectual and manual action, whereas the latter refers to structured and formatted data-sets that remain passive until used by those with the knowledge needed to interpret and process them (David and Foray, 2003). Information, thus, is a passive resource given meaning by the application of knowledge, which facilitates action. While information can easily be reproduced and transferred, the reproduction and transfer of knowledge is a more difficult process, owing to the fact that only a part of the available knowledge becomes codified. Codification refers to translating knowledge into symbolic representations so that it can be stored on a particular medium. Only *codified knowledge* can be acquired through the reading of books and articles, or through the use of ICT and the Internet.

An important share of knowledge, however, remains tacit. *Tacit knowledge* refers to that component of knowledge which defies codification or articulation, either because the actor is not fully aware of it or because the codes of language are not developed well enough to permit codification (Breschi and Lissoni, 2001a). By stating 'we can know more than we can tell' Michael Polanyi (1966: 4) circumscribed this tacit dimension of knowledge. It refers to skills such as swimming, playing music, cooking, performing art or a handicraft. Master–apprentice relations and learning by doing are important mechanisms for acquiring these skills (Gertler, 2003). But tacit knowledge is also an important dimension for cognitive processes and research activities. It refers to those cognitive frames, understandings and preconceptions which are not made explicit but are used as implicit filters for the selection of information and for interpreting data and messages. In fact, the acquisition and use of codified knowledge to some extent always need tacit forms of knowledge as complements.

The generation and application of knowledge requires the combination and articulation of those various knowledge types in order to become effective. Nonaka and Takeuchi (1995) and the OECD (1996) see this as a dynamic process, a knowledge spiral, where individual implicit (tacit) knowledge (1), is made explicit (codified) (2), and then shared with others (socialized) (3). Once knowledge becomes routinized it may become embedded into collective rules and habits (4), which are taken up and learned by individuals (1). Through such a revolving process of externalization and internalization, as well as through interaction and the socialization of knowledge, both the individual and collective knowledge bases are continuously enhanced.

As has been pointed out already in the introductory chapter, it is important to bear in mind that knowledge is more complex than these binary relations between implicit and explicit kinds of knowing suggest. Tacit and
codified knowledge may be linked in a number of ways. For example, tacit
knowledge may act as a facilitator for the transmission of codified knowledge and the availability of a skilled workforce also might play an important role in localized dissemination of knowledge (Zitt et al., 2003). Cooke
(2005a) has referred to *complicit knowledge* as another relevant knowledge
type in this context. It is interjected into the processes by which tacit
(implicit) knowledge becomes codified (explicit). Thus places may 'monop-
olize' knowledge in certain fields where specific, frequently implicit, exploration knowledge assets exist in geographical proximity. Such implicit
knowledge requires extraction in order to be applied and appropriated. In
this space complicit intermediary firms and other actors mediate the 'trans-
lation' of such implicit, exploration knowledge by examination of it for
validity, reliability and safety, and so on. It then may become codified as
explicit, appropriable knowledge in the form of, for example, a copyright,
trademark or patent. Supporting knowledge capabilities are necessary in
order to achieve this. These are often provided in regional innovation
systems whose innovation institutions focus on network interactions assisting translational research for commercialization of new knowledge.

Knowledge tends to be exchanged in 'communities' where people group
together in an intensive effort to produce and exchange new knowledge,
often with the help of ICT. Such communities may exist within companies
or organizations (research centres, government agencies, and so on) extend-
ing usually across various departments or locations of firms. More often,
however, knowledge communities cut across the boundaries of conventional organizations, leading to inter-organizational links or temporary
organizations at various spatial levels. Amin and Cohendet (2004) in
this context consider innovative regions or cities as places where various
knowledge communities intersect. Such regions, thus, become points of
exchange of various types and kinds of knowledge or locations where
widely distributed knowledge bases can easily be integrated.

PROXIMITY AND THE SPATIAL CONCENTRATION OF THE KNOWLEDGE ECONOMY

Proximity and geography matter in the knowledge and innovation process,
although often in a more differentiated way, as has been suggested. Unlike
the views of Cairncross (1997), we find no 'death of distance' through the
use of ICT and the Internet but a strong role of proximity in knowledge

generation and application. The literal meaning of *proximity* includes geographical connotations such as nearness, closeness and contiguity. However, it includes also geographically unconfined meanings involving nearness in context, domain or opinion (Cooke, 2005b). A theoretical analysis of the relations between innovation capability and varieties of proximity is presented in Boschma (2005). He argues that proximity may be geographically, organizationally or institutionally defined. Zeller (2004) investigating the dependence of Swiss 'big pharma' on innovative biotechnology clusters elsewhere, arrives at a more differentiated typology of proximity. Besides geographical proximity, he refers to institutional (for example, national laws), cultural (for example, communities of practice), relational (for example, social capital), technological (for example, Linux software users), virtual (for example, a multinational) and internal and external proximity (for example, firm supply chain management). Few of these types of proximity feature prominently in his empirical analysis and those that do are utilized assertively rather than analytically. Geographical proximity, thus, turns out as still being highly relevant.

The *role of geographical proximity* for tacit knowledge exchange has been pointed out extensively in the literature on high-technology districts (Saxenian, 1994; Storper, 1997; Keeble and Wilkinson, 2000) and innovative milieux (Maillat et al., 1996; Ratti et al., 1997; Capello, 1999; Crevoisier, 2001). The argument is that the exchange of innovation-relevant knowledge requires trust, a common language and understanding, and is strongly favoured by face-to-face interaction. Ideally, this has been observed for specific regions such as Silicon Valley, Boston, Cambridge or Munich. Gertler (2003) has argued in this context that the cognitive and cultural aspects of such milieux are not enough to understand the innovation process. He has emphasized that in particular the role of institutions, that is, common habits, routines, prevailing practices in companies, labour markets and financial organizations, are relevant for the knowledge and innovation process. This, in fact, links up to the concepts of innovation systems and of business systems, which have been taken up in the Introduction (Chapter 1) and in other chapters of this volume.

What these various approaches have in common is the finding that *knowledge economies are unevenly distributed in space* and give rise to various kinds of disparities: knowledge organizations such as universities and research centres as well as business innovation activities such as R&D, patenting, and product innovations are usually concentrated in a few specific regions or urban areas and are not evenly spread across geographical space. Many studies have shown that R&D activities and high-tech sectors are strongly concentrated in space, much more so than employment or GDP (Gehrke and Legler, 2001; EC, 2003b; Feldman, 2000; Cooke,

2002; Laafia, 2002). This in turn gives rise to disparities in competences, productivity, incomes and career opportunities.

We find, thus, that *large cities and urban places hold key positions* in the emerging knowledge economy (Brower et al., 1999; Simmie, 2003). Geographical proximity seems to be a means to achieving many of the other kinds of proximity. Thus while the expectation of a death of distance or end of geography was wrong in respect of the idea of ubiquitous access to the full range of production and other factors over space, it was correct in postulating globally networked information flows. This rests on the observation that globalization proceeds through networks linking nodes of economic power, mainly cities. Such nodes can be regarded as results of increasing returns to urban agglomeration (Sternberg and Litzenberger, 2004). In the emerging knowledge economy these are derived from a variety of spillovers, especially knowledge spillovers, that tend to concentrate in cities as will be shown in Chapter 3.

There is still much debate, however, regarding the question as to which type of agglomeration or cluster is providing the best results and performance. One view argues along the line of Alfred Marshall (1918) that it is *specialized* agglomerations or clusters which give rise to sectoral knowledge spillovers as a key to innovation (Glaeser et al., 1992; Baptista and Swann, 1998; Henderson, 2003). Another view originates with Jane Jacobs (1996) and has been supported by Feldman and Audretsch (1999) showing that sectoral *diversity* is most strongly associated with regional innovativeness. Advantages of specialization and of diversification have been brought into discussion much earlier in the context of (mainly static) 'externalities' with the concept of localization versus agglomeration economies (for an overview see Maier and Tödtling, 2005), whereas more recently they were discussed from a dynamic perspective in the context of the cluster approach (Porter, 1990; Steiner, 1998; Baptista and Swann, 1998). Indeed, there seems to be a trade-off in terms of short-term versus long-term advantages (Tichy, 2001): in the short term or in the early phases of cluster development specialized clusters seem to be well able to reap benefits from localization economies and spillovers of specialized knowledge. In the longer term, however, there is the danger of cognitive, functional and political lock-ins as Grabher (1993) and others have demonstrated (Hassink and Shin, 2005). Such findings have been supported recently by Henderson (2003) who showed specialization effects on knowledge spillovers to have strong but short-lived impact in high-technology industry, while diversification effects persist far longer. Boschma (2005) resolved many of these specialization–diversification dilemmas with the evolutionary economic geography concept of *related variety* (see Chapter 10).

IS THERE AN EMERGING KNOWLEDGE ECONOMY? SOME EVIDENCE FOR EUROPE AND THE USA

To what extent can we observe a move towards a knowledge economy, and what differences can we find between countries in Europe? In the following we will look at some indicators describing investments or inputs into the knowledge economy as well as performance indicators. Basically we will compare the EU with its main competitor, the USA, and we will also characterize the two countries which are analysed in greater detail in the present volume, Austria and the UK. We will use evidence from the EC (2003a; 2005) as well as from other sources. The aspect of regional concentration and of clustering of knowledge economies will be taken up in later chapters of the book, focusing on the UK and Austria.

The European Commission (EC, 2005) has applied two composite indicators in its attempt to describe the complex and multidimensional nature of the knowledge economy. The first refers to investments in the knowledge economy and includes R&D effort, investment in highly skilled human capital (researchers and PhDs), the capacity and quality of education systems (education spending and lifelong learning), purchase of new capital equipment and modernization of public services (e-government). The second composite indicator tries to capture the performance of knowledge-based economies and includes overall labour productivity, scientific and technological performance, usage of ICT (e-commerce) and the effectiveness of the education system (schooling success rates). Most of the data refer to the period from 1997 until 2002–03.

Within Europe we can identify a marked North–South divide regarding *investments in the knowledge economy* (see Figure 2.1). The leading countries are Sweden, Denmark and Finland with clearly above average investment levels, and in the case of Denmark also growth rates. Laggards in terms of investment levels are Greece, Italy, Spain and Portugal. In particular, Greece has been catching up, from a very low position however, between 1997 and 2002. Together with Belgium and France, the UK and Austria belong to a middle group of countries above the EU average in terms of investment levels. From the latter countries, the UK and Belgium have shown stronger overall growth of investment in the knowledge economy, between 1997 and 2002.

From Figure 2.2 we can see that there is a positive correlation between investment levels and *performance in the knowledge economy*. In countries such as the UK, Belgium, France and Spain performance seems to meet quite well their level of investments. However, there are also interesting deviations. The leading countries in terms of investment levels

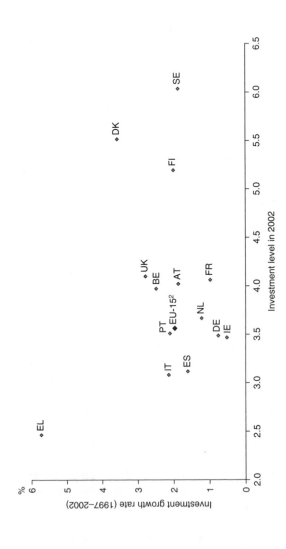

Notes:
1 All 7 sub-indicators were included for the investment level 2002 (horizontal axis), but the indicator on e-government could not be included in the comparison of the growth rates (no data available for 1997).
2 EU-15 does not include LU.

Source: DG Research, Key Figures 2005.
Data: Eurostat, OECD, DG Information Society.

Figure 2.1   Composite indicator of investment in the knowledge-based economy: relative country positions in 2002 and annual growth rate, 1997–2002[1]

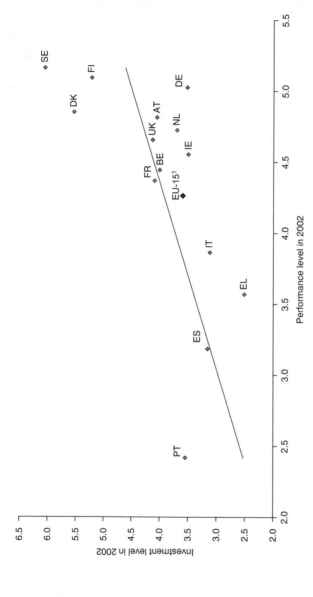

Note: 1 EU-15 does not include LU.

Source: DG Research, Key Figures 2005.
Data: Eurostat, OECD, DG Information Society.

Figure 2.2 Investment versus performance in the knowledge-based economy: relative country positions in 2002

(Sweden, Finland and Denmark) are also high performers, but they stay below their expectations (indicated by the correlation line). The same is true at the other end of the distribution, where Portugal's performance clearly stays behind its relatively high investment levels. Better in terms of performance, compared with their investments, are Germany, the Netherlands, Ireland and Austria. Obviously, in these countries there are also other factors at work, having an impact on the knowledge economy performance, besides the investments. These can be a higher efficiency of individual actors (firms, knowledge providers), or a better interplay and coherence of the overall innovation system (Lundvall, 1992; Edquist, 2005).

In the following, we are going to look at some selected indicators of investment into, and performance of, knowledge economies. One of the key indicators regarding knowledge generation is the *R&D intensity* of an economy (GERD (Gross Expenditure on R&D) as a percentage of GDP: see Figure 2.3). For the year 2003 we can observe that the EU-25 with 1.93 clearly lags behind the USA (2.59) and Japan (3.15). Only the Nordic countries (Sweden, Finland and Denmark) have higher or similar R&D levels in comparison to the USA and Japan. Austria is with 2.37 below the USA but above EU-25, whereas the UK stays below the EU-25 average. Austria's R&D intensity has grown, at 5 per cent per year, quite strongly in the period 1997–2003, whereas the UK's R&D intensity grew only around 1 and 2 per cent per year from 1997 to 2003 (EC, 2005: 24).

Another relevant knowledge economy indicator refers to the *share of highly qualified scientific and technical workers* as a percentage of the total labour force (see Figure 2.4). The highest shares (more than 20 per cent) are again to be found in the Nordic countries (Denmark, Sweden and Finland). The UK with 16.2 per cent also has a high share of such highly qualified workers, whereas Austria with a share of 9.4 per cent is well below the EU-25 average of 13.8 per cent. It is surpassed even by many new EU entrants such as Estonia, Slovenia or Poland, and by cohesion countries such as Greece and Spain in this respect. A major weakness of the Austrian knowledge and innovation system, thus, seems to be the lack of such higher qualified scientific and technical workers.

Also with regard to *high-tech venture capital* (per 1000 GDP) we can identify a European as well as an Austrian weakness (see Figure 2.5). The USA is well ahead of the European countries, and in particular venture capital for the expansion phase of (new) companies is abundant. Within Europe, the UK follows the three Nordic countries, whereas Austria's venture capital is rather weak in international comparison. In particular, venture capital for the expansion phase seems to be rare in Austria.

Investments and inputs are important preconditions for moving towards a knowledge economy. However, of key interest is the performance of

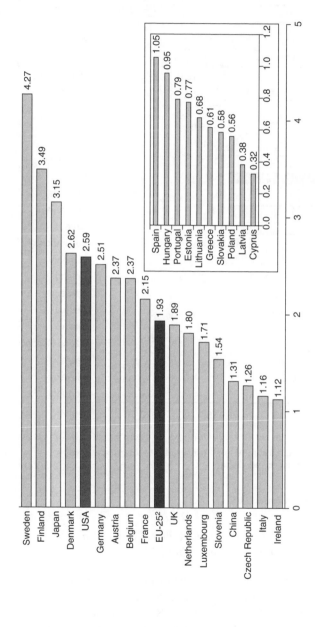

Notes:
1 LU: 2000; SE: 2001; IE, IT, NL: 2002; BE: 2004; AT: 2005.
2 EU-25 was estimated by DG Research and does not include LU and MT.

Source: DG Research, Key Figures 2005.
Data: Eurostat, OECD.

Figure 2.3   R&D intensity (GERD as a percentage of GDP), 2003[1]

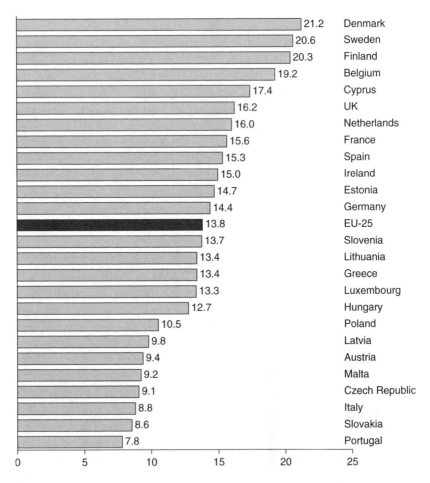

21.2	Denmark
20.6	Sweden
20.3	Finland
19.2	Belgium
17.4	Cyprus
16.2	UK
16.0	Netherlands
15.6	France
15.3	Spain
15.0	Ireland
14.7	Estonia
14.4	Germany
13.8	EU-25
13.7	Slovenia
13.4	Lithuania
13.4	Greece
13.3	Luxembourg
12.7	Hungary
10.5	Poland
9.8	Latvia
9.4	Austria
9.2	Malta
9.1	Czech Republic
8.8	Italy
8.6	Slovakia
7.8	Portugal

0   5   10   15   20   25

Notes:
1 Highly-qualified scientific and technical workers refer here to the group of people educated *and* employed in Science and Technology.
2 NL: 2002.

Source: DG Research, Key Figures 2005.
Data: Eurostat.

Figure 2.4   Highly-qualified scientific and technical workers[1] as a percentage of the total labour force, 2003[2]

economies in terms of knowledge output, innovation and competitiveness. Relevant indicators in this respect are publications, patents, high-tech manu-facturing exports and productivity growth. Regarding *publications* per million of population (Figure 2.6) we again find that the USA is publishing

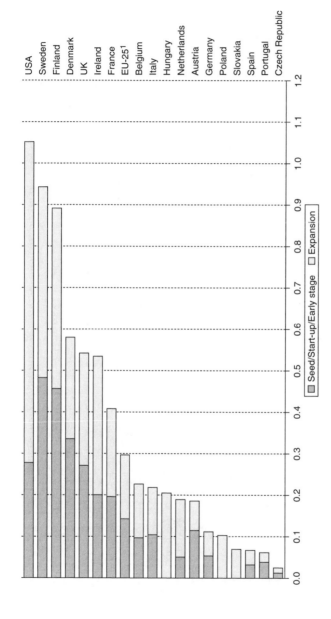

Note: 1 EU-25 does not include EE, LU, CY, LV, LT, LV, MT and SI.

Source: DG Research, Key Figures 2005.
Data: PriceWaterhouseCoopers (Moneytree Survey, Money for Growth 2004).

Figure 2.5 High-tech venture capital by stage per 1000 GDP, 2003

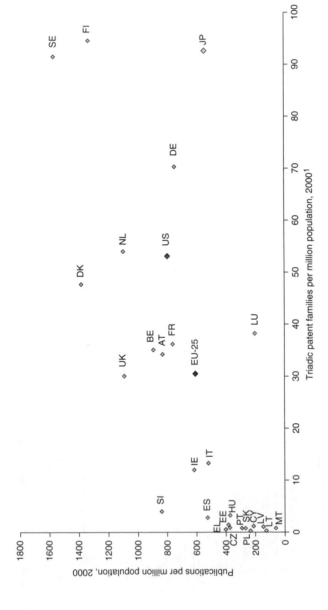

Note: 1 Data by earliest priority date and country of residence of the inventors.

Source: DG Research, Key Figures 2005.
Data: Eurostat, OECD, Thomson Scientific/CWTS, Leiden University.

Figure 2.6 Triadic patent families and publications per million population, 2000

relatively more than Europe. However, there are a number of European countries (the Nordic countries, the Netherlands, the UK and Belgium) which are more active than the USA. Austria, together with Germany and France, has about the US level of publications per 1000 of population.

Triadic patents (per million of population) are also shown for the year 2000 in Figure 2.6. There are some interesting differences compared to the publication activity: Japan and Germany, and to a lesser extent the USA, seem to favour knowledge application. Their patenting activity is clearly better than their publication record. On the other hand, the UK and Denmark seem to follow a more basic science approach: they are patenting fewer than their publication record would suggest. Austria has fewer patents than the leading countries, but in relative terms more than the UK and the EU-25.

An important indicator, although disputed (see Smith, 2002), of how far a country has moved towards a knowledge-based economy[1] is the share and performance of knowledge-based sectors, in particular *high-tech manufacturing*. In terms of the employment share of high- and medium-tech industries we find for the year 2001 surprisingly many Eastern European countries on top of the ranking (Czech Republic 9.3 per cent, Slovenia 8.6 per cent, Hungary 7.1 per cent). This probably results both from foreign direct investment (FDI) inflows and endogenous sectors in those countries. From the old EU, leading countries are Germany (9.25 per cent) and Sweden (8.3 per cent) as well as Finland (6.9 per cent) and Ireland (6.9 per cent: see EC, 2003a: 80). Interestingly, both Austria (5.2 per cent) and the UK (5.1 per cent) have below average employment shares of those industries.

Despite these low employment shares, the *export performance of high-tech industries* seems to be better in the UK and Austria (Figure 2.7). In the UK the share of high-tech exports reaches 27 per cent of all manufacturing exports and is above the EU-25 average of 20 per cent. Austria holds an intermediate position with 15 per cent of high-tech exports. In fact, from 1997 to 2002 Austria saw a strong growth of its world market share in high-tech exports (annual growth of 11.6 per cent), whereas the UK has had a slight decrease in the same indicator (−0.6 per cent: see Figure 2.8). It is interesting to observe that world market shares of high-tech exports grew in particular in countries of Eastern and Southern Europe, as well as in Ireland between 1997 and 2002, whereas the established countries in those industries, like the USA, Sweden and Japan, lost world market shares of exports from −4.5 per cent to −6.2 per cent annually.

We find a different picture for the shares of *knowledge-intensive services*. These are broadly defined, since they reach 37 per cent in the EU-15 (EC, 2003a). Here, the UK has a clear lead among the EU-15 countries

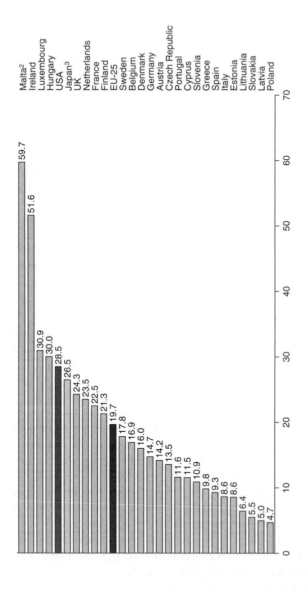

Notes:
1 The value for EU-25 does not include intra-EU-25 exports.
2 Data for MT refer to 2001.
3 Data for JP refer to 2002.

Source: DG Research, Key Figures 2005.
Data: Eurostat (Comext), UN (Comtrade).

Figure 2.7 High-tech manufacturing industries: exports as a percentage of total manufacturing exports, 2003[1]

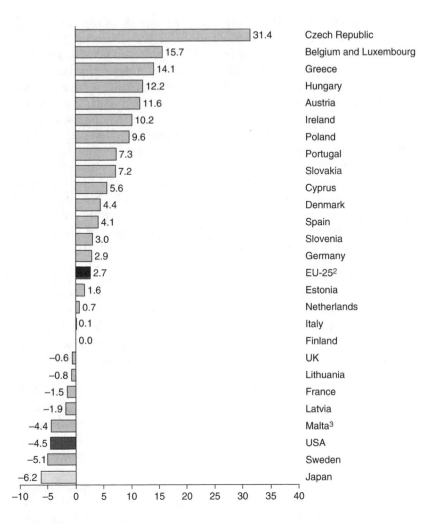

31.4	Czech Republic
15.7	Belgium and Luxembourg
14.1	Greece
12.2	Hungary
11.6	Austria
10.2	Ireland
9.6	Poland
7.3	Portugal
7.2	Slovakia
5.6	Cyprus
4.4	Denmark
4.1	Spain
3.0	Slovenia
2.9	Germany
2.7	EU-25[2]
1.6	Estonia
0.7	Netherlands
0.1	Italy
0.0	Finland
−0.6	UK
−0.8	Lithuania
−1.5	France
−1.9	Latvia
−4.4	Malta[3]
−4.5	USA
−5.1	Sweden
−6.2	Japan

Notes:
1 All data include intra-EU-25 high-tech exports and the world market refers to total world high-tech exports including intra-EU-25 exports.
2 EU-25 does not include MT.
3 Data for MT refer to 1996–2001.

Source: DG Research, Key Figures 2005.
Data: Eurostat (Comext), UN (Comtrade).

Figure 2.8 *High-tech manufacturing industries: world market shares of exports, average annual growth (percentage), 1997–2002[1]*

regarding share of value-added (61.3 per cent) and also a top rank regarding the employment share (48.8 per cent: see EC, 2003a: 82f). Austria, however, is at the bottom of the league (23.7 per cent for value-added; 27 per cent for employment), with only Italy and Portugal lying further behind. While the UK, thus, has strengths in knowledge-intensive services, Austria so far has a rather small high-technology sector, in particular in services.

The share and growth of knowledge-intensive sectors is a relevant aspect of a country moving towards a knowledge economy. However, as Smith (2002) has observed, it does not capture knowledge and innovation processes in the rest of the economy. In this respect, *productivity changes* in the overall economy, as well as in ICT-related sectors, are of interest. Table 2.1 shows data for the periods 1990–95 and 1996–2002. In the first phase, Sweden, Finland and Italy were the leaders, with productivity growth rates between 2.65 per cent and 2.95 per cent. The UK (2.2 per cent) and Austria (2.3 per cent) also had a good performance compared to countries such as France, Spain or the Netherlands. Most European countries, in fact, had a higher productivity growth than the USA (1.12 per cent) in this period. The picture changed for the period from 1995–2002 as shown in Chapter 1. The US's productivity growth rate, owing to the New Economy boom, went up to 1.74 per cent, whereas in most European countries except Ireland, Sweden and Finland, the growth rate declined strongly. This was also the case in the UK (1.08 per cent), but less so in Austria (1.73 per cent).

Regarding the role of ICT for productivity changes Pilat and Wölfl (2004) have calculated the respective contributions (see Table 2.1). We can observe that Ireland, Finland and Sweden benefited considerably from productivity growth in ICT-producing manufacturing, in particular in the period from 1995 to 2002 (between 0.5 and 0.9 percentage points). To a lesser extent this is also true for the USA (0.45 per cent). Information and communication technology producing services demonstrated strong productivity growth in Germany, Finland and the UK from 1995 to 2002 (between 0.24 and 0.46 percentage points), whereas in the USA this factor was negligible. Strong productivity growth in ICT-using services occurred in the UK (0.85 percentage points), Ireland (0.73 per cent), Sweden (0.60 per cent) as well as in Austria (0.51 per cent).

By contrast with Sweden, Finland and Ireland, overall productivity growth in the UK and in Austria is not strongly related to the ICT manufacturing sector. The UK, however, clearly has strengths in the ICT producing and using services. Austria's relatively good overall productivity performance results less from ICT production in manufacturing or services. To some extent it results from ICT use in a variety of sectors, also in services and from other factors.

Table 2.1 Sectoral contribution to labour productivity growth in selected EU countries and the USA, 1990–95[1] and 1996–2002[2] (total economy, value-added per person employed, contribution in percentage points)

	Total economy		ICT – producing manufacturing[3]		ICT – producing services[4]		ICT – using services[5]	
	1990–1995	1996–2002	1990–1995	1996–2002	1990–1995	1996–2002	1990–1995	1996–2002
Belgium	1.90	0.78	0.03	0.13	0.12	0.05	0.47	0.17
Denmark	1.99	1.45	0.09	0.09	0.27	0.13	0.18	0.37
Germany	2.11	1.38	0.17	0.09	0.18	0.46	0.17	0.12
Spain	1.22	0.28	0.14	0.01	0.09	0.16	−0.17	−0.03
France	1.13	1.00	0.20	0.21	0.02	0.14	0.01	−0.17
Ireland	2.39	3.76	0.43	0.89	0.10	0.28	0.15	0.73
Italy	2.83	0.56	0.09	0.02	0.12	0.20	0.88	0.14
Luxembourg	2.08	0.51	−0.03	−0.01	0.74	0.32	1.13	−0.20
Netherlands	0.63	0.77	0.10	0.03	0.09	0.17	0.25	0.28
Austria	2.32	1.73	0.12	0.11	0.15	0.13	0.59	0.51
Finland	2.65	2.02	0.20	0.82	0.13	0.36	0.10	0.22
Sweden	2.95	2.67	0.27	0.51	0.24	0.22	0.45	0.60
UK	2.20	1.08	0.19	0.12	0.18	0.24	0.37	0.85
USA	1.12	1.74	0.33	0.45	0.14	0.16	0.24	1.29

Notes:
1 DE: 1991–95; FR, IT: 1992–96.
2 SE: 1996–98; ES: 1996–99; IE: 1996–2000; FR, UK, USA: 1996–2001.
3 ISIC Rev 3 30–33.
4 ISIC Rev 3 64 and 72.
5 ISIC Rev 3 71–74.

Source: Pilat and Wölfl (2004).
Data: OECD.

SUMMARY

In this chapter we have tried to understand and conceptualize the knowledge economy. We have argued that there are various views and perspectives on the emerging knowledge economy. It can be characterized as an economy where knowledge inputs are key, and where knowledge as a product (knowledge-intensive services, high-tech products) is central. Also, an increasing trend towards codification and ICT development and use are relevant features. At the core, however, is Castells's perspective that it is the 'action of knowledge upon itself', that is, the exploitation of existing and new knowledge in order to create more new knowledge.

Then we have dealt with the nature and types of knowledge. We have argued that, different from 'passive' information as unstructured messages, knowledge should be understood as a cognitive capacity which empowers its possessors for various kinds of actions. We have also pointed out that the knowledge process is a social process involving a dynamic interplay and exchange of implicit (tacit) and explicit (codified) forms of knowledge among various actors. The transformation of implicit to explicit knowledge, however, often needs complicit knowledge and respective competences as interfaces in order to make knowledge generation, application and exploitation effective.

Knowledge is a social product since it is typically exchanged in 'communities' of various kinds. Proximity is a key feature of such communities providing nearness, a common understanding and possibly a sense of neighbourhood or identity. The forms of proximity may be quite diverse, however. In addition to geographical proximity these may be of organizational, institutional, cultural or virtual kind. Geographical proximity, thus, is not alone in creating a common understanding for tacit knowledge exchange as has often been claimed in the literature. However, as the literature shows, it is still highly important in this respect. This is reflected in the highly uneven and polarized character of knowledge generation, application and innovation in geographical space. Innovative milieux, high-tech regions and knowledge-based city regions are expressions of such asymmetries and knowledge monopolies. These regions and areas should not be understood, however, as a result of only geographical proximity and regional features, but in the sense of Amin and Cohendet (2004) as nodes where various kinds of knowledge communities and networks intersect. Large cities and clusters at the same time have to be seen as local concentrations of knowledge providers such as universities, research organizations, and firms from which various kinds of knowledge spillovers and knowledge links emanate.

Looking at recent empirical evidence provided by the European Commission and the OECD, we find that Europe in many aspects lags behind the USA and Japan in its development towards a knowledge economy. The gap is particularly large in terms of R&D activities in the business sector as well as in terms of venture capital in high-technology, specifically for the expansion phase of new companies. The gap is smaller, but can also be observed in terms of publications and triadic patents (relative to population) and in the share of high-tech manufacturing exports. The stronger investment into the knowledge economy in the USA seems to be reflected also in a better productivity performance since 1995.

Within Europe we can see a marked North–South divide in terms of the knowledge economy. Sweden, Finland and Denmark are clearly the leading countries both in terms of investment and performance. It is interesting to observe, however, that some of the countries with moderate investment in the knowledge economy, as, for example, Austria, have a relatively good performance in some of the indicators.

Austria has considerably improved its position with respect to R&D intensity since 1997, but it still has weaknesses regarding the share of highly qualified scientific and technical workers, the venture capital situation and the share of knowledge-intensive sectors, both in manufacturing and in services. Despite these weaknesses it does remarkably well in terms of scientific publications, patents and productivity growth. Clearly, its economy benefits from knowledge application and innovation in many sectors, not just high-technology, as well as from the use of ICT in the services and other areas.

The UK in some respects seems to have moved more towards a knowledge economy. The most obvious strengths are in knowledge-intensive services and in the ICT producing and using services. It has also relatively high shares of highly qualified scientific and technical workers and a comparatively good venture capital situation. On the other hand, it has a surprisingly low R&D intensity in the business sector, a relatively weak patent performance and a rather strong decline in productivity growth.

NOTE

1. For the difference between 'knowledge economy' and 'knowledge-based economy' see Cooke and Leydesdorff (2006).

REFERENCES

Amin, A. and Cohendet, P. (2004), *Architectures of Knowledge – Firms, Capabilities, and Communities*, Oxford: Oxford University Press.

Archibugi, D. and Lundvall, B.-A. (eds) (2002), *The Globalizing Learning Economy*, Oxford: Oxford University Press.

Audretsch, D. (1998), 'Agglomeration and the location of innovative activity', *Oxford Review of Economic Policy*, **14**, 18–29.

Baptista, R. and Swann, P. (1998), 'Do firms in clusters innovate more?', *Research Policy*, **27**, 525–40.

Boschma, R. (2005), 'Proximity and innovation: a critical assessment', *Regional Studies*, **39**, 61–74.

Breschi, S.L. and Lissoni, F. (2001a), 'Knowledge spillovers and local innovation systems: a critical survey', *Industrial and Corporate Change*, **10**(4), 975–1005.

Breschi, S. and Lissoni, F. (2001b), 'Localised knowledge spillovers versus innovative milieux: knowledge "tacitness" reconsidered', *Papers in Regional Science*, **80**, 255–73.

Brower, E., Budil-Nadvornikowa, H. and Kleinknecht, A. (1999), 'Are urban agglomerations a better breeding place for product innovation? An analysis of new product announcements', *Regional Studies*, **33**, 541–9.

Cairncross, F. (1997), *The Death of Distance: How Communications Revolutions will Change our Lives*, Boston, MA: Harvard Business School Press.

Capello, R. (1999), 'SME clustering and factor productivity: a milieu production function model', *European Planning Studies*, **7**, 719–35.

Castells, M. (1996), *The Rise of the Network Society*, Oxford: Blackwell.

Cooke, P. (2002), *Knowledge Economies*, London: Routledge.

Cooke, P. (2004), 'Regional knowledge capabilities, embeddedness of firms and industry organisation: bioscience megacentres and economic geography', *European Planning Studies*, **12**(5), 625–41.

Cooke, P. (2005a), 'Global bioregional networks: a new economic geography of bioscientific networks', paper presented at the Kiel Institute for World Economics, April.

Cooke, P. (2005b), 'Regionally asymmetric knowledge capabilities and open innovation: exploring "Globalisation 2" – a new model of industry organisation', *Research Policy*, **34**, 1128–49.

Cooke, P. (2007), *Growth and Cultures: the Globalisation of Bioregions*, London: Routledge.

Cooke, P. and Leydesdorff, L. (2006), 'Regional development in the knowledge-based economy: constructing regional advantage', *The Journal of Technology Transfer*, **31**(1), 5–15.

Cowan, R.C., David, P. and Foray, D. (2000), 'The explicit economics of knowledge codification and tacitness', *Industrial and Corporate Change*, **9**(2): 212–53.

Crevoisier, O. (2001), 'Der Ansatz des kreativen Milieus', *Zeitschrift für Wirtschaftsgeographie*, **45**, 246–56.

David, P. and Foray, D. (2003), 'Economic fundamentals of the knowledge society', *Policy Futures in Education*, **1**, 20–49.

Drucker, P. (1998), 'From capitalism to knowledge society', in D. Neef (ed.), *The Knowledge Economy*, Woburn, MA: Butterworth.

Dunning, J. (ed.) (2000), *Regions, Globalization, and the Knowledge-Based Economy*, New York: Oxford University Press.

Edquist, C. (2005), 'Systems of innovation – perspectives and challenges', in J. Fagerberg, D. Mowery and R. Nelson (eds), *The Oxford Handbook of Innovation*, Oxford: Oxford University Press, pp. 181–208.

European Commission (EC) (2003a), *Towards a European Research Area, Key Figures 2003–2004, Science, Technology and Innovation*, Brussels: DG Research.

European Commission (2003b), 'European innovation scoreboard: technical paper no 3, regional innovation performances', European Commission, Brussels.

European Commission (2005), *Key Figures 2005 on Science, Technology and Innovation, Towards a European Knowledge Area*, Brussels: DG Research.

Feldman, M. (2000), 'Location and innovation: the new economic geography of innovation, spillovers, and agglomeration', in G. Clark, M. Feldman and M. Gertler (eds), *The Oxford Handbook of Economic Geography*, Oxford: Oxford University Press, pp. 373–94.

Feldman, M. and Audretsch, D. (1999), 'Innovation in cities: science-based diversity, specialisation and localised competition', *European Economic Review*, **43**, 409–29.

Gehrke, B. and Legler, H. (2001), *Innovationspotenziale deutscher Regionen im europäischen Vergleich*, Berlin: Duncker and Humblot.

Gertler, M. (2003), 'Tacit knowledge and the economic geography of context or the undefinable tacitness of being (there)', *Journal of Economic Geography*, **3**, 75–99.

Glaeser, E., Kallall, H., Scheinkman, J. and Shleifer, A. (1992), 'Growth in cities', *Journal of Political Economy*, **100**, 1126–52.

Grabher, G. (1993), 'The weakness of strong ties: the lock-in of regional development in the Ruhr-area', in G. Grabher (ed.), *The Embedded Firm: On the Socioeconomics of Industrial Networks*, London: T.J. Press, pp. 255–78.

Hassink, R. and Shin, D.-O. (2005), Guest editorial: 'The restructuring of old industrial areas in Europe and Asia', *Environment and Planning A*, **37**, 571–80.

Henderson, V. (2003), 'Marshall's scale economies', *Journal of Urban Economics*, **53**, 1–28.

Johnson, B., Lorenz, E. and Lundvall, B.-A. (2002), 'Why all this fuss about codified and tacit knowledge?', *Industrial and Corporate Change*, **11**, 245–62.

Kaufmann, A., Lehner, P. and Tödtling, F. (2003), 'Effects of the internet on the spatial structure of innovation networks', *Information Economics and Policy*, **15**, 402–24.

Keeble, D. and Wilkinson, F. (eds) (2000), *High-Technology Clusters, Networking and Collective Learning in Europe*, Aldershot: Ashgate.

Laafia, I. (2002), 'National and regional employment in high tech and knowledge intensive sector in the EU – 1995–2000', Statistics in Focus, Science and Technology, Eurostat.

Lundvall, B.-A. (ed.) (1992), *National Systems of Innovation: Towards a Theory of Innovation and Interactive Learning*, London: Pinter.

Lundvall, B.-A. and Borrás, S. (1999), *The Globalising Learning Economy: Implications for Innovation Policy*, Luxembourg: Office for Official Publications of the European Communities.

Lundvall, B.-A. and Johnson, B. (1994), 'The learning economy', *Journal of Industry Studies*, **1**, 23–42.

Maier, G. and Tödtling, F. (2005), *Regional- und Stadtökonomik 1 – Standorttheorie und Raumstruktur*, 5th edition, Wien: Springer Verlag.

Maillat, D., Léchot, G., Lecoq, B. and Pfister, M. (1996), 'Comparative analysis of the structural development of milieux: the example of the watch industry in the Swiss and French Jura Arc', Working Paper 96-07, Institut de recherches économiques et régionales, Université de Neuchatel, Neuchatel.

Marshall, A. (1918), *Industry and Trade*, London: Macmillan.

Morgan, K. (2004), 'The exaggerated death of geography', *Journal of Economic Geography*, **4**, 3–21.

Nonaka, I. and Takeuchi, H. (1995), *The Knowledge-Creating Company*, Oxford: Oxford University Press.

Organisation for Economic Co-operation and Development (OECD) (1996), *The Knowledge-Based Economy*, Paris: OECD.

Organisation for Economic Co-operation and Development (OECD) (2001), *OECD Science, Technology and Industry Scoreboard. Towards a Knowledge-Based Economy*, Paris: OECD. Also available at www1.oecd.org/publications/e-book/92-2001-04-1-2987/.

Pilat, D. and Wölfl, A. (2004), 'ICT production and ICT use: what role in aggregate productivity growth', in OECD, *The Economic Impact of ICT*, Paris: OECD.

Polanyi, M. (1966), *The Tacit Dimension*, London: Routledge.

Porter, M. (1990), *The Competitive Advantage of Nations*, New York: Free Press.

Ratti, R., Bramanti, A. and Gordon, R. (eds) (1997), *The Dynamics of Innovative Regions: The GREMI Approach*, Aldershot: Ashgate.

Saxenian, A. (1994), *Regional Advantage: Culture and Competition in Silicon Valley and Route 128*, Cambridge, MA: Harvard University Press.

Simmie, J. (2003), 'Innovation and urban regions and national and international nodes for the transfer and sharing of knowledge', *Regional Studies*, **37**, 607–20.

Smith, K. (2002), 'What is the "Knowledge Economy"? Knowledge intensity and distributed knowledge bases', United Nations University, Institute for New Technologies, Discussion Paper Series, Maastricht.

Soete, L. (2002), 'The new economy: a European perspective', in D. Archibugi and B.-A. Lundvall (eds), *The Globalizing Learning Economy*, Oxford: Oxford University Press.

Steiner, M. (ed.) (1998), *Clusters and Regional Specialisation*, London: Pion.

Sternberg, R. and Litzenberger, T. (2004), 'Regional clusters in Germany – their geography and relevance for entrepreneurial activities', *European Planning Studies*, **12**, 767–91.

Storper, M. (1997), *The Regional World*, New York: Guilford Press.

Storper, M. (2002), 'Institutions of the learning economy', in M. Gertler and D. Wolfe (eds), *Innovation and Social Learning: Institutional Adaption in an Era of Technological Change*, Basingstoke: Palgrave.

Tichy, G. (2001), 'Regionale Kompetenzzyklen – Zur Bedeutung von Produktlebenszyklus- und Clusteransätzen im regionalen Kontext', *Zeitschrift für Wirtschaftsgeographie*, **45**, 181–201.

Tödtling, F., Lehner, P. and Trippl, M. (2006), 'Innovation in knowledge intensive industries – the nature and geography of knowledge links', *European Planning Studies*, **14**(8), 1035–58.

Zeller, C. (2004), 'North Atlantic innovative relations of Swiss pharmaceuticals and the proximities with regional biotech areas', *Economic Geography*, **80**, 83–111.

Zitt, M., Ramanana-Mahary, S., Bassecoulard, E. and Laville, F. (2003), 'Potential science-technology spillovers in regions: an insight on geographical co-location activities in the EU', *Scientometrics*, **57**, 295–320.

3. Knowledge-based sectors: key drivers of innovation and modes of knowledge exchange

INTRODUCTION

In the previous chapter we have investigated what the knowledge economy is, pointing out that it combines an increasing importance of knowledge as input, an increasing share of sectors where knowledge constitutes an output, and a stronger role of codified knowledge as well as of ICT in the process of generation and application of knowledge. In a dynamic way the knowledge economy can also be characterized by the exploitation of existing and new knowledge in order to create more new knowledge. The empirical evidence that is available so far has demonstrated that the European countries, in fact, have been moving towards knowledge economies in terms of inputs, outputs and the use and application of ICT. However, such a move has occurred with different speed within Europe and in most countries it is slower than in the USA. There are also considerable regional imbalances involved since knowledge-based industries tend to cluster in specific regions, cities or clusters.

In this chapter we look more closely at those sectors where knowledge is both an important input and output. We first deal with the key actors and drivers for the development of knowledge-based sectors and of innovation. An important role is held by the science sector and universities, but of course ultimately the companies are key actors responsible for the application and commercialization of knowledge. These may be large and global firms with large R&D budgets and competencies, but increasingly they are innovative SMEs or spin-off companies, often supported by venture capitalists and other services. Also, the state has a relevant and partly new role in the knowledge economy in support of knowledge activities. We then deal with differences between synthetic, analytic and symbolic knowledge bases and the implications for the knowledge process in respective industries. The different types of knowledge base imply specific kinds of knowledge and its sources as well as different ways of innovating. Focusing on knowledge-based industries we subsequently deal with the commercialization of scientific knowledge, university–industry links and academic spin-offs, pointing out supporting factors for such activities. In the final section of this chapter we look at the prevailing types of innovation interactions of firms, classifying them according to their static/dynamic and formal/informal nature. We differentiate between market links and networks, knowledge spillovers and milieu, and we discuss the geographical levels of those links.

KEY DRIVERS OF INNOVATION AND THE KNOWLEDGE ECONOMY

Since the late nineteenth century the perspective on key actors and drivers of knowledge generation and application as well as of innovation has considerably changed. For the early Schumpeter (Mark I) the *'heroic' entrepreneur* was the main driving force both for innovation and for economic development (Schumpeter, 1911). Schumpeterian entrepreneurs were looking for pioneer profits by introducing new combinations of production factors such as new products, processes and organizations, or by opening up new markets. Owing to such innovations, economic development was stimulated through new and growing markets and productivity improvements as well as through sectoral and organizational change. For Schumpeter such a process was, however, in no way smooth and beneficial for all participants. To the contrary, he saw it as a process of 'creative construction' where older firms or sectors were out-competed by innovators and their new technologies. Unlike Alfred Marshall (1918), who saw positive effects of local industry concentrations for the efficiency of firms and for knowledge transmission, Schumpeter did not analyse the social or spatial conditions for innovation and entrepreneurship but was more interested in the role of outstanding individual entrepreneurs for economic development.

In the first half of the twentieth century and with the growth of the large firm, the perspective on the drivers of innovation has changed. These were often identified with vertically integrated *large firms*, such as Ford, which gave the name to the accumulation regime of 'Fordism' as deployed by the 'regulation school'. Key drivers of the knowledge and innovation process at that time, thus, were considered to be large corporations, for example, in the automotive, chemical and electrical goods industries, and their R&D departments. This was pointed out by, for example, the 'late' Schumpeter (Mark II: Schumpeter, 1942) and by Chandler (1962). In these firms there was generally a clear distinction and separation of conception (that is knowledge-related activities) from execution. This implied a strict separation of management, product development and production as well as the organization of the work process according to the principles of 'scientific management' (Taylorism). The linear innovation model was based on the strong role of R&D in large firms and R&D organizations, viewing them as the most important starting points and drivers for innovation and for the introduction of new products or processes.

In studies on industrial districts (Beccatini, 1991; Asheim, 1996), innovative mileux (Aydalot and Keeble, 1988; Maillat, 1998) and technology districts (Storper, 1995; 1997), undertaken since the 1970s, small firms have moved to centre stage again – this time not the individual entrepreneurs but ensembles or *networks of SMEs*, embedded into the social fabric of particular regions or localities. According to the underlying *interactive* innovation model, the sources of new ideas and knowledge became more diverse and reached beyond R&D. Innovations were regarded as a result of interactive learning jointly with customers, suppliers, service companies and external knowledge providers. The drivers of innovation, thus, were to be found in networks of firms both at local and higher levels (Camagni, 1991; Asheim, 1996; Cooke and Morgan, 1998). Whereas the earlier work on districts and milieux largely focused on networks of SMEs, later studies pointed out that networks may differ considerably, both in terms of firm size (large and small firms) and in terms of governance modes (egalitarian/hierarchical governance: De Bresson and Amesse, 1991; Storper and Harrison, 1991; Powell, 1998). Egalitarian networks are often regarded as advantageous for innovation in the long run, owing to a more diverse knowledge base (higher redundancy) and a lower likelihood for 'lock in' compared with networks which are dominated by large focal firms (Storper and Harrison, 1991; Grabher, 1993; Tichy, 2001).

More recently, the literature on *innovation systems* (IS), brought the role of the state and other public institutions into attention. The detailed picture of relevant institutions and organizations has varied, however, between sectoral, national or regional variants of the innovation systems approach (Lundvall, 1992; Edquist, 1997; 2005; Cooke et al., 2000; Doloreux, 2002; Cooke et al., 2004). The actors of the knowledge and innovation process are more diverse than in the earlier approaches. They are to be found as core elements in the subsystem of knowledge generation and diffusion (universities and research organizations, complemented by transfer agencies and intermediaries, schools and training organizations) as well as in the subsystem of knowledge application and commercialization. Despite the institutional focus of the IS approach, the latter subsystem is considered the key driver by some observers such as Etzkowitz and Leydesdorff (2000). It is made up by the main clusters of firms of a region or country, consisting of leading firms in a particular industry, vertically and horizontally related companies, service providers and supporting institutions (see Figure 3.1).

According to the regional innovation system (RIS) approach various support organizations and policy actions may promote learning and innovation at the regional level. The setting up or expansion of universities and research institutions, science parks, innovation centres, technology transfer agencies and educational institutions can stimulate and enhance the production, diffusion and application of knowledge. An important precondition is, however, that these organizations develop dense links to the firms of the respective regions and beyond. Other important organizations supporting innovation-based growth include venture capital firms, business angels, standard-setting bodies and development agencies. The regional innovation system approach highlights that regional authorities can shape local learning and innovation processes in a significant way by providing R&D infrastructure and educational infrastructure, supporting academic spin-offs, enhancing human capital and encouraging the formation of social capital (Cooke et al., 2000; Doloreux, 2002; Tödtling and Trippl, 2005).

The key role of *science, universities and public research organizations* in the emerging knowledge economy has been pointed out in particular by Gibbons et al. (1994) and by Etzkowitz et al. (2000). For Gibbons and his collegues the process of knowledge generation has been changing fundamentally from a more traditional and disciplinary model (Mode 1), where knowledge was produced in university settings, to a more recent Mode 2 where knowledge sources are widely distributed (government laboratories, industries, think tanks) and new knowledge is generated in trans-disciplinary contexts of application. Research organizations, firms and users of knowledge, thus, are involved in an interactive process of knowledge generation and application. Etzkowitz et al. (2000) also demonstrate a new role of universities in their 'triple helix' approach. Universities in the knowledge economy, they argue, are increasingly moving towards entrepreneurial activities, encompassing a 'third mission' in addition to research and teaching. According to the 'triple helix' model, universities are translating research into economic development through various forms of technology transfer, and they assist also the modernization of low- and mid-tech firms.

In reality, however, such knowledge flows are far from smooth, since there are substantial barriers to technology transfer, in particular as regards SMEs in low-tech industries (Cooke et al., 2000; Tödtling and Kaufmann, 2001; Asheim et al., 2003; Trippl, 2004). These barriers result from differences in focus among universities and firms (basic versus applied research), motivation and incentives (publication versus private appropriation and protection of inventions) as well as of concepts and language

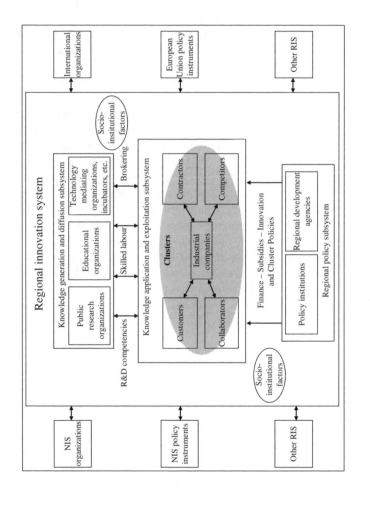

Note: NIS – national innovation system.

Figure 3.1   Diagrammatic representation of a regional innovation system

(scientific versus practical). As it turns out, science and business operate as distinct systems according to different principles and rules, and they apply different concepts and ways of finding solutions.

Despite such interaction barriers between universities and specific firm segments, we are indeed observing an increasing overlap between teaching, research and business in recent years, as Chapter 2 and other empirical work has shown. There are more science–industry links, R&D cooperations, mobility of researchers and qualified staff, and spin-offs nowadays than in the past. In a knowledge economy, universities are assuming a more important and partly new role as knowledge producing and disseminating institutions. They become key elements both as human capital providers and as seedbeds for new firms (see the section on spin-offs below).

The development towards entrepreneurial universities is not uncontested, however. There are also voices and arguments saying that universities should focus more on their original tasks, namely, teaching, qualification and doing critical research instead of getting too involved in the setting up of new ventures and in intensive collaborations with corporations. According to this view it is exactly the difference between science and business systems which might stimulate innovations of a more radical nature. Still, differences between science and business systems can only become productive if those different worlds are able to 'meet' and if 'boundary crossing' occurs (Kaufmann and Tödtling, 2001). In the emerging knowledge economy we can observe an increasing variety of such boundary crossings and interfaces, such as mobile labour and talent, technology licensing, R&D cooperations, spin-off companies or informal knowledge transfer and networks as pointed out further below.

TYPES OF KNOWLEDGE BASE

The findings and arguments so far suggest that the knowledge and innovation process has in recent years become increasingly complex. There is a larger variety of knowledge sources and inputs to be used by organizations and firms, and knowledge bases have become more distributed (Smith, 2002). There are more and partly new players on stage, and older players such as universities have taken on new roles and activities. There is in general more interdependence and division of labour among actors (individuals, companies, universities, state and other organizations). The process of knowledge exploration and exploitation requires a dynamic interplay and transformation of tacit and codified forms of knowledge as well as a strong interaction of people within organizations and among them, as Nonaka and Takeuchi (1995), Lundvall and Borrás (1999),

Gertler (2003) and Amin and Cohendet (2004) have pointed out. The knowledge process, thus, becomes increasingly inserted into various forms of 'complicit' communities and networks, as pointed out further below. Partly these have a territorial nature in the form of local industry clusters or regional and national innovation systems. Partly they are shaped by other principles, as is the case in intra-corporate or professional groups or in virtual communities.

Although there is a generic trend towards increased diversity and inter-dependence in the knowledge process, we argue here that the innovation process of firms is also strongly shaped by their *specific* knowledge base, depending on the industry in question. In accordance with Laestadius (1998), Tödtling et al. (2004) and Asheim and Gertler (2005), we distin-guish between 'synthetic', 'analytical' and 'symbolic' types of knowledge base. These imply different mixes of tacit and codified forms of knowledge, codification possibilities and limits, qualifications and skills required, organizations and institutions involved (networks) as well as innovation challenges and pressures.

Many traditional sectors such as engineering, industrial machinery, special vehicles and instruments are characterized by a *synthetic knowledge base*. In such settings the application of existing knowledge or the new com-bination of available knowledge may lead to innovations. This often occurs in the need to solve actual problems on the shop floor or in interaction with key customers or users and suppliers. Research and development are gen-erally less important than in analytical knowledge base settings, and it is of an applied character, aiming at the development of products or processes rather than at new basic knowledge or radical new solutions. University–industry links are generally less frequent and are focused more on actual problems and on the development of products or processes rather than on basic research.

Accordingly, knowledge is created less through deductive approaches, abstraction or theoretical work, but more often in an inductive process of testing, experimentation, or through practical work. Much of the relevant knowledge is in tacit form, in particular due to the fact that it often results from experience gained at the workplace, and through learning by doing, using and interacting. Increasingly, firms in industries such as engineering and machinery also rely on codified knowledge and ICT which allows them to access and combine available pieces of knowledge in order to create new or customised solutions. Still, the workforce in such industries requires less theoretical and abstract knowledge but, rather, specific know-how, crafts and practical skills which are often provided by professional and polytechnic schools or by on-the-job training. Firms innovate mainly in an incremental way by modifying existing products and processes. Such improvements often target the efficiency and reliability of new solutions, or the practical utility and user-friendliness of products instead of big technological advances examination, through trialing and testing, is of crucial importance.

In economies characterized by the entrepreneurial model, such as the UK or the USA, interaction between users and producers in industries with a synthetic knowledge base tends to be shorter term, with market links predominating. Under the associative model, such as in Germany and Austria, in contrast, close user–producer interaction between partners is more common. These often result from a long, stable history of interactions, rooted in shared conventions, norms, attitudes and values, are often dense technology transfer networks, supported by industrial associations and governments, which enhance the mediating knowledge capabilities associated with complicit knowledge. This is complemented by a well-developed system for general education and vocational training.

This contrasts quite strongly with industrial settings where an *analytical knowledge base* is dominant, such as in genetics, biotechnology and information technology. Here, scientific knowledge and access to respective sources is key. Knowledge exploration is more often based on cognitive and rational processes, analytical techniques or formal models. Basic and applied research as well as systematic development of technologies are among the core activities of firms. Companies, therefore, typically have their own R&D departments, but they also rely on the research results of universities and other research organizations to bring forward innovations. University–industry links and networks to science, as well as academic spin-offs are, consequently, more common than in the synthetic of symbolic knowledge bases. Knowledge inputs and outputs are codified to a greater extent than in the other types, but tacit knowledge is often needed in order to interpret, understand and work with codified knowledge in an appropriate way, as was pointed out in Chapter 2 (Nonaka et al., 2000; Johnson et al., 2002). There are several reasons for the importance of codification: knowledge generation is based on reviews of existing studies and the application of scientific principles and methods. Knowledge processes, furthermore, are rather formally organized (for example, in R&D departments) and outcomes tend to be documented in scientific papers, reports, electronic files or patent descriptions. These activities require specific capabilities such as analytical skills, abstraction, theory building and testing, and documentation. The workforce, therefore, needs university training or research experience to a greater extent. Scientific discoveries and technological inventions are the aims of R&D activities which may lead to patents and licensing activities. New products or processes in such industries tend to be of a more radical type than in the other knowledge bases. They may give rise to technology-based start-ups and spin-off companies, which are frequently supported by universities and other incubators.

The entrepreneurial model as represented by the USA and the UK seems to have a clear advantage for such knowledge-based industries. Elite universities and education institutions, such as Harvard, the Massachusetts Institute of Technology (MIT), Stanford in the USA or Oxford and Cambridge in the UK, provide strengths in R&D, knowledge generation, inventions, and radical innovations. Close university–industry links, academic spin-offs, technology licensing and a high mobility of researchers promote the transfer and application of scientific knowledge. By contrast, the associative model, as represented, for example, by Austria or Germany, tends to give more attention to traditional sectors (which often have a synthetic knowledge base) leading to a slower growth of newly emerging knowledge industries such as biotechnology and ICT. In recent years, however, there were deliberate attempts in countries such as Austria and Germany to catch up in knowledge-intensive sectors both on central and regional levels (see Chapter 8 on Austria). These policies have partly tried to copy features of the entrepreneurial model such as provision of risk capital and support of entrepreneurship. There were also new competitive programmes like the BioRegio Programme in Germany (Dohse, 2000; 2003; Cooke, 2002; Eickelparsch and Fritsch, 2005) or the LISA programme and the Genome Research Programme (GEN-AU) in Austria (Tödtling and Trippl, 2005).

In cultural industries such as media (film making, publishing, music), advertising, design or fashion (Scott, 1997; 2001) we find other types of knowledge to be relevant, which can be characterized by a *symbolic knowledge base*. Knowledge in these industries is related to the aesthetic attributes of products, the creation of designs and images, and the economic use of cultural artefacts. These industries can be regarded as design-intensive since a crucial share of work is dedicated to the creation of new ideas, images and fashions and less to the actual physical production process. Competition shifts from the 'use-value' of products to the 'sign-value' of brands (Lash and Urry, 1994: 122). 'Branding' is therefore a core activity. The knowledge involved is typically incorporated and transmitted in aesthetic symbols, images, (de)signs, artefacts, sounds and narratives, and thus codified. At the same time, however, it is tied to a deep understanding of the habits and norms and 'everyday culture' of specific social groupings and, thus, is culturally embedded. As in the synthetic knowledge base, there is a strong tacit component. Cultural industries rely on aesthetic rather than cognitive inputs, therefore abilities in symbol interpretation rather than information processing and hard data analysis are required. Such creative, imaginative and interpretive skills are acquired less through formal qualifications and university degrees and more through practical experience in the creative process. In cultural industries knowledge of potential collaborators with complementary capabilities and skills (that is, 'Know Who') is often more important than 'Know How'. This is due to the fact that production in such industries, like film production, is often organized in temporary projects (Grabher, 2001; Sydow and Windeler, 2004). These allow the bringing together of a diverse spectrum of professional cultures ranging from the artistic world to the commercial world of business services for the duration of the project. Often, productive tensions and creative conflicts, triggering innovation, characterize such projects (Grabher, 2002).

These different types of knowledge bases, thus, imply specific kinds of knowledge used and respective sources, types of innovation and of innovation relations as well as ways of commercializing new technologies, as argued below. Also the geography of innovation links and the spatial concentration tendency of industries are influenced by their respective knowledge base, as demonstrated in later sections of this chapter and in the empirical chapters.

COMMERCIALIZATION OF KNOWLEDGE AND ACADEMIC SPIN-OFFS

In the following section we take a closer look at the commercialization of knowledge in industries operating on an analytical knowledge base. As stated above, we find here a science-driven innovation model, with a strong role for universities and public research organizations (PROs). There are also new channels of technology transfer and ways of commercializing knowledge. Traditionally, licensing has been the dominant route for the commercialization of new technologies invented by universities or public research organizations. This has created a steady but modest flow of income for those organizations active in technology licensing. More recently the formation of university-based spin-off companies has become an important, yet underexploited option (Keeble and Wilkinson, 2000; Cooke, 2002; Lockett et al., 2005). Academic spin-offs can often, but not exclusively, be observed in industries with an analytical knowledge base, such as biotechnology, ICT, software or computing, since they are based on particular findings or inventions coming from scientific research. Frequently, universities or public research organizations take an equity stake in such start-ups and they may subsequently benefit from future value increase of the firms if new products are successful and the firm grows. In particular in the USA, new ventures spun off from universities and other public research organizations were seen as a way of transferring technology and as a source of income for these organizations. Within Europe, university spin-offs were until the 1990s a less common phenomenon, but more recently European countries and regions have caught up in this respect (Keeble and Wilkinson, 2000).

Academic spin-offs combine a variety of knowledge links according to our classification developed in the following section. To some extent they can be considered as 'knowledge spillovers' since knowledge generated in universities or PROs is, usually without full compensation, put to private use and exploited in the form of a new company. However, since universities or PROs often take equity shares, and with academics being members of the company boards, there are also strong features of formalized 'networks'. Finally, there are also elements of a milieu involved, since these start-ups are frequently located in incubators or in local industry clusters close to the university where there is informal knowledge exchange and collective learning among participants.

Lockett et al. (2005) state that there are several process issues involved in the development of spin-offs at PROs. These are opportunity recognition, the decision to commercialize and due diligence, the choice between licensing and spinning off, the time period over which technology transfer offices (TTOs) are involved in spin-offs, and accessing resources and knowledge. Opportunity search is, in the USA and the UK context, primarily undertaken by the academic or the (complicit) TTO, occasionally also by (complicit) external entrepreneurial actors. Commercialization typically occurs when the technology is commercially viable and it implies the clarification of the intellectual property right issue, that is, verifying patent ownership. Regarding the choice between licensing and spinning off, key factors are the estimated financial return and the willingness of the entrepreneur to exploit the invention. However, as Lockett et al. find, there are considerable differences among US and UK universities in their ability to carry out such an assessment in a systematic way. Access to resources, technology, human capital and finance are key. At a very early stage, before firms are set up, universities or public funds are typically the sole sources of financial capital, whereas in later phases private financial capital from friends and family, business angels and venture capitalists is most important. Furthermore, there is public funding in early phases of spin-off development such as the University Challenge Fund in the UK or the Small Business Innovation Research (SBIR) programme in the USA. Also in Austria several public initiatives in this field have been launched recently, including the programmes 'LISA preseed' and 'uni venture' (see Chapter 8).

There are, meanwhile, a number of studies investigating the determinants of spin-off activity of universities. For the USA, O'Shea et al. (2005) in an empirical study examined 141 research universities regarding their spin-off activity. They found that in addition to 'history', that is, the university's previous success in technology transfer, faculty quality, commercial capability, and the extent of federal science and engineering funding were significant determinants of university start-up rates. Lockett and Wright (2005) in a similar investigation of UK research universities found that the number of spin-out companies created were significantly positively associated with expenditure on intellectual property protection, the business development capabilities of technology transfer offices and the royalty regime of the universities. In contrast they did not find significant relations to the number of TTO staff, or the age of the respective TTO. According to the authors 'the results suggest that universities and policymakers need to devote attention to the training and recruitment of technology officers with the broad base of commercial skills' (Lockett and Wright, 2005: 1043).

In some other studies the linkages of spin-offs to universities were analysed. Rothaermel and Thursby (2005) investigated 79 start-ups from the Georgia Institute of Technology finding that a new venture's university linkages through a Georgia Tech licence and/or through a Georgia Tech professor on the firm's management senior team significantly reduced the new venture's probability of failure, but it also retarded the firm's graduation from the incubator to a significant extent. Ensley and Hmieleski (2005) compared 102 high-technology start-ups from university incubators with 154 ventures unaffiliated with such facilities. They found that university-affiliated start-ups had more homogenous but less dynamic management teams than their unaffiliated counterparts. Also, university-affiliated start-ups had a significantly lower performance in terms of net cash flow and revenue growth than the unaffiliated new firms. Audretsch et al. (2005) examined the role of location of high-technology start-ups for accessing knowledge spillovers from universities. Analysing data from 281 German start-ups listed on Neuer Markt between 1997 and 2002 they found that those start-ups in high-technology sectors were influenced not only by the traditional regional and economic characteristics but also by the opportunity to access knowledge from universities.

Summing up, we find that academic spin-offs have become an important but still underexploited route to technology transfer in addition to licensing for universities and other public research organizations. Spin-off activity is influenced by various factors such as resources for opportunity search and intellectual property protection, as well as capabilities for business development of TTOs. However, they also depend on the entrepreneurial climate, the innovative milieu of the region and the density and strength of university–firm links. Regarding the last it is interesting to observe that 'strong' links between universities and start-ups tend to reduce the probability of failure but also might lead to a slower development of new firms retarding their graduation from university incubators. Spatial proximity of incubators and spin-offs in general favours knowledge flows, a finding which is in line with earlier work on knowledge spillovers.

A key factor for the development of start-ups in addition to knowledge and technology is finance. For the very early phases (opportunity search, patenting) finance often comes from universities or public research organizations; in the later phases public funding or private venture capital typically steps in. Major gaps in conditions for venture capital are certainly an obstacle for the development of spin-offs and high-technology start-ups, as can be observed for many countries in Europe and in particular for Austria (see Chapter 2). Venture capital firms fulfil important functions for the development of technology start-ups beyond the mere provision of capital (Tödtling, 1994; Cooke, 2002). They are usually competent in selecting promising technologies and ventures, supporting the formation of management teams. They monitor the development of new firms helping them if there are emerging difficulties. In some cases they may directly provide consultancy or help to find reliable partners in this respect.

MODES OF KNOWLEDGE EXCHANGE AND THEIR GEOGRAPHY

As was argued in Chapter 2, there is some evidence (Feldman and Audretsch, 1999; Gehrke and Legler, 2001; Carrincazeaux, 2002; Cooke, 2002), that knowledge-based firms and activities exhibit a strong tendency to concentrate in particular regions, cities or clusters. In particular for high-technology industries, notably biotechnology and information and communication technologies, spatial clustering seems to be a striking feature (Prevezer, 1997; Swann et al., 1998; Baptista and Swann, 1999; Bresnahan et al., 2001; Cooke, 2002; Fuchs, 2003). The industry or cluster life-cycle hypothesis (Swann, 1998; Tichy, 2001) argues in this context that in particular in the early stages of industry development, where many start-ups and spin-offs occur, geographical proximity is vital, whereas in the latter stages, when the industry matures, economic activities become more geographically dispersed. Proximity to universities and research organizations and access to (often tacit) knowledge is important in the early phases in the innovation process. Since tacit knowledge is best transmitted via face-to-face contacts and through direct interaction, it is a key factor to explain spatial clustering in knowledge-based sectors (Gertler, 2003). Malmberg and Maskell (2002) noted that co-located firms undertaking similar activities benefit from 'monitoring advantages', that is, they can observe competitors and imitate superior solutions. They may also benefit from a specialized labour market and from labour mobility.

In many studies on high-tech clusters the existence of knowledge spillovers and other relations has been claimed. What is often missing, however, is a clear differentiation of knowledge interactions, both conceptually and empirically. In the following, we present a typology of knowledge links which are also applied in the empirical chapters below. We distinguish between two dimensions, the first refering to Storper's (1995; 1997) distinction between traded and untraded interdependencies in the innovation process. Storper has argued that it is in particular the untraded, often informal relations which might explain the spatial concentration of innovative industries and activities rather than the traded, more formalized interactions among firms. The second dimension refers to the static versus dynamic aspects of knowledge exchange and innovation interactions, and was brought into discussion by authors of the milieu approach (Camagni, 1991; Maillat, 1998; Capello, 1999). Static knowledge exchange implies the transfer of 'ready' pieces of information or knowledge from one actor to the other. Examples are the licensing of a specific technology, the reading of a patent description of another firm or the observation and imitation of other firms. Dynamic knowledge exchange refers to interactive learning among actors through, for example, cooperation or other joint activities as described by Lundvall (1992), Capello (1999) and Lawson (2000). In this case the collective stock of knowledge is increased through the interaction. Along this line of reasoning we arrive at the four main types of relations in Figure 3.2.

These four types of relations constitute 'ideal types' which in real situations can rarely be observed in pure form. It has been argued that market

	Static (knowledge transfer)	Dynamic (collective learning)
Formal/ traded relation	(1) Market relations	(3) Cooperation/ formal networks
Informal/ untraded relation	(2) Knowledge externalities and spillovers	(4) Milieu informal networks

Figure 3.2   Types of knowledge exchange in the innovation process

relations and networks cannot clearly be separated, but are positioned rather along a continuum between these ideal types. This has been stated, for example, for relations to suppliers and customers by Dosi (1988) and von Hippel (1988) which often show a more durable and interactive nature than pure market links. Then, there are considerable overlaps between knowledge spillovers and milieux, making it difficult to differentiate these categories in real situations. Also the distinction between milieux and networks has remained unclear in the literature, since an innovative milieu is often identified with networks. We follow here Camagni (1991) who has argued that the milieu is characterized by informal links and a high degree of tacit knowledge exchange among actors, whereas networks are constituted by formal cooperations or agreements connecting the regional milieu with external knowledge sources.

Market relations (1) in the present context refer to the buying of 'embodied' technology or knowledge in codified forms (Scherer, 1992). This includes, for example, the buying of machinery, ICT equipment or software, or the buying of licences. Since the traded technology or knowledge is more or less in a 'ready' form, we consider this as a static relation vis-à-vis knowledge transfer. As we have mentioned already above, in reality the relations between the buyers and suppliers of machinery and equipment are often more durable and interactive, moving this type of relation more towards networks (von Hippel, 1988; Gertler, 1993). Nevertheless, we classify these relations as market relations, since, in principle, trade partners could be changed swiftly and the level of interaction is in many cases rather low. Under certain conditions there may be transaction cost savings and other advantages of *regional* trade links in specific cases. Regional user–producer relations may be relevant, for example, for early phases of the product cycle (Tichy, 2001), new firms or industrial districts of SMEs (Asheim, 1996). More often, however, trade relations are at higher spatial levels, reaching clearly beyond the region (Storper, 1997; Sternberg, 2000). Feldman (2000) considers interregional and international trade links as one of the most important mechanisms of technology transfer.

There exists a considerable body of literature, pointing out that markets are far from perfect with respect to knowledge generation and exchange. A number of authors have demonstrated through econometric methods that there are considerable *local knowledge externalities or spillovers* (2) from universities and research organizations to firms. Unlike market links, there is no contract or formal compensation for the acquired knowledge. Jaffe (1989), Audretsch and Feldman (1996), Anselin et al. (1997) and Bottazi and Peri (2003) have investigated and identified such local knowledge spillovers applying a knowledge production function approach. Jaffe et al. (1993) found considerable proximity effects with respect to patent citations.

It is argued that such local knowledge spillovers might result from various mechanisms such as mobile labour or informal contacts. The same authors also found a time decay with local knowledge spillovers: the most frequent citations of local patents were in the first few years after the patents had been granted. A geographical distance decay of knowledge spillovers was also shown by Anselin et al. (1997) for the USA and by Bottazi and Peri (2003) for Europe. Knowledge spillovers, thus, have been verified for a number of North American and European cities.[1]

Networks and milieux are conceptually different from the above categories. They are based on evolutionary or sociological approaches while the arguments and reasoning go beyond the transaction costs logic. Compared to market links, *networks* (3) are more durable and interactive relations between specific partners in the innovation process. There is not just an exchange of a given technology or piece of knowledge but a collective further development and enhancement of the respective knowledge base. This constitutes a dynamic process of collective learning (Lundvall and Johnson, 1994; Lundvall and Borrás, 1999). Innovation networks may take different forms (De Bresson and Amesse, 1991; Powell and Grodal, 2005). Some are based on formal agreements or contracts (R&D co-operations, R&D alliances, research consortia) including formal statements on the sharing of tasks, cost, benefits and revenues. These types of networks often, but not exclusively, include large and international firms, specialized technology companies or major research organizations. The partners are selected according to their strategic importance for the firms or their complementary competences. As a consequence, these formal innovation networks are often established at an international or even global level (Archibugi and Iammarino, 1999). Hagedoorn (2002) noted that at the end of the 1990s the share of international partnerships was about 50 per cent of all newly made R&D partnerships between firms, whereas high-tech sectors were less internationalized in their R&D partnering. Furthermore, he showed that in recent years over 80 per cent of the newly made R&D partnerships were found in information technology and pharmaceutical industries. Innovation networks were investigated in particular for knowledge-based industries such as ICT and biotechnology (Powell et al., 1996; Powell, 1998; Godoe, 2000; Matuschewski and Zoche, 2001; Cooke, 2002; McKelvey et al., 2003). In the latter studies it was demonstrated that there is networking also at the regional and national levels, often involving local universities, venture capital and smaller companies (Cooke, 2002; Powell and Grodal, 2005). However, more frequently the networks identified in these knowledge-based industries were among international partners. This could be observed in particular for small open economies such as Sweden (McKelvey et al., 2003) or Austria (Schartinger et al., 2000).

Innovation networks may also include informal links and collaborations among companies and organizations, such as those in industrial districts (Asheim, 1996) and in high-tech regions (Saxenian, 1994). These are often based on trust, and a shared understanding of problems and objectives, and the acceptance of common rules and behavioural norms. In the literature this is referred to as social capital (Putnam, 1993; Wolfe, 2002) or a shared culture leading to a specific *innovative milieu* (4) (Camagni, 1991; Maillat, 1998; Ratti et al., 1997). The swift exchange of ideas and knowledge are key to an innovative milieu, but as in the case of networks there is a dynamic element of collective enhancement of the local knowledge base through continuous innovation interactions, that is, collective learning (Camagni, 1991; Capello, 1999; Lawson, 2000). Collective learning processes in innovative milieux were investigated for a number of regions in Europe by the GREMI group (Aydalot and Keeble, 1988; Camagni, 1991; Ratti et al., 1997) as well as by Keeble and Wilkinson (2000). These studies demonstrated the importance of the region for such knowledge interactions. An innovative milieu is tied to a specific locality or region since it is based on personal relations and face-to-face interaction, common rules and a shared understanding. These often result from interactions in a specific local/regional production system or cluster. We pointed out in Chapter 2 that a shared understanding is not confined to a local milieu, but may also be established through organizational or institutional proximity or through the virtual exchange and discussion groups. Amin and Cohendet (2004) have referred to 'communities of practice' in this context, which can be established, for example, through the Internet on a global scale (Kaufmann et al., 2003).

A general finding seems to be that local and territorial milieux have to be complemented by global networks in order to provide access to complementary knowledge and resources and to avoid 'lock in' (Camagni, 1991; Tödtling, 1994). While not neglecting that geographical concentration can provide enormous opportunities for the transmission of sticky, non-articulated forms of knowledge between firms (from the advantages of 'being there'), Bathelt et al. (2004) as well as Cooke (2004) also emphasize the importance of 'global links through which access to codified external knowledge is secured'. In this context Owen-Smith and Powell (2004) have argued that there is a difference between 'channels' which have a more open character and 'pipelines' which are of a more closed nature. The former offer more opportunity for knowledge capability enhancement since they are more 'leaky' and give rise to knowledge spillovers. Pipelines, which can also be local, offer more confidential, contractual means of proprietary knowledge transfer. This may occur locally or over great geographical distances based on contractual agreements. These are less 'leaky' because they are closed rather than open. In their analyses of research and patenting practices in the Boston regional biotechnology cluster they found that network pipelines offered reliable and excludable information transfer at the cost of fixity, and thus were more appropriate to a stable environment, whereas permeable channels rich in spillovers are responsive and more suitable for variable environments (Owen-Smith and Powell, 2004).

SUMMARY

In this chapter we investigated key drivers in the emerging knowledge economy. Unlike earlier periods, these are to be found less in individual firms or actors but rather in more complex innovation systems of regional, national or sectoral nature. The knowledge base of companies has become more 'distributed', both in terms of actors and geography. Innovation systems are made up of networks of knowledge providers such as universities and research organizations, clusters of large and small firms underpinned by venture capital, education, technology transfer and other supporting institutions. Owing to the partly sticky nature of knowledge, territorial innovation systems are playing an increasing role in the knowledge economy, since they help to integrate knowledge which is distributed in many different sources.

Key actors in the emerging knowledge economy are universities, and their role and the way they relate to regional or national economies has been changing considerably in recent years. Universities have become more entrepreneurial, taking over a third role in addition to teaching and research. They increasingly engage in links with industry, spin-offs and other kinds of knowledge application. Through these activities their traditional functions of teaching and research are also subject to change. There are many more overlaps between these functions in the sense that research and teaching are undertaken in a context of application, described as 'Mode 2' by Gibbons et al. (1994).

We have argued that there are different types of knowledge base which imply considerably differences in the knowledge and innovation process of various industries. Firms in traditional industries such as engineering or machinery often innovate on a 'synthetic knowledge base', that is, they combine various forms of existing and new knowledge in order to bring forward innovations. These tend to be incremental, strongly based on tacit knowledge exchanged with key customers and suppliers. Cultural industries such as new media, advertising and entertainment rely on a 'symbolic knowledge base'. Here, knowledge is represented in cultural artefacts such as images, designs, or music brought forward by individuals in a creative process. These are often brought together in projects (temporary organizations) where professions from the artistic and commercial worlds meet. 'Know Who', thus, tends to be more important than 'Know How' in such industries. Firms in knowledge-intensive industries, in contrast, innovate on an analytical knowledge base. Here we find a strong role of science, university–industry links and spin-offs. Knowledge is more often codified, relations tend to be of a contractual nature and innovations are more radical.

The commercialization of scientific knowledge has traditionally occurred through technology licensing but more recently also through academic spin-offs. We have found that spin-offs have become an important but, in particular in Europe, still underexploited route to technology transfer in addition to licensing. The number and success of spin-offs depends on various factors such as previous experience of universities in technology transfer, faculty quality, commercial capability and business development capabilities of TTOs. Spin-off activity also depends on the entrepreneurial climate of the region and the strength of university–firm links. Spatial proximity to the university in general favours knowledge flows. There is a trade-off involved, since strong university links tend to reduce the probability of failure but may lead to slower growth of the start-ups.

In the final section we presented a typology of knowledge interactions based on the formal/informal nature of links and the static/dynamic character of interactions. We distinguished between market links, networks, spillovers and milieux. Through these different types of links specific kinds of knowledge are exchanged. Tacit knowledge seems to be more easily exchanged through informal networks or milieux, whereas codified knowledge exchange often takes the route of market links or of formal networks. Furthermore, the importance of these knowledge links varies between synthetic, analytic and symbolic knowledge bases, presented above. For knowledge-intensive industries all four types are relevant, since these industries rely on a broad variety of knowledge sources, as demonstrated in the empirical chapters. These types of links also have different geographies, as spillovers and milieux are tied to localities and regions, whereas market links and networks constitute the local interregional and international 'pipelines' through which specific complementary knowledge assets, usually in codified form, are accessed.

NOTE

1. For the USA see furthermore Glaeser et al. (1992) and Florida (2002). For Canada, see Polese (2002). For the UK, see Begg (2002). For Europe, see Cheshire and Magrini (2002).

REFERENCES

Amin, A. and Cohendet, P. (2004), *Architectures of Knowledge – Firms, Capabilities and Communities*, Oxford: Oxford University Press.

Anselin, L., Varga, A. and Acs, Z. (1997), 'Local geographic spillovers between university research and high technology innovations', *Journal of Urban Economics*, **42**, 422–48.

Archibugi, D. and Iammarino, S. (1999), 'The policy implications of the globalisation of innovation', in D. Archibugi, J. Howells and J. Michie (eds), *Innovation Policy in a Global Economy*, Cambridge: Cambridge University Press, pp. 242–71.

Asheim, B. (1996), 'Industrial districts as "learning regions": a condition for prosperity', *European Planning Studies*, **4**, 379–400.

Asheim, B. and Gertler, M. (2005), 'Regional innovation systems and the geographical foundations of innovation', in J. Fagerberg, D. Mowery and R. Nelson (eds), *The Oxford Handbook of Innovation*, Oxford: Oxford University Press, pp. 291–317.

Asheim, B., Isaksen, A., Nauwelaers, C. and Tödtling, F. (eds) (2003), *Regional Innovation Policy for Small-Medium Enterprises*, Cheltenham, UK and Northampton, MA, USA: Edward Elgar.

Audretsch, D. and Feldman, M. (1996), 'Innovative clusters and the industry life cycle', *Review of Industrial Organisation*, **11**, 253–73.

Audretsch, D., Lehmann, E. and Warning, S. (2005), 'University spillover and new firm location', *Research Policy*, **34**(7), 1113–22.

Aydalot, P. and Keeble, D. (eds) (1988), *High Technology Industry and Innovative Environments: The European Experience*, London: Routledge.

Baptista, R. and Swann, P. (1999), 'A comparison of clustering dynamics in the US and UK computer industries', *Journal of Evolutionary Economics*, **9**, 373–99.

Bathelt, H., Malmberg, A. and Maskell, P. (2004), 'Clusters and knowledge: local buzz, global pipelines and the process of knowledge creation', *Progress in Human Geography*, **28**, 31–56.

Becattini, G. (1991), 'Italian industrial districts: problems and perspectives', *International Studies of Management and Organisation*, **21**, 83–9.

Begg, I. (ed.) (2002), *Urban Competitiveness*, Bristol: Policy Press.

Bottazzi, L. and Peri, G. (2003), 'Innovation and spillovers in regions: evidence from European patent data', *European Economic Review*, **47**, 687–710.

Breshnahan, T., Gambardella, A. and Saxenian, A. (2001), ' "Old Economy" inputs for "New Economy" outcomes: cluster formation in the new Silicon Valleys', *Industrial and Corporate Change*, **10**, 835–60.

Camagni, R. (1991), 'Local "milieu", uncertainty and innovation networks: towards a new dynamic theory of economic space', in R. Camagni (ed.), *Innovation Networks*, London: Belhaven Press, pp. 121–44.

Capello, R. (1999), 'SME clustering and factor productivity: a milieu production function model', *European Planning Studies*, **7**, 719–35.

Carrincazeaux, C. (2002), 'The role of geographical proximity in the organisation of industrial R&D', in M. Feldman and N. Massard (eds), *Institutions and Systems in the Geography of Innovation*, Boston, MA: Kluwer Adacemic, pp. 145–79.

Chandler, A.D. (1962), *Strategy and Structure: Chapters in the History of Enterprise*, Cambridge, MA: Harvard University Press.

Cheshire, P. and Magrini, S. (2002), 'The distinctive determinants of European urban growth: does one size fit all?', *Research Papers in Spatial and Environmental Analysis*, **73**, Department of Geography, London School of Economics.

Cooke, P. (2002), *Knowledge Economies: Clusters, Learning and Cooperative Advantage*, London: Routledge.

Cooke, P. (2004), 'Regional knowledge capabilities, embeddedness of firms and industry organisation: bioscience megacentres and economic geography', *European Planning Studies*, **12**, 625–41.

Cooke, P. and Morgan, K. (1998), *The Associational Economy – Firms, Regions and Innovation*, New York: Oxford University Press.

Cooke, P., Boekholt, P. and Tödtling, F. (2000), *The Governance of Innovation in Europe – Regional Perspectives on Global Competitiveness*, London and New York: Pinter.

Cooke, P., Heidenreich, M. and Braczyk, H.-J. (eds) (2004), *Regional Innovation Systems*, 2nd edition, London: UCL Press.

De Bresson, C. and Amesse, F. (1991), 'Networks of innovators: a review and introduction to the issue', *Research Policy*, **20**, 363–79.

Dohse, D. (2000), 'Technology policy and the regions – the case of the BioRegio contest', *Research Policy*, **29**, 1111–33.

Dohse, D. (2003), 'Taking regions seriously: recent innovations in German technology policy', in J. Bröcker, D. Dohse and R. Soltwedel (eds), *Innovation Clusters and Interregional Competition*, Berlin, Heidelberg, New York: Springer, pp. 372–94.

Doloreux, D. (2002), 'What we should know about regional systems of innovation', *Technology in Society*, **24**, 243–63.

Dosi, G. (1988), 'The nature of the innovative process', in G. Dosi, C. Freeman, R. Nelson, G. Silverberg and L. Soete (eds), *Technical Change and Economic Theory*, London: Pinter, pp. 221–38.

Edquist, C. (1997), 'Introduction: systems of innovation approaches – their emergence, and characteristics', in C. Edquist (ed.), *Systems of Innovation: Technologies, Institutions and Organizations*, London: Pinter, pp. 1–35.

Edquist, C. (2005), 'Systems of innovation – perspectives and challenges', in J. Fagerberg, D. Mowery and R. Nelson (eds), *The Oxford Handbook of Innovation*, Oxford: Oxford University Press, pp. 181–208.

Eickelparsch, A. and Fritsch, M. (2005), 'Contests for cooperation – a new approach in German innovation policy', *Research Policy*, **34**(8), 1269–82.

Ensley, M.D. and Hmielski, K.M. (2005), 'A comparative study of new venture top management team composition, dynamics and performance between university-based and independent start-ups', *Research Policy*, **34**(7), 1091–105.

Etzkowitz, H. and Leydesdorff, L. (2000), 'The dynamics of innovation: from national systems and "Mode 2" to a triple helix of university–industry–government relations', *Research Policy*, **29**, 109–23.

Etzkowitz, H., Webster, A., Gebhardt, C. and Cantisano Terra, B.R. (2000), 'The future of the university and the university of the future: evolution of ivory tower to entrepreneurial paradigm', *Research Policy*, **29**, 313–50.

Feldman, M. (2000), 'Location and innovation: the new economic geography of innovation, spillovers, and agglomeration', in G. Clark, M. Feldman and M. Gertler (eds), *The Oxford Handbook of Economic Geography*, Oxford: Oxford University Press, pp. 373–94.

Florida, R. (2002), *The Rise of the Creative Class*, New York: Basic Books.

Fuchs, G. (ed.) (2003), *Biotechnology in Comparative Perspective*, London: Routledge.

Gehrke, B. and Legler, H. (2001), *Innovationspotenziale deutscher Regionen im europäischen Vergleich*, Berlin: Duncker and Humblot.

Gertler, M. (1993), 'Implementing advanced manufacturing technologies in mature industrial regions: towards a social model of technology production', *Regional Studies*, **27**, 665–80.

Gertler, M. (2003), 'Tacit knowledge and the economic geography context, or the undefinable tacitness of being there', *Journal of Economic Geography*, **3**, 75–99.

Gibbons, M., Limoges, C., Novotny, H., Schwartzman, S., Scott, P. and Trow, M. (1994), *The New Production of Knowledge – The Dynamics of Science and Research in Contemporary Societies*, London: Sage, reprinted 1999.

Glaeser, E., Kallall, H., Scheinkman, J. and Shleifer, A. (1992), 'Growth in cities', *Journal of Political Economy*, **100**, 1126–52.

Godoe, H. (2000), 'Innovation regimes, R&D and radical innovations in telecommunications', *Research Policy*, **29**, 1033–46.

Grabher, G. (1993), 'The weakness of strong ties: the lock-in of regional development in the Ruhr-area', in G. Grabher (ed.), *The Embedded Firm: on the Socioeconomics of Industrial Networks*, London: T.J. Press, pp. 255–78.

Grabher, G. (2001), 'Ecologies of creativity: the village, the group, and the heterarchic organisation of the British advertising industry', *Environment and Planning A*, **33**, 351–74.

Grabher, G. (2002), 'Cool projects, boring institutions: temporary collaboration in social context', editorial to special issue, *Regional Studies*, **35**.

Hagedoorn, J. (2002), 'Inter-firm R&D partnerships: an overview of major trends and patterns since 1960', *Research Policy*, **31**, 477–92.

Jaffe, A. (1989), 'Real effects of academic research', *American Economic Review*, **79**, 957–70.

Jaffe, A., Trattenberg, M. and Henderson, R. (1993), 'Geographic localization of knowledge spillovers as evidenced by patent citations', *Quarterly Journal of Economics*, **79**, 577–98.

Johnson, B., Lorenz, E. and Lundvall, B.-A. (2002), 'Why all this fuss about codified and tacit knowledge?', *Industrial and Corporate Change*, **11**, 245–62.

Kaufmann, A., Lehner, P. and Tödtling, F. (2003), 'Effects of the Internet on the spatial structure of innovation networks', *Information Economics and Policy*, **15**, 402–24.

Keeble, D. and Wilkinson, F. (eds) (2000), *High-Technology Clusters, Networking and Collective Learning in Europe*, Aldershot: Ashgate.

Laestadius, S. (1998), 'Technology level, knowledge formation and industrial competence in paper manufacturing', in G. Eliasson, C. Green and C.R. McCann (eds), *Microfoundations of Economic Growth: A Schumpeterian Perspective*, Ann Arbor, MI: University of Michigan Press.

Lash, S. and Urry, J. (1994), *Economies of Signs and Space*, London: Sage.

Lawson, C. (2000), 'Collective learning, system competences and epistemically significant moments', in D. Keeble and F. Wilkinson (eds), *High-Technology Clusters, Networking and Collective Learning*, Aldershot: Ashgate, pp. 182–98.

Lockett, A. and Wright, M. (2005), 'Resources, capabilities, risk capital and the creation of university spin-out companies', *Research Policy*, **34**(7), 1043–57.

Lockett, A., Siegel, D., Wright, M. and Ensley, M.D. (2005), 'The creation of spin-off firms at public research institutions: managerial and policy implications', *Research Policy*, **34**, 981–93.

Lundvall, B.-A. (ed.) (1992), *National Systems of Innovation: Towards a Theory of Innovation and Interactive Learning*, London: Pinter.

Lundvall, B.-A. and Borrás, S. (1999), 'The globalising learning economy: implications for innovation policy', report to DG XII, Commission of European Union, Brussels.

Lundvall, B.-A. and Johnson, B. (1994), 'The learning economy', *Journal of Industry Studies*, **1**, 23–42.

Maillat, D. (1998), 'Vom "Industrial District" zum innovativen Milieu: ein Beitrag zur Analyse der lokalisierten Produktionssysteme', *Geographische Zeitschrift*, **86**, 1–15.

Malmberg, A. and Maskell, P. (2002), 'The elusive concept of localization economies: towards a knowledge-based theory of spatial clustering', *Environment and Planning A*, **34**, 429–49.

Marshall, A. (1918), *Industry and Trade*, London: Macmillan.

Matuschewiki, A. and Zoche, P. (2001), 'Regionale Verankerung von Informations- und Kommunikationstechnologie-Unternehmen. Eine Fallstudie der TechnologieRegion Karlsruhe', *RuR*, **2–3**, 154–65.

McKelvey, M., Alm, H. and Riccaboni, M. (2003), 'Does co-location matter for formal knowledge collaboration in the Swedish biotechnology-pharmaceutical sector?', *Research Policy*, **32**, 483–501.

Nonaka, I. and Takeuchi, H. (1995), *The Knowledge Creating Company*, Oxford and New York: Oxford University Press.

Nonaka, I., Toyama, R. and Konno, N. (2000), 'SECI, Ba and leadership: a unified model of dynamic knowledge creation', *Long Range Planning*, **33**, 5–34.

O'Shea, R.P., Allen, T.J., Chevalier, A. and Roche, F. (2005), 'Entrepreneureial orientation, technology transfer and spinoff performance of U.S. universities', *Research Policy*, **34**, 994–1009.

Owen-Smith, J. and Powell, W. (2004), 'Knowledge networks as channels and conduits: the effects spillovers in the Boston biotechnology community', *Organization Science*, **15**, 5–21.

Polese, M. (2002), 'The periphery in the knowledge economy', www.inrs-ucs.uquebec.ca.

Powell, W. (1998), 'Learning from collaboration: knowledge and networks in the biotechnology and pharmaceutical industries', *California Management Review*, **40**, 228–40.

Powell, W. and Grodal, S. (2005), 'Networks of innovators', in J. Fagerberg, D. Mowery and R. Nelson (eds), *The Oxford Handbook of Innovation*, Oxford: Oxford University Press, pp. 56–85.

Powell, W., Koput, K. and Smith-Doerr, L. (1996), 'Interorganizational collaboration and the locus of innovation: networks of learning in biotechnology', *Administrative Science Quarterly*, **41**, 116–45.

Prevezer, M. (1997), 'The dynamics of industrial clustering in biotechnology', *Small Business Economics*, **9**, 255–71.

Putnam, R. (1993), *Making Democracy Work: Civic Traditions in Modern Italy*, Princeton, NJ: Princeton University Press.

Ratti, R., Bramanti, A. and Gordon, R. (eds) (1997), *The Dynamics of Innovative Regions: The GREMI Approach*, Aldershot: Ashgate.

Rothaermel, F.T. and Thursby, M. (2005), 'Incubator firm failure or graduation? The role of university linkages', *Research Policy*, **34**, 1076–90.

Saxenian, A. (1994), *Regional Advantage: Culture and Competition in Silicon Valley and Route 128*, Cambridge, MA: Harvard University Press.

Schartinger, D., Gassler, H. and Schibany, A. (2000), *Benchmarking Industry – Science Relations, National Report – Austria*, Seibersdorf: Austrian Research Centers.

Scherer, F. (1992), *International High-Technology Competition*, Cambridge, MA: Harvard University Press.

Schumpeter, J. (1911), *Theorie der wirtschaftlichen Entwicklung – Eine Untersuchung über Unternehmergewinn, Kapital, Kredit, Zins und den Konjunkturzyklus*, München und Leipzig: Duncker and Humblodt.

Schumpeter, J. (1942), *Capitalism, Socialism and Democracy*, New York, Harper and Brothers.

Scott, A.J. (1997), 'The cultural economy of cities', *International Journal of Urban and Regional Research*, **2**, 323–39.

Scott, A.J. (2001), 'Capitalism, cities, and the production of symbolic forms', *Transactions of the Institute of British Geographers*, **26**, 11–23.

Smith, K. (2002), 'What is the "knowledge economy"? Knowledge intensity and distributed knowledge bases', United Nations University, Institute for New Technologies, Discussion Paper Series, Maastricht.

Sternberg, R. (2000), 'Innovation networks and regional development – evidence from the European Regional Innovation Survey (ERIS): theoretical concepts, methodological approach, empirical basis and introduction to the theme issue', *European Planning Studies*, **8**, 389–407.

Storper, M. (1995), 'The resurgence of regional economies, ten years later: the region as a nexus of untraded interdependencies', *European Urban and Regional Studies*, **2**, 191–221.

Storper, M. (1997), *The Regional World: Territorial Development in a Global Economy*, New York and London: Guilford Press.

Storper, M. and Harrison, B. (1991), 'Flexibility, hierarchy and regional development: the changing form of industrial production systems and their form of governance in the 1990s', *Research Policy*, **20**(5), 407–22.

Swann, P. (1998), 'Towards a model of clustering in high-technology industries', in P. Swann, M. Prevezer and D. Stout (eds), *The Dynamics of Industrial Clustering*, Oxford: Oxford University Press, pp. 52–76.

Swann, P., Prevezer, M. and Stout, D. (eds) (1998), *The Dynamics of Industrial Clustering*, Oxford: Oxford University Press.

Sydow, J. and Windeler, A. (eds) (2004), *Organisation der Content-Produktion*, Wiesbaden: VS Verlag.

Tichy, G. (2001), 'Regionale Kompetenzzyklen – Zur Bedeutung von Produktlebenszyklus- und Clusteransätzen im regionalen Kontext', *Zeitschrift für Wirtschaftsgeographie*, **45**, 181–201.

Tödtling, F. (1994), 'Regional networks of high-technology firms – the case of the Greater Boston region', *Technovation*, **14**, 323–43.

Tödtling, F. and Kaufmann, A. (2001), 'The role of the region for innovation activities of SMEs', *European Urban and Regional Studies*, **8**, 203–15.

Tödtling, F. and Trippl, M. (2005), 'Knowledge links in high-technology industries: markets, network or milieu? The case of the Vienna biotech cluster', paper presented at DRUID 10th Anniversary Summer Conference on Dynamics of

Industry and Innovation: Organisations, Networks and Systems, Copenhagen, 27–29 June.

Tödtling, F., Lehner, P. and Trippl, M. (2006), 'Innovation in knowledge intensive industries – the nature and geography of knowledge links', *European Planning Studies*, **14**(8), 1035–58.

Trippl, M. (2004), 'Das Verhältnis von Wissenschaft und Wirtschaft aus systemtheoretischer Perspektive', in D. Rehfeld (ed.), *Arbeiten an der Quadratur des Kreises: Erfahrungen an der Schnittstelle zwischen Wissenschaft und Praxis*, München: Hampp.

Von Hippel, E. (1988), *The Sources of Innovation*, New York: Oxford University Press.

Wolfe, D. (2002), 'Social capital and cluster development in learning regions', in J. Holbrook and D. Wolfe (eds), *Knowledge, Clusters and Regional Innovation*, Montreal: McGill-Queen's University Press, pp. 11–38.

Zeller, C. (2002), 'Project-teams as means for restructuring research and development in the pharmaceutical industry', *Regional Studies*, **36**(3), 275–89.

4.   Local clusters and global networks

INTRODUCTION

In the first part of this chapter, we move deeper following our review of the elements of knowledge economy interactions and milieux to investigate the configurations of knowledge-based clusters. In the second part, the final three sections explore more the global network linkages in varieties of interaction with such milieux. The key aim of this chapter is to show how processes of knowledge exchange between these clusters exert influence upon global economic geography. We think this has occurred to the extent that we propose globalization itself to be changing direction because of the rise of the knowledge economy. The basic thesis is that until approximately the 1990s what we term 'Globalization 1' prevailed. This order was animated principally by global corporations, the multinationals, who sought cheap labour zones in which to locate routine production, much of which was sold in their home markets, principally the EU and the USA. This rather straightforward quest for low unit labour costs and heightened productivity was subject to elaborations. Global value chains emerged as multinationals became more sophisticated regarding which elements of the production chain were best located where, and such locations also contributed to the emergence of global value networks requiring delicate management, both strategically and on a day-to-day basis (Gereffi, 1999; Henderson et al., 2002). This in turn presaged the quarrying by multinationals of industry clusters present in developing countries and, furthermore, global linkages forming between clusters on an international basis (see, for example, Cooke and Memedovic, 2003).

However, simultaneously a newer kind of knowledge-based rather than labour or componentry quest was under way. It was typified in the late-1990s technology boom through the search by software firms for offshore knowledge and technological capabilities. Software engineers had been in short supply in the EU and the USA for many years. Numerous studies (for example, Saxenian, 2000) had begun to show how, in the years leading to the boom, Silicon Valley's workforce was 25 per cent foreign born and that in 1990 one-third of higher qualified workers (engineers and scientists) were born overseas. Of these, two-thirds were Chinese (including Taiwanese) and Indian. Moreover, many were 'knowledge entrepreneurs': whereas in

1980 only 12 per cent of Silicon Valley's high-technology start-up firms were Indian or Chinese, by 1998 29 per cent were of those origins, two-thirds of them Chinese. Moreover, these entrepreneurs represented a new kind of so called 'to-and-fro migration' rather than a simple 'brain drain' (Balasubramanyam and Balasubramanyam, 2000). That is, they frequently returned home and, in India especially, created new software and technology businesses as well as venture capital funds that gave rise to dynamic cluster growth in such places as Bangalore and Hyderabad. Global information technology (IT) outsourcing was estimated to be around $39.6 billion in 2004, with India's share amounting to nearly $17.2 billion while China attracted only $1.9 billion of outsourcing revenues. More recently such knowledge entrepreneurship has extended to R&D outsourcing contributing to an Indian R&D outsourcing value for IT services as a whole of $2 billion in 2004 (Sarma, 2005). This quest for capabilities by multinationals marks what we hypothesize to be a transition to 'Globalization 2'. This is characterized by a new kind of capability deficiency (in varieties of knowledge) on the part of multinationals who are forced to seek out global talent pools and knowledge assets that are extremely asymmetrically distributed around the globe.

Hence, this chapter proceeds by, first, reviewing varieties of knowledge-based clustering, mainly ICT and biotechnology, to pin down, as far as possible, the rationale behind it. In doing this theoretical exegesis will range from market-driven explanations of the clustering phenomenon to newer monopoly, notably knowledge monopoly and quasi-monopoly practices on the part of incumbent firms (see also, Cooke, 2002). Of central importance in the analysis of this kind of clustering compared to, say, the remarkably successful clustering that generated so many employment opportunities in traditional Italian industries as discussed in Chapter 1, is the role of the science and research base. Moreover, its key components such as the talent available, the novelty and quality of the 'research industry' in specific knowledge 'hotspots' and the opportunities for 'open science' and even 'open innovation' are pronounced (Chesbrough, 2003). These are often found in proximity in the form of clusters, many warranting a post-cluster 'megacentre' designation since they contain major public or non-profit facilities like universities, hospitals, research laboratories, and government research institutes as anchors as well as firms, the more common element in business clusters after Porter (1998). Within them are numerous intermediaries that are masters of many kinds of knowledge from exploration to exploitation and in between (March, 1991); analytic, synthetic and symbolic in science, engineering and creative production; and tacit, codified and, as we suggested in the introductory chapter, something again in between that we term 'complicit' knowledge (Cooke, 2004).

Other dimensions of knowledge interaction concerning distant networks or proximity as key factors, and specialized or diversified kinds of knowledge spatially nearby or not, are reviewed. Thereafter, attention is devoted to the role of global firms in relation to such knowledgeable clusters in an attempt to get a clear perspective on the following issue: to what extent are knowledge asymmetries of the kind referred to fatal for global firms, marking the rise to prominence of clustered firms with global knowledge networks? Going a step further, to what extent is there evidence that successful global firms of the future are already practising new kinds of control through their knowledge network management capabilities?

VARIETIES OF KNOWLEDGE-BASED CLUSTERING: THE SCIENCE AND RESEARCH BASE

The point has been made that what distinguishes, say, a biotechnology cluster from an Italian industrial district is not knowledge per se but the nature of knowledge utilized in the innovation process. In the first case, the cluster is saturated in the formalized analytical theories, methodologies and testing processes of science and research – much of it codified, some still tacit and more somewhere in between. In the second category this will certainly not be the case, although there will be synthetic knowledge concerning production engineering and materials processing, methodologies and skills – many tacit but many also codified and fewer in between, although intermediary *functions* and knowledges are common. Much more symbolic knowledge of the aesthetic kind will saturate any design-intensive cognitive input to the production process. Some of this will be pre-tacit, self-transcending knowledge, some tacit, but in regard to inspirational 'exemplars' and design 'blueprints' there will be a strong explicit knowledge dimension and some degree of 'complicit' translational skill also present. So a key contrast is revealed in the knowledge emphasis between the two, the one more scientific, the other more symbolic – yet both with origins in the pre-tacit realm of creative thought, analogy, inspiration and dissatisfaction with the status quo. Importantly, this occurs in both cases in contexts where these provocations can be expressed in discourse with others, possibly in teams or groups 'working on the project', therefore having varying degrees of collective consciousness of how to 'tackle the problem'.

Two cases illustrate this: the first from the genes industry, the second from the jeans industry. Since the earliest days of biotechnology, the leading DBFs were all associated with leading scientists from Massachusetts (MA) and California, specifically Cambridge, San Francisco and San Diego.

Alongside University of California San Francisco (UCSF) medical school's Herb Boyer with Genentech were Walter Gilbert of Harvard with Biogen, Ivor Royston of University of California San Diego with Hybritech, Mark Ptashne of Harvard with the Genetics Insitute, and William Rutter of UCSF with Chiron. In the 1980s Nobel Laureate David Baltimore[1] (MIT) founded SyStemix, Malcolm Gefter of MIT founded ImmuLogic, and Jonas Salk, Salk Institute San Diego founded Immune Response. After leaving Cambridge, Lasker Prize winner and founder of molecular studies of gene regulation Dr Mark Ptashne became Ludwig Chair of Molecular Biology at the Sloan-Kettering Institute in New York. Tom Maniatis was co-founder of the Genetics Institute, one of the first biotechnology companies. The Genetics Institute was acquired for $3 billion by American Home Products (AHP) in 1996. American Home Products subsequently changed its name to Wyeth and the Genetics Institute no longer operates as a separate business unit, although Ptashne remained with Wyeth Research.

In February, 2004 it was announced that Ptashne and Maniatis had formed a new company, Acceleron Pharma of Cambridge, Massachusetts. The newly formed company raised $25 million in venture capital to open its initial laboratory in December 2003. Founding investor Polaris of Waltham MA invested $250 000 in initial seed funding. Advanced Technology Ventures of Palo Alto, California and Waltham MA, Flagship Ventures of Cambridge and Alameda, California, and Venrock Associates of Menlo Park, California, New York and Cambridge co-led the round. Sutter Hills Ventures in Palo Alto also contributed financing. Acceleron used the money to develop protein-based treatments for a range of diseases from osteoporosis to obesity and muscle-wasting illnesses such as Lou Gehrig's disease. After he left the Genetics Institute, Harvard professor Maniatis went on to found ProScript in Cambridge, which developed the drug Velcade[2] to treat multiple myeloma, a rare form of leukaemia. That company eventually became part of Millennium Pharmaceuticals Inc. in Cambridge after a series of acquisitions, and Velcade is now Millennium's signature drug. Acceleron co-founder Wylie Vale himself launched Neurocrine Biosciences Inc. in San Diego, a $2 billion company where he was a founding board member. Millennium bought LeukoSite from Cambridge, MA and COR Therapeutics from San Francisco with $76 million earned by selling its DNA sequencing technology to Eli Lilly and Monsanto. This enabled the company to fulfil Millennium's founders' vision of moving from platform technologies into therapeutic treatments. Millennium also merged with Cambridge Discovery Chemistry in 2000 giving it a UK presence and adding 100 chemists to a firm that was only founded in 1993 and which, by 2000 was valued at $11 billion and employing 1530 people. These are highly illustrative details of the strong local and distant network linkages among biotechnology firms, intermediaries and individuals over a 30-year period in these global bioregions.

While formalized scientific research and knowledge entrepreneurship of the most advanced kind characterizes the cluster practices of the US scientists referred to, the alternative illustrative type is a classic case of contemporary value-adding activity overwhelmingly reliant upon advanced and sophisticated exploitation of symbolic knowledge. One of north-central Italy's celebrated but less researched industrial district regions is Marche, to the east of Tuscany and Umbria, fronting the Adriatic shore. It has four provinces – Pesaro, Ancona, Macerata and Ascoli Piceno. Although each has a mix of industrial districts or 'clusters', Pesaro has a roughly equal and quite extensive focus on clothing and furniture, Ancona specializes to a limited degree in both, Macerata has roughly equal amounts of clothing, furniture and footwear districts and in Ascoli Piceno it is (luxury, for example, Tod's, and high-tech, for example, Finproject, specialist soles for hiking boots) footwear and clothing. On the Abruzzo border the clothing industry traditionally specialized in workwear, and over time this transmuted into denim jeans.

A leading customer is Miss Sixty,[3] a local fashion chain with global reach specializing in fashion clothing for girls from their teens to their twenties. Local firm Wash has Miss Sixty as a majority shareholder, an acquisition that saved the firm from bankruptcy given the decline of its former markets. In 2005 Wash was bulk buying indigo denim jeans (and jackets) from Romania and China for €1 each and then exposing them to a multi-stage division of labour to 'distress' them. This Schumpeterian apotheosis of creative destruction began with, in some cases, special enzyme-tolerant coatings being applied to the fabric, in other cases immediate application of high-pressure powdered glass spraying, to 'wear' and 'age' the appearance of the cloth. Thereafter, hand sanding with emery boards 'added value' by further distressing the fabric at strategic points. Workers from Islamic countries, mostly North Africa and Pakistan, performed these unhealthy tasks. Next, in some cases, paint marks of the kind that do-it-yourself (DIY) might leave, or grafitti, were applied before being pressure-bonded into the fabric. Thereafter rips were strategically made to prominent parts of the product. The final production stage involved the distressed garments being transported a few metres to a neighbouring Chinese-owned factory, employing 30 Chinese seamstresses, artfully to sew up the rips and tears arising from the previous distressing. The resulting product was then sold to Miss Sixty for €10 who marketed it in its retail outlets for €80–€100.

The fashion element in this value-adding distressing process was, of course, not accidental but design-intensive. The pioneer and inspiration for this kind of distressing was innovator Renzo Rosso's anti-haute couture firm, Diesel, based near Vicenza in Veneto region. Nevertheless, whereas Karl Lagerfeld contributes to Diesel designs, Miss Sixty retains a design agency based in Rome to innovate and advise on its designs. The source of 'self-transcending' inspiration for this is the actual fashion practices of Roman teenagers in the edgier nightclubs of the capital, where, in 2005 slashing, smearing and generally distressing their jeans was an ultimate fashion statement with global reach. Of special interest here are the following features of this highly innovative exemplar of the commercial exploitation of symbolic knowledge. First and foremost, the importance of self-transcending, pre-tacit knowledge in informing designer judgement in a highly uncertain and fast-moving sector. Second, the manifold iterations by which the application of designer tacit knowledge to retailer market research and producer tacit knowledge of the manufacturability of the innovated design occurs. This includes search for and sourcing labour willing and able to work on sometimes unpleasant processes. The whole process is a study in complexity management in itself. Finally, notable too is the presence, in hitherto culturally rather homogeneous Italy, of several ethnic minority communities that enable such design-intensive added-value to continue to be extracted. Naturally the multidimensional local–global dynamics of designing, producing and marketing such 'quick fashion' products in such generally high-wage contexts as Italy strikes the uninitiated observer as remarkable.

The role, for example, of Chinese entrepreneurs and Chinese and Islamic workers in Italy's industrial districts is recent and in process of being researched and written about.[4] As many as 23 000 Chinese entrepreneurs and employees are now found in Prato, one of Italy's traditionally important textile manufacturing districts near Florence. The town has a total population of some 150 000. Carpi, near Bologna and other clothing towns also have sizeable Chinese populations that settled mainly in the 1990s. There is an interesting history to this involuted aspect of globalization in such locales. In the 1980s, as the Italian industrial district model came into focus internationally, those Italian regions with communist governments exported, through consultants, advisers and study visits, the industrial district model to communist China. The three top clothing regions of Jiangsu, Zhejiang and especially Guangdong benefited from Deng Xiaoping's 'open door policy' and the Special Economic Zones. Although some 'specialized industrial towns' existed in Guangdong, for example, many new ones were encouraged to follow the small firm, collaborative industrial district developmental model, stimulated by the emergence of small-scale private enterprise. Thus Chinese entrepreneurs find it easy to adapt to Italian industrial district conventions (Bellandi and Di Tommaso, 2005). Accordingly, as such Italian districts adapt by moving into higher value-added segments (for example Prato from textiles to clothing; Ascoli Piceno from workwear to designer denim) such entrepreneurs have capabilities to supply missing skills like sewing, button attachment and production of accessories. It is thus the case that without the entry of Chinese entrepreneurs and cheap machinists, possibly in some cases undocumented, some Italian industrial districts would be less competitive than they are, given the intensity of global competition in such industries.

These two extremes in the spectrum of industrial cluster categories fit reasonably neatly into the analysis of such phenomena performed by Bottazzi et al. (2002) who developed a five-category cluster typology. It compares with more variation yet some overlaps with Markusen's (1996) earlier, more US-derived analysis discussed subsequently below, but there is no evidence the former knew of Markusen's work. The categories are:

- horizontally diversified – 'Made in Italy' districts with small firms producing high-quality, design-intensive fashion products in traditional sectors like clothing, ceramics and jewellery
- vertically disintegrated – a variant on the 'Made in Italy' type with a 'Smithian' division of labour in localized supply chains, specialized with local input–output linkages and user–producer knowledge exchange; shoes, textiles and some clothing are produced in such districts
- local hierarchical – an 'oligopolistic core' connected to subcontracting networks, as in transport equipment and white goods, for example
- knowledge complementarities – science and engineering driven, as in Silicon Valley and biotechnology clusters. Absent in Italy
- path dependent – spatially inert agglomerations without any particular advantage from agglomeration in itself, for example, Detroit as described by Klepper (2002).

They point out that different types of agglomeration suggest different drivers for agglomeration itself taking into account also different sectoral specificities. To test this out they conduct a stochastic econometric test of various agglomeration hypotheses. Using Italian data this confirms a statistically significant advantage, measured in export performance, for the industrial district categories of agglomeration over the rest. The authors ascribe this advantage to two phenomena theoretically and empirically pronounced in such districts: sectoral patterns of knowledge accumulation, and localized knowledge spillovers of various kinds, from innovation to labour. As will be obvious, our first exemplary cluster paradigm focused on biotechnology fits the penultimate 'knowledge complementarities' category well. The second illustration is obviously a category two type of industrial cluster with internal and external 'Smithian divisions of labour'.

More recently, and building on such insights, contributions by Paniccia (2006) and Belussi (2006) further typologize and taxonomize industrial districts and clusters. The difference is typologies are *ex ante* and taxonomies *ex post* empirical analyses. Both deploy wide ranges of criteria by which they differentiate categories and both draw, to a greater or lesser extent, upon the preceding analysis of Markusen but not that of their Italian colleagues. For Paniccia the sixfold typology involves:

- semi-canonical – small family firms, egalitarian, 'star structure' of dense networks
- craft-based or urban – a strong nucleus of final firms subcontracting to localized supply chains
- satellite platform or hub and spoke – oligopolistic internal or external to the district
- co-location areas – a *filière* of reasonably large, vertically integrated firms producing different products in the same sector
- concentrated or evolutionary – 'world' or 'regional product mandate', as with the Belluno spectacles cluster with lead firms overseeing the whole production process from R&D to retailing, but outsourcing specific technology-intensive functions to local specialists
- science-based – universities, large firms and small firms interact to commercialize sometimes short-lived discoveries from basic research in a cooperative, high social capital but ultimately highly marketized competitive context.

Belussi takes four of these categories for her taxonomy of industrial districts and clusters, correctly for her purposes conflating the first two and excluding 'co-location areas' in a tight conceptual model drawing on the many studies already conducted. She argues strongly for the geographic proximity connotation of clusters, thus moving the field forward conceptually from the period of Porterian confusion regarding that central issue. The task both for this taxonomy and Paniccia's typology is to provide the theoretical groundwork for hypothesis-testing research to determine the relative performance, efficiency and market effectiveness of the categories.

A first approximation from Paniccia is that those clusters with high firm density will *ceteris paribus* outperform the others, something, it will be recalled, that is consistent with the Bottazzi et al. (2002) conclusions. The primary empirical parts of this book in Chapters 7 and 8 take this into account in explaining firm performance in clustered – unfortunately only 'science-based' – against non-clustered settings. Prefiguratively speaking, it can be intimated at this point that our results are not quite as clear-cut as the Paniccia–Bottazzi et al. hypothesis proposes. In other words, there are crucial differences within science-based industry regarding performance in relation to clustering – one of the key findings of this book. Nevertheless, the Paniccia–Belussi elaboration of Bottazzi et al. is helpful in that it captures in the new category of 'concentrated evolutionary' *global product mandate* cluster, key essentials of the 'Miss Sixty' interaction with its (distressed denim) supply chain.

Hence, the analytical precision regarding varieties of clusters has evolved markedly since the pioneering intervention of Markusen (1996) who identified five types of industrial district:

- Marshallian – small firms, localized investment links, preferred suppliers, labour market loyalty, flexible work regime
- Marshallian (Italianate variant) – with added cooperation, design-intensive work and collective institutions plus local government support
- hub and spoke – Hollywood-like oligopolies plus suppliers, high intra-district trade, low cooperation among oligopolies
- satellite platform – externally owned oligopolies, low cooperation, low social capital, strong government support
- state anchored – government installations, scale economies, low local investment, strong external linkages.

Thus there is no science-based categorization and while Marche's distressed denim can easily be placed in the Marshallian (Italianate variant) category that is far too ill-fitting and roomy a category to give much explanatory purchase. In truth, this typology has relatively little relevance to the differentiation of clusters, except to remind us of the variety of ways even large firms may utilize geographical proximity for reasons of history, policy or comparative advantage. Thus New Jersey's 'Medicine Chest of the World' hosts the US headquarters of many global pharmaceuticals firms, but they scarcely interact, subcontract locally or share facilities. They simply appropriated tax breaks and reduced rents across the Hudson River from New York. Meanwhile, bulk chemicals complexes host many large, global petrochemical plants as a result of 1960s estuarine growth pole policy, whereby governments subsidized infrastructure costs for capital-intensive production in specific oil-importing locations. Nowadays those firms are, in some cases, diversifying into 'biologics' as petrochemicals production migrates to the originating oil-producing countries. Finally, military bases and capital cities come to mind as recipients of enormous taxpayer largesse with notable service, income and consumption spillovers for the luckily state-anchored districts, but not without vulnerabilities when land rents rise or military policies shift.

THE TALENT BASE: EXAMINATION KNOWLEDGE

It is clear we are mainly focusing upon Bottazzi et al.'s (2002) *knowledge complementarities* and Paniccia's *science-based* cluster category. The latter usefully draws attention to institutional variety where university research and training interact with the needs of firms large and small involving elements of collaboration rather than only arm's length exchange. Moreover, the 'science and engineering driven' designation of the former, while useful in reminding us that *analytical* and *synthetic* knowledge types relate, nevertheless has a certain 'linear' feel to it that trivializes the undoubted interactive complexities involved. So, from now on, we will refer to the kind of knowledge-based clustering which informs an important part of our analyses as *science-based clustering* for convenience, bearing in mind two things: first *science* also embraces *engineering* in practical terms while the overall analysis continues to be interested in non-clustered, non-agglomerated firms in its comparative assessment of firm-performance. A further key point is that we are interested in forms of knowledge and their interactions in the processes being analysed. Table 4.1 captures these key knowledge forms of interest, with stylized exemplars.

We noted earlier the important distinction made by March (1991) at the organizational level between exploration and exploitation as corporate functions. In a project analysing regional knowledge economies this

Table 4.1   Forms and stages of knowledge production

	Analytical	Synthetic	Symbolic
Exploration	Mathematical reasoning	e.g. Gene therapy	Experimental art work
Examination	Theorems to test	Clinical trials	Art exhibition
Exploitation	e.g. Penrose tiles/patterns[1]	Therapeutic treatment	Gallery sale

Note:
1 Penrose 'tiles' are patented mathematical formulae for producing computer graphic software patterns used in various electronic media. They were innovated by UK physicist Sir Roger Penrose.

Source: Cooke (2006).

analysis must extend beyond the firm as well as inquiring about firm practices. Complexity rises accordingly, but in any case we feel justified in introducing the new category of *examination* knowledge given the high premium placed upon safety and efficacy of products and services, as well as producers of goods and services by new regulatory rules and conventions. Our focus on the biotechnology sector naturally alerts us to this conspectus of testing and trialling activities. Since the thalidomide disaster, drugs trialling is the major time constraint upon the emergence of approved pharmaceutical products. But in other industries, notably agro-food, automotives and manufacturing more generally, testing in both real time and by structural calculation of finite elements algorithms on computers has grown in importance. Although attention is not much focused on *symbolic* knowledge here, it is clear that artists have traditionally tested the product through exhibitions, not to mention rehearsals for theatre and music, while authors have their proposed outputs reviewed. Indeed, it may be thought surprising that economics and other social sciences pay virtually no attention to this important element of economic activity.

Hence varieties of knowledge have economic value. Innovative knowledge-based clusters possess these distinctive *talent* bases as a consequence of recruiting and retaining scientific, technological, entrepreneurial and risk investment talent, as noted in our introductory chapter. Furthermore, in contrast to the view that knowledge-based clustering activity was best understood as a mainly market-related activity, we also suggested in the Introduction that it is by no means outlandish to conceive that it occurs also as a means to control knowledge access.[5] This involves protection of valuable knowledge by researchers and/or their clients from competitors, and communication of such knowledge formally or informally only with the trusted few who earn that position by virtue of possessing complementary and valuable knowledge of their own. In other words, it is as valid to hypothesize clustering in proximity of this kind to be quasi-monopolistic practice[6] as it is to argue it is primarily an open market practice.[7] For why, it has to be asked, would firms willingly pay the extra costs in land rent, wages, congestion diseconomies and so on denoted by spatial concentration, if knowledge freely flowed outwards in ubiquitous open market fashion? They could operate through distant networks or arm's length exchange. Of course, for some firms and some sectors that is precisely what happens. It depends on the kind of knowledge in question. Thus, biotechnology firms often cluster near the exploration knowledge base because, as Zucker et al. (1998) show, academic researchers wish to control the knowledge advantage they have, and tend to outsource examination and exploitation to trusted spin-out firms (in which they may have a financial interest) who must locate in proximity to conduct joint, for example, patent-related, or commercialization work. But this also involves *complicit* actors with specialized talent in knowledge of patient trialling, patent law, commercial due diligence and venture capital for an innovation to have a chance of ensuing. However, such conditions of perfect implicit–complicit–explicit knowledge interaction occur globally in relatively few places.[8]

Contrastingly, we suggested earlier that ICT firms are likely to cluster elsewhere despite also being high-technology businesses.[9] As Chapter 7 will show they find location in proximity to customers and suppliers far more important than proximity to exploration knowledge. Unlike biotechnology, the expertise and equipment sophistication of universities are insufficient to their requirements as 'synthetic' knowledge specialists themselves outsourcing or being outsourced to meet customized client requirements. Being located in proximity to such contractual value chain opportunities enables them to exert control over the company's destiny since anticipatory knowledge concerning contracts is frequently 'in the air' as Marshall famously put it.[10]

So what are the key talent-formation mechanisms? Florida (2002) identifies the following. Utilizing a composite index he calls the Creative Index based on cosmopolitanism, high-tech industry, innovation and share of creative workers in the labour force, Florida shows how they combine in a statistically significant way to predict the strongest, most innovative high-tech clusters in the USA. Research based on the Florida method in Europe[11] shows somewhat comparable results. Namely, the locations with the greatest spread of the listed characteristics combine creativity in the symbolic knowledge sense (art, music and culture industries generally) with synthetic knowledge (high-tech industry) and analytical knowledge (abundant university education and research opportunities). In the USA , leading locations combining creativity, high-tech and innovation include Boston, Austin (Texas), Raleigh-Durham and San Francisco while European equivalents are London, Helsinki, Stockholm and Oslo. Unsurprisingly, cities and their satellite metro-areas overwhelmingly dominate – usually, as the above selection shows, state or country capitals, excepting San Francisco. These places thus score highly on the 'three Ts' index measuring technology, talent and tolerance. They attract top students because they are perceived as liberal, interesting, lively places with opportunities for employment in growing and challenging business sectors. Thus in the knowledge economy, places like these that have a strong talent base attract and create business opportunities, whereas in the industrial economy places with employment opportunities attracted talent. Nowadays the huge recruitment of national engineer or chemistry cohorts to unprepossessing places with large concentrations of jobs for such graduates – ICI Teesside and

Merseyside or Siemens and Erlangen come to mind – no longer hold sway. Hand in hand with this, of course, goes the massive, globally widespread revaluation upwards of tertiary education, academic entrepreneurship, and knowledge-based employment among 'varieties of capitalism' and by governments of all stripes.

INTERACTIVE KNOWLEDGE EXCHANGE: IMPLICIT, COMPLICIT, EXPLICIT

The ICT and biotechnology sectors are widely perceived to be of key importance to many economies.[12] Some, including the UK and Austria, are strong in exploration knowledge but see problems in exploitation or commercialization capabilities. Accordingly, the leading research universities may have strong industrial research collaborations with leading global corporations but not with indigenous SMEs. German biotechnology was shown to be like this and as elsewhere, steps were taken to create biotechnology clusters. Also in Germany the InnoRegio programme has sought to clone Silicon Valley in Dresden with not a little success, thanks to former Siemens semiconductor divestiture Infineon. Where the innovation system is often least 'joined-up' is with its indigenous SME population comprised of firms in many sectors, including advanced technologies, that were not academic spin-outs.

Throughout Europe, improving the exploitation of research excellence is seen as the priority, with integration of SMEs a more complex challenge to be faced later. To that end, the key problem has been identified as the inability of entrepreneurial talent to innovate commercial products and services. Arguably, the transfer of implicit into explicit knowledge is trivialized in Europe. The rather 'one-club' strategy of valuing spin-outs over other forms of university research engagement externally, while simultaneously underestimating the contribution of large firms as sources of such knowledge entrepreneurship is bewildering. It is thus in part a developmental problem involving too little *complicit* knowledge in the market, a problem shared in many regions compared to leading US high-tech 'megacentres' such as Boston and San Francisco.

To anticipate a key finding of the UK research by way of a brief illustration, a key innovation system instrument introduced in Scotland to boost knowledge entrepreneurship and clustering is the Intermediary Technology Institute (ITI). Three of these, in Life Sciences, ICT and Energy were announced in 2003.[13] They exist to compensate for perceived market failure in knowledge commercialization. The manner in which this is achieved differs somewhat according to the focus of each ITI. With respect to the 'TechMedia ITI,' the colloquialism for ITI ICT and Digital Media, there are three innovative dimensions to organizational strategy. First, it adopts a thematic approach to its technology promotion activities. This means that the focus is upon 'platform technologies' having qualities of 'ubiquity'. Thus they are likely to be pervasive in their impact rather than being confined to a sector. Hence Broadband Wireless Technologies, Content Creation Tools, Human–Computer Interfaces, and Networked Sensor Technologies have wide potential application. Networked Sensor Technologies, for example, allow remote monitoring of animals, wind farms or even personal health care.

A second innovative aspect in ITI approach is to take seriously the difficult task pf *cluster-building*, a high-tech policy commitment in Scotland since the early 1990s. Basically, in policy terms, the cluster idea involves taking policy actions that enable businesses to gain from complementarities, collaborations and knowledge spillovers, especially where related firms operate in geographical proximity. As we have hypothesized, this may be perceived by firms in terms of exclusivity or even construction of quasi-monopoly conditions where a group of firms enjoy 'club' benefits. But building the cluster may, in fact, mean disrupting such equilibrium conditions by, for example, introducing a competitor. Intermediary Technology Institute Life Sciences concurred with Scottish Enterprise facilitating this, causing anxieties on the part of incumbents. It was justified in terms of 'building the cluster' rather than protecting established interests. The cluster perspective is thus seen from a market viewpoint (which forms the leadership of these specific intermediary organizations) as a policy retro-model. This means it is based upon 'a compartmentalized 1980s picture' of sectors rather than being a future-shaping, disruptively innovative vision. We easily forget that the desire for innovative high-tech clusters comes principally from policy-makers. Their thinking is backward from the ITI perspective. For example, software is seen by firms as a pervasive platform technology. But the policy action-line in Scotland subsumed it under 'e-learning'. This has been a (possibly fading) obsession of politicians and policy-makers since the 1990s but such an equation is seen as ludicrous in ICT where 'e-learning' is something that from a *market* perspective has almost no profile.

Finally, 'complicit' intermediaries must overcome barriers arising from the existence of 'communities of practice'. Here, for example, SMEs without university traditions either in respect of management training or innovation frequently fail to interact significantly with such knowledge centres for innovation purposes. Similarly 'triple helix' models that privilege senior manager 'communities of practice' of universities, industry and governance bodies may exclude the academics actually conducting research. Accordingly, decoding of tacit research knowledge and its relevance to large, small or spin-out businesses, may be inadequate. Equally, interaction with academics to the point where arcane university conventions concerning such technical matters as intellectual property rights (IPR) licensing, probably means an administrative functionary is best suited to conduct decoding of administrative conventions. Or a senior organization figure needs to be complicit with the aims of ITIs and their market-facing responsibilities to effect action in cutting red tape. In whichever dimension, the creation of some modest, 'weak ties' social capital in order to push forward the innovation agenda is necessary if innovation is to be seen as 'systemic'. In their first 20 months the ITIs have performed valuable network-building functions. This comes about in two main ways: formally, by creating membership clubs; informally by breaking red tape and increasing regional trust-building.

THE DIFFERENCE SPACE MAKES: LOCALIZED, GLOBALIZED, PROXIMATE AND VIRTUAL KNOWLEDGE EXCHANGE

In recent years, with the rise of the knowledge economy, science-based clusters, interactive innovation, and creative, tolerant and talented concentrations of politically desirable economic growth, much attention has been devoted to the fascinations of the idea of *proximity*. Foremost are its attendant opportunities for tacit, complicit, codified face-to-face and tactile contact among and between communities of practice as the *ne plus ultra* of modern economic development. It is, of course, astonishing that the role of proximity in human life could have been so occluded by modern economics and other social sciences. It was once said rather frivolously by an anonymous economist that 'economics is about houses and cars': that is, we have to spend on accommodation (and subsistence) and we will spend what it takes to get close as swiftly as possible to others comparably minded to make life worthwhile. Marx[14] put it more succinctly when he wrote about 'capitalist expansion that annihilates space with time'. Were a historical audit of commercialized new knowledge in the form of patented innovations in recent time conducted, it is probable that a majority would in some way or other be concerned with 'annihilating space with time' to facilitate the achievement of proximity. All the great Kondratieff bursts of innovation from steam trains, to ship technologies, to air transport, the automobile, communications technologies, television and the Internet were predicated on enhancing proximity. The main contending family of innovations is that focused upon health care, an equally restless field, a much-imitated candidate motto for which was innovated by the British Columbia health service in 1987 as 'adding years to life, and life to years'.[15]

The death of distance and the end of geography were rumours much exaggerated upon the implementation of all these innovations, most recently and erroneously with the advent of the Internet (Morgan, 2004). However, what happened as a consequence of these innovations is that *proximity*, the literal meaning of which includes *inter alia* nearness, closeness, contiguity and propinquity, all with traditionally geographical connotations, has evolved elaborated and geographically unconfined meanings involving nearness in context, domain and even opinion. Thus chat rooms are quite neighbourly places in virtual space. A multinational company displays characteristics of organizational proximity in all its global operations because of its common rules, conventions and resources from job titles to the commonalities of its intranet. Zeller (2004) in an interesting article tracing the dependence of Swiss 'big pharma' on innovative biotechnology clusters elsewhere, lists, as well as geographical proximity, the following: institutional (for example, national laws); cultural (for example, communities of practice); relational (for example, social capital); technological (for example, Linux software users); virtual (for example, a multinational); and internal and external (for example, firm supply chain management). Actually, as noted in Chapter 2, few feature prominently in his empirical analysis and those that do are utilized assertively rather than analytically with the most obviously useful, contrarily, proving to be *geographical* proximity. This is because Novartis is shown to be obsessed with quarrying American biotechnology clusters due to its own knowledge asymmetries, resulting in that multinational now settling its leading R&D headquarters in the San Diego and Cambridge, Massachusetts, biotechnology 'megacentres'. Nevertheless, the point has been made that, although for words not to become wholly meaningless concepts should not be overstretched, the notion of proximity, unlike propinquity, never had any expressly spatial 'nearness in place' meaning. Thus Zeller (2004) performed a useful service in this sense.[16]

However, it is hard to escape if not an 'iron law' then a reinforced concrete one that dictates that past and contemporary economic development tends to become increasingly urbanized. Indeed, while the emphasis in the viewpoint that foresaw the end of geography/death of distance happening was wrong in respect of the idea of ubiquitous access to the full range of necessities for economic activity over space, it was correct in postulating globally networked information flows. However, it missed the point that such nodes would be the result of increasing returns to urban agglomeration. By and large this has meant increasing returns derived from a variety of spillovers, especially knowledge spillovers, that tend to concentrate in cities. This is true for North American and European (including Israeli) cities for which the required analysis has been performed.[17] Clearly, such a wide array of city settings means the growth process is by no means identical in all cases. Moreover, as Begg (2002) and others show, the competitiveness of cities often accompanies social polarization. However, this is also a by-product of growth where in-migrants are attracted because of perceived economic opportunities absent in their location of origin. Thus in liberal market economies urban sprawl into neighbouring non-metropolitan municipalities is usual, whereas in coordinated market economies this is less the case.

A pattern in all these settings, underlined by Begg (2002), is that proximity to knowledge spillovers is crucial to city growth in the era of the knowledge-based economy. This harks back to the initial contention of Glaeser et al. (1992) that human capital and scarce skills are significant factors in a city's capability to retain and augment its economic growth. This is thus something of a progenitor of Florida's (2002) talent-led analysis of US city growth in the contemporary era. However, much of the finer detail of variations within growth trajectories is lost in these analyses, not least because of definitional, and even unit of data analysis complexities. One interesting differentiation first hypothesized from a *static* analysis of major concentrations of European knowledge economy sectoral activities, derived from European Union city and region-level data on high-technology manufacturing and knowledge-intensive business services (KIBS), was that major cities, sometimes also capital cities, accreted much of the KIBS employment. Contrariwise, more specialist satellite cities concentrated high-technology manufacturing employment to a greater extent. Live instances of that modern urbanization process would include, for example, Cambridge and numerous lesser high-tech satellites of Boston such as Waltham, Worcester, Woburn and Andover; San Francisco vis-à-vis many such places in Silicon Valley, London in relation to Cambridge, Oxford and the Thames Valley; Stockholm and Uppsala; Helsinki and Espoo; and Copenhagen in cross-border relationship to Lund, the so-called Medicon Valley, traversed by the Øresund bridge.

In a test of this proposition, Cooke and Schwartz (2003) collected appropriate longitudinal statistics for Israel and found support for the proximity to the major city thesis. Statistically significant relationships were found for the growth over time of such knowledge-based urbanization. But in Israel specialist satellites were growing in employment around the main KIBS city of Tel Aviv rather than the capital Jerusalem. Yet even Jerusalem, for all its political troubles, managed to evolve the country's main biotechnology cluster. Nevertheless, in such proximate satellite locations as Rehovot, Herzliah and Ramla higher than average employment growth rates in high-technology manufacturing occurred from 1995 to 2002. There was even a biotechnology emphasis to Rehovot's growth and ICT emphases in the other two instances. This suggests that in countries where the main financial centre is not the capital city the former will exert the stronger proximity effect but that where, as in the UK and Austria, the capital is also the leading financial services centre, a strong spatial monopoly or more accurately quasi-monopoly proximity effect is exerted. This is the classic result modelled by Krugman (1995) in applying increasing returns to scale theory, under conditions of imperfect knowledge, to two hypothetically competing candidate cities with the consequence that one always ended up monopolizing space. Contemporary city growth theory places knowledge spillovers from (geographical) proximity at the forefront of the explanation for these observed tendencies.

To repeat, this is not to say that geographical proximity determines economic activity to an overwhelming degree. If anything, the implications of what has been concluded here is that the defining feature of knowledge spillovers from geographical proximity is qualitative and quantitative in equal measure. That is, an incumbent located proximately and actively in relation to multiple and varied sources of high-grade intelligence, creativity and connectivity is in principle at an advantage compared to a competitor who is not. However, connectivity to other appropriate knowledge nodes elsewhere in the relevant global knowledge networks is likely to be quantitatively less intensive, albeit of qualitative equivalence or even superiority. In their discussion of precisely this geographically proximate as against virtually proximate relationship, Owen-Smith and Powell (2004) argued for the superiority of geographical proximity along the following lines. Key processes by which dynamic proximity capabilities are expressed interactively in research or *exploration knowledge* transfer and commercialization or *exploitation knowledge* transfer include the following:

- There is a difference between 'channels' (open) and 'pipelines' (closed). The former offer more opportunity for knowledge capability enhancement since they are more 'leaky' and 'irrigate' more geographically proximately. Pipelines offer more confidential, contractual means of proprietary knowledge transfer. This may occur locally or over great geographical distances based on contractual agreements. These are less 'leaky' because they are closed rather than open.
- In high-tech fields, research centres may be a magnet for firms because they operate an 'open science' policy, promising spillover innovation opportunities. These are possible sources of productivity improvement, greater firm competitiveness, accordingly proximate, localized economic growth.

- Such open science conventions influence interfirm innovation network interactions. Although researchers may not remain the main intermediaries for long as successful firms grow through patenting and commercialization, they experience greater gains through the combination of proximity and conventions, than through either proximity alone or conventions alone.

These propositions each receive strong support from statistical analyses of research and patenting practices in the Boston regional biotechnology cluster. Thus:

> Transparent modes of information transfer will trump more opaque or sealed mechanisms when a significant proportion of participants exhibit limited concern with policing the accessibility of network pipelines . . . closed conduits offer reliable and excludable information transfer at the cost of fixity, and thus are more appropriate to a stable environment. In contrast, permeable channels rich in spillovers are responsive and may be more suitable for variable environments. In a stable world, or one where change is largely incremental, such channels represent excess capacity. (Owen-Smith and Powell, 2004: 15)

Finally, though, leaky channels rather than closed pipelines represent also an opportunity for unscrupulous convention-breakers to sow misinformation among competitors. However, the strength of the 'open science' convention means that so long as research institutes remain a presence, as in science-driven contexts they often do, such 'negative social capital' practices are punishable by exclusion from interaction, reputational degrading or even, at the extreme, convention shift, in rare occurrences, towards more confidentiality agreements and spillover-limiting 'pipeline' legal contracts. We noted in the introduction how open science conventions attract, in further evolutionary rounds, 'open innovation' to such knowledgeable clusters when it might otherwise be assumed openness should mean knowledge advantage erosion. But likely gains are perceived to outweigh losses by customers taking the plunge. This is a major factor in proximity-based economic growth since knowledge supplier firms garner a substantial share of their income from, especially, R&D outsourcing by larger customer firms.

SPECIALIZATION AND DIVERSIFICATION IN INNOVATION CONTEXTS

The localized knowledge spillovers and proximity questions, which we feel we have shown not to be based in fantasy, relate in interesting and complex ways to issues of specialization versus diversification as wellsprings of innovative capability. At issue in the former debate is whether such localized knowledge spillovers as those discussed, reside in the combination of effects that gives local innovation proactivity to the region or locale itself. Or does the locale derive its extra proximity capabilities from the combined effects of the firms that typify it? The latter is argued by Breschi and Lissoni (2001) who see no convincing evidence that non-pecuniary spillovers, the apotheosis of localized knowledge spillovers, have displaced Marshall's classic definition of 'external economies' in terms of pecuniary advantages.

However, the existence of and contribution to an entrepreneurial social infrastructure of *localized knowledge spillovers* is advocated by most of the academic community interested in the explanation for clustering and, at a higher scale, the formation of regional innovation systems. These are of course highly germane to the interest of this book and it is currently an area of hot debate, the outlines of which are explored by Caniëls and Romijn (2003). The proponents of the view that proximity offers innovation advantages in itself begins in relatively recent times with Jaffe et al. (1993). The argument here was that R&D in particular constitutes a public good in locations where it concentrates and that this is sufficient to cause firms to concentrate in proximity to such knowledge spillover opportunities to access them as free goods in advance of competitors. This, of course, is entirely consistent with the attractions of 'open science' and subsequently 'open innovation' in proximity to (informal) knowledge spillovers discussed in the previous section of this chapter. Thereafter the likes of Audretsch and Feldman (1996) and Malmberg and Maskell (2002) also argued in favour of the power of localized knowledge spillovers as drivers of innovation, especially in knowledge-based clusters. However the 'market' dimension of such growth impulses may, as we have suggested, be overstated since, if free markets in such knowledge really operated, proximity would not be necessary. In this sense we hypothesize clusters as more like clubs than commodity exchanges.

Interestingly both sides argue their cases in respect of the meso-level of analysis. Thus the Caniëls and Romijn critique is of both sides for ascribing too much influence to regional *milieu* and too little to firm *capabilities* or what may also be referred to as *entrepreneurship* and specifically in this vein *knowledge entrepreneurship*. The compromise position suggested as far as the treatment of such externalities (static or dynamic) is concerned is to establish a more penetrative analysis of the firm-level contribution to regional *capabilities*. For now, further work is required on types of agglomeration advantage, ranging from static to dynamic spillovers, pecuniary to non-pecuniary, and pure versus impure knowledge spillovers at the firm level but aggregated up to at least the regional level. This means, of course, stepping further than firms alone in the analysis, since regional capabilities, including knowledge entrepreneurship capabilities, are seldom reducible to those of firms alone. One of the puzzles of knowledge economies is how cognitive boundary-crossing among 'epistemic communities' (Haas, 1992) such as entrepreneurs and researchers, usually representing the private and public sectors respectively, actually occurs. Epistemic communities are - *professionally* distinctive associations, gaining identity and economic status precisely from their distinctiveness.

This is excellent candidate territory for the exploration of *complicit* knowledge spillovers of the kind we referred to in the Introduction. Authors such as Galison (1997) refer to the means of explicit–implicit interchange as 'contact languages' and even talk of 'pidgins' as means of communication. In summary, it is arguable that concentrations of scarce and valuable capabilities among incumbents and intermediaries such as varieties of knowledge to explore an idea, test or examine it, and then exploit it as a commercial innovation, may be a kind of quasi-monopoly of knowledge it is in the interests of incumbents to protect, as far as they are able, from outsiders. 'Openness' in innovation may clearly have certain 'club' characteristics, as we have argued already a number of times.

In connection with this, there is also a further and evolving debate between two perspectives regarding the most propitious setting for the stimulation of innovation. One says capabilities from sectoral *specialization* produce the best results, the other says *diversification*. The former position is associated with Glaeser et al. (1992) and Griliches (1992) who, as we have seen, envisage specialized knowledge 'spillovers' as key to innovation. The latter view begins with Jane Jacobs (1969) and is supported by, for example, Feldman and Audretsch (1999) who show sectoral diversity is most strongly associated with regional innovativeness. The first tend to emphasize *markets* while the second give greater weight to the *milieu* of institutional infrastructure (innovation support systems) and microeconomic linkages across agents and firms (networks). So we have a comparable issue as that which cropped up regarding localized knowledge spillovers. However, a slightly different resolution arose when, subsequently, Henderson (2003) showed *specialization* effects on knowledge spillovers to have strong but short-lived impact in high-technology industry while *diversification* effects persist far longer. This suggests that as they evolve such clusters first specialize then later diversify. This makes sense in that science-based clusters are demonstrably formed in historical terms around a single innovation, which swiftly transmutes into families of related products – and processes. One only has to think of the semiconductor that was the trigger for Silicon Valley firms later spawning the integrated circuit, graphic interface, mouse, personal computer (PC), and so on. Similarly, both the San Francisco and Cambridge, Massachusetts, biotechnology clusters began with the production by, respectively, Genentech and Biogen, of human insulin.

It is incumbent upon us to try to place these debates in good order in so far as that ambition is feasible in rapidly evolving, empirically informed theoretical fields. But before that attempt, one more, long-standing, binary opposition of relevance to the foregoing discussion requires integration into the debate. This concerns the relationship of Marshall's (1918) contrast between *urbanization* and *localization* economies to the foregoing, not least because important elements of the contemporary debate have their origins in Marshall's insights of a century ago. These, of course, were fully informed by his clinical observation of the industrial districts of Britain, which reached their apogee shortly before the outbreak of the 1914–18 war. Urbanization economies are somewhat cognate with Jacobs's notion of the innovative buoyancy of industry diversification, which is naturally far more pronounced in urban than non-urban areas. Localization economies are more specialized and their externalities equivalently so. Thus Marshall speaks of the manner in which British (Yorkshire) woollen towns specialized in specific kinds of smooth (for example, worsted) or rough (tweed-type) *woollens* and (Lancashire) cotton towns Blackburn, Nelson, Oldham and Preston displayed different types of weaving of cotton and 'the constant intercommunication of ideas between machine makers and machine users'[18] that underpinned the evolved specializations in specific localities. These dynamic, partly non-pecuniary localization externalities enabled firms to specialize while maximizing equipment utilization. This contrasts with equivalent static, pecuniary externalities derived from access to a highly specialized, localized skills base. It is conceivable, but only just, that such ruminations possibly followed by specifications could be conducted in virtual proximity. But as Sapsed et al. (2005) show, tactile meetings are still required for project design even within single distributed corporations engaged in complex projects. Hence we come to Table 4.2 and an effort to juxtapose the master concepts regarding varieties of knowledge-based clustering. This is as a prelude to the second, somewhat briefer analysis of contemporary global knowledge flows. To explain the content of Table 4.2 we distinguish between firm and milieu characteristics, between diversification/urbanization and specialization/localization innovation contexts and types of externality. We further distinguish static and dynamic externalities associated with urbanization versus localization economies. Finally, we categorize the *main* mode by which knowledge spillovers are accessed. Unfortunately this cannot be done in terms of pure/impure or pecuniary/non-pecuniary spillovers in the interests of limiting the complexity of an already three-dimensional graphic space. Table 4.2 is intended to be illustrative of theoretical nuances arising from the juxtaposition of

*Table 4.2   Firm/milieux by innovative context, externalities and knowledge
            spillovers*

	Knowledge spillovers			
	Firm		Milieux	
Diversification/ urbanization	*Free rider (recombinant)*	*Proximate networks (incremental/ disruptive)*	*Industry agglom- eration*	*Modern industrial district*
	Static	Dynamic	Static	Dynamic
Innovation context		Externalities		
Specialization/ localization	*Classical industrial districts (incremental)*	*Virtual networks (radical)*	*Company town or oligopoly*	*Innovative business cluster*

conceptual categories discriminating between types of knowledge and types of industry organization.

For the *firm*, the taker of static externalities in a diversified, urbanized innovation context is not 'networked' but may still be innovative, possibly as described in Cooke (2005b), as a *recombinant* innovator, finding diverse customers for incremental innovations. The *milieu* most associated with low-networking propensity because of its static externalities is the industry agglomeration. The firm exploiting dynamic urbanization externalities in a diversified innovation context without significant 'distant (knowledge) networking' is a stylized description of the modern, particularly Italian, industrial district firm, capable of incremental innovation, occasionally innovating disruptively, but seldom radically.[19] Hence its typical milieu is a modern industrial district. On the contrary the firm seeking only static knowledge spillover externalities in a localized, specialized innovation context is perhaps 'protectionist', path dependent and capable only of incremental innovation at best – it is reminiscent of firms in mature, classical industrial districts such as the silk or ceramics districts of Great Britain that died from this kind of protectionist disposition (Cooke, 2002). They equate to the oligopolistic cluster, a category that resonates well with the Bottazzi et al. (2002) categorization discussed above. However, we would not necessarily expect only decaying oligopolies to operate in such clusters. In Finland, for example, Nokia operates as a highly innovative company in such contexts, often with its R&D laboratories embedded in university science parks with clusters of spin-out suppliers. Finally we propose the virtually networked firm to be the likeliest candidate for radical innovation, particularly when also engaged in the similarly dynamic knowledge spill-over settings of diversified, urbanized economic space. Much research shows such innovators to be highly networked globally and locally (Gans and Stern, 2003). Accordingly, they are likely to be located in innovative business clusters where open science and open innovation opportunities abound.

THE ROLE OF GLOBAL FIRMS AND NETWORKS: GLOBAL FIRMS AS CLUSTER FLAGSHIPS AND FINANCIERS

We turn in this second part of Chapter 4 to global knowledge and asset flows to open up analysis of globalization processes under the changed conditions we referred to as 'Globalization 2' occasioned, we argue, by the rise of the knowledge economy as we have defined it in Chapter 2. Recall that the key explanatory point behind this shift is the *metamorphosis*, as Penrose (1995) referred to it, engendered by the need for the growing firm to master international knowledge networks as a fundamental capability beyond the basic ones of efficient and effective resource and administrative management. More specifically, even multinationals do not control the knowledge exploration process internally as they used to. Rather, by various means to be described, they increasingly outsource that function to knowledgeable firms and research institutes, often found in innovative business clusters, increasingly overseas and in some developing countries such as China and India. To that extent global companies are supplicants rather than sovereign in the knowledge-acquisition processes of their core business. Thus the economic geography of power over knowledge has, as argued already, tilted towards university towns often, like the Cambridges, satellites of large cities where dynamic knowledge spillovers from a diversified innovation context are accessible.

In this section we examine the role of global firms as in some sense the most visible leaders of innovative business clusters, and particularly their still-dominant function as financiers. They are investors in the smaller incumbents on whom they depend for knowledge but who depend upon them for financial resources. Thereafter we consider the other main vehicle of global knowledge transfer in what we call the *research industry*, composed of knowledge entrepreneurs in and outside universities and other redoubts of knowledge exploration, such as specialist research institutes. Notable in the Penrosian *metamorphosis* in which we are interested is the extent to which public funding and infrastructure play a key role regarding production of research and talent. This is a key factor in the changing economic geography of 'knowledge capital' (Burton-Jones, 1999). It also resonates with the varieties of capitalism framework in that it might be anticipated that coordinated market economies would be skilled at managing the public–private interface complexities implied by the exploitation of public research. But the reality is that they are no better and actually in most cases worse than the liberal market economies in doing this. Even so, liberal market UK is only marginally better than its European partners while it is the USA, and to a lesser extent Canada, that perform best in respect of realized commercial innovations arising from scientific research.

Of key importance in this state of affairs is venture capital and risk management by accomplished entrepreneurs. This is especially evident in economic, and specifically high-tech, downturns when risk investment dries up because of the decline in anticipated returns. Thus, after the end of the dot.com bubble in 2000, US venture capital investment, most of it directed towards ICT and biotechnology, declined by some 80 per cent from $106 billion in 2000 to $18 billion in 2003.[20] In the EU the decline was from €19.6 billion to €9.8 billion from 2000 to 2002. Moreover, the percentage shares of smaller tranche investments rose while the larger tranche investments declined. Thus, taking biotechnology as a case in point: in the UK, the share of investments valued at less than €5 million rose from 40 per cent in 2001 to 70 per cent in 2003. In the rest of Europe it moved from 45 per cent to 52 per cent, while in the USA it rose from 21 per cent to 33 per cent over the same period. There were corresponding declines in shares taken by deals involving investments of €20–€50 million (DTI, 2005). Because of the intimate interaction of venture capital with knowledge entrepreneurship (Feldman et al., 2005) a severe disequilibrium occurs between the rate of candidate firms forthcoming from academic entrepreneurship and business incubation, and the rate of investment by venture capitalists at the take-off stage, which is among the riskier stages of firm evolution. That is, the 'system' by which small-scale, mainly public support for start-up businesses operates, continues at a pace that is relatively unaffected by stock market fluctuations, only for those firms with IPR and early-stage seed capital invested to then face a drought at the point they seek to move their product or service towards becoming a commercial innovation.

This actually brings into question the issue of whether venture capital is a necessary evil which entrepreneurship would be best advised to avoid. Thus, why do entrepreneurial businesses not bypass venture capitalists and move straight into contractual negotiations with the penultimate users of commercially valuable innovations, namely, larger firms? Unsurprisingly, there is plenty of evidence that this happens, especially in ICT where firms like Cisco Systems, Intel and Microsoft engage in corporate venturing and, notoriously in the case of Cisco and Microsoft, conduct innovation primarily by acquisition of such firms once they have demonstrated a viable product or service. There is evidence that this is happening increasingly in biotechnology too as firms seek suitors from outside the venture capital community that since the downturn has increasingly invested in such relatively safe havens as utilities and retail. Thus, French venture capital firm Sofinova reported that numerous venture capital firms of their acquaintance were missing good investment opportunities because pharmaceuticals corporations, desperate to replenish their drying drug pipelines, had become more aggressive and agile identifiers of valuable spin-out investment opportunities.[21]

Nevertheless, by 2004 venture capital was beginning to return to biotechnology, at least in the USA where the total for 2003 rose slightly to $2.8 billion compared with the $2.7 billion invested in 2002. Nevertheless, in the USA this was not the most popular means of raising investment capital as Table 4.3 shows. In the USA, most popular was so-called convertible debt, where a firm sells shares at below market price as an investment incentive. In the EU however, on far smaller overall investment magnitudes, venture capital remained the clear favourite. Venture capital in ICT in the USA continues to be overwhelmingly (two-thirds) devoted to Internet firms, a category that is not within the scope of this book since it is arguable that they belong to the creative industries. In the European Union ICT-related venture capital investments fell from $9.6 billion (0.12 per cent of GDP) to $4.1 billion (0.05 per cent of GDP) between 2000 and 2002 while for biotechnology (and health) the decline was a more modest $3.2 billion to $2.5 billion (down 0.04 per cent to 0.03 per cent of GDP). The comparison

Table 4.3   Types of risk investment in US and EU biotechnology, 2003

Equity type	Quantity (€billion)	
	USA	European Union
PIPEs[1]	1.6	0.04
Venture capital	2.8	0.36
Other private equity	0.4	0.05
Public equity	3.8	0.29
Total	8.6	0.74

Note: 1   Private investment in public equity.

Source: DTI (2005).

for ICT in the USA was \$65.4 billion in 2000 declining to \$12.0 billion in 2002 (0.67 per cent down to 0.12 per cent of GDP) with health and biotechnology declining from \$7.8 billion in 2000 to \$5.0 billion in 2002 (0.08 per cent to 0.05 per cent of GDP). Approximately half of EU venture capital expenditure is accounted for by the UK (OECD, 2004).

Hence three points of interest arise here. First the 'liberal market' economies account for over three-quarters of the combined venture capital expenditure of the EU and the USA, indicating the extent to which knowledge entrepreneurship of the advanced kind on which this book focuses is dependent on private risk capital investment. Second, the ICT and health-care biotechnology sectors absorb very large shares of that investment, although much goes to Internet businesses that may as easily be in the retail or gambling sectors as ICT where in the USA they are accounted statistically. Third, it is noticeable that the difference in levels of investment in biotechnology and health between the USA and the EU were not that great at the peak of the technology boom and they converged thereafter, at least until 2002.

PIPELINE INTERACTIONS AND EXPLOITATION NETWORKS

We noted earlier that when venture capitalists lose interest in risk investment in high-tech businesses, the load is taken up by other vehicles to some extent, although such opportunities must be well along the innovation pipeline. In such cases, particularly but not exclusively in biotechnology, larger firms are nowadays less likely than they once were to acquire such firms but rather take a majority shareholding if it is a reasonably mature firm or make 'milestone' payments against IPR as a firm progresses past agreed development and trialling targets with a view ultimately to licensing the technology in question. They may also offer to buy the technology rather than the licensed service, something Chesbrough (2003) portrays in relation to Monsanto and Eli Lilly vis-à-vis Millennium Pharmaceuticals' DNA sequencing software, the sale of which generated sufficient revenue to buy three smaller biopharmaceutical companies to become the drug manufacturer its owners had always sought to be. This is nowadays a major means of knowledge transfer in biotechnology and ICT through varieties of often triangular 'partnership' among a large customer, a smart technology SME and, sometimes, a university or institute researcher or team responsible for the discovery or invention.

Table 4.4 is instructive as an illustrative indication of this for biotechnology in the single post-boom year of 2003. There we see a major effort

Table 4.4   *Top ten biotechnology partnerships, 2003*

Biotechnology firm	Partner	Transaction	Value (€million)
Millennium Pharmaceuticals (USA)	Johnson & Johnson (USA)	Global marketing rights for oncology drug	425
Biovitrum (Sweden)	Amgen (USA)	Develop and commercialize Phase 2 diabetes drug	414
Regeneron Pharmaceuticals (USA)	Aventis (France)	Develop and commercialize Phase 1 cancer drug	405
Regeneron Pharmaceuticals (USA)	Novartis (Switzerland)	Develop Phase 2 rheumatoid arthritis drug	278
Actelion (Switzerland)	Merck (USA)	Marketing heart, kidney and hyper-tension inhibitors	216
Idenix Pharmaceuticals (USA)	Novartis (Switzerland)	Licensing hepatitis B drug. Co-develop Phase 1/2 compound	199
Corgentech (USA)	Bristol-Myers Squibb (USA)	Co-develop Phase 3 vein graft failure drug	199
Medivir (Sweden)	Boehringer Ingelheim (Germany)	Out-license of Phase 2 HIV therapy	158
Lexicon Genetics (USA)	Bristol-Myers Squibb (USA)	Neuroscience drug R&D, commercialization	152
Flamel Technologies (France)	Bristol-Myers Squibb (USA)	Licence and commercialize insulin product	131

Source: DTI (2005).

by US pharmaceuticals corporations to transfer-in knowledge from European and US dedicated biotechnology firms by forming a variety of partnership agreements with them. Many of these firms might otherwise have been targets for venture capital houses but it is clear that by 2003 pharmaceuticals firms were willing to invest substantial sums in products that in some cases were only at early-stage trialling. It is noteworthy too, that many such deals focused on the end-point of the drug development process, namely, commercialization. For the foreseeable future the large scale of retained earnings continuing to be enjoyed by pharmaceuticals companies means they maintain a commercialization advantage in respect of distribution and marketing staffs (the latter often in the tens of thousands) that DBFs do not have, and possibly would not want. Hence the successful 'pharma' firm of the future is likely to be drawn from those who realize their strengths and weaknesses and act as, fundamentally, bankers and sales engines for the products of DBF and university research power.

Before completing this section on 'pipeline' interactions where the pipeline is the time line of different stages in the evolution of a potential product as it moves from laboratory bench or *in silico* representation as computer graphics, software or design, to innovation on the market, the other main means of global knowledge transfer involving *research* deserves mention.

In published papers such as Cooke (2004) evidence is provided of the main global co-publishing patterns among biotechnology 'star' scientists in prestigious research institutes (Figure 4.1). It is largely pre-private sector entry research and largely funded by research councils, governments and non-profit foundations. It reveals the strength at the centre of the hierarchy of US clusters in California, Massachusetts and New York. In Europe university towns are often the heart of the excellence network. It is thus no coincidence that the locational preferences of recent pharmaceuticals industry openings has been influenced by the existence of these concentrations of accomplishment. We are not in possession of the research to enable us to say to what extent a comparable pattern of co-publication is found in ICT, but the strong probability is that there is some overlap as at Cambridge (UK) and in Silicon Valley for example.

Finally, closer to the point of exploitation when a large pharmaceutical firm and other investors, including venture capital will likely be interested in investing, is the patented outcome of the exploration and examination research stages of the pipeline. Again, we have conducted research into biotechnology co-patenting activity by 'stars' in comparably accomplished contexts and these network hierarchies are comparably structured, though in a more condensed way (Figure 4.2). This reinforces the commercial interest of such clusters for large commercializing companies whose

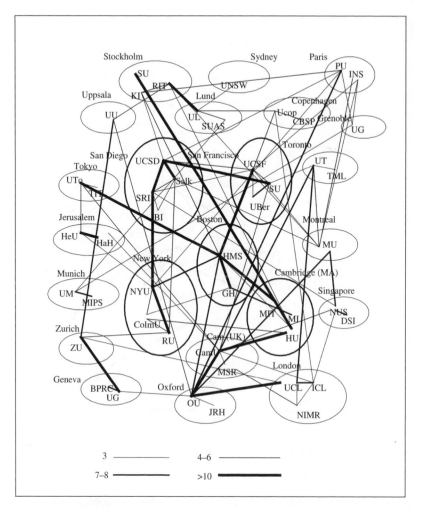

Note: Represented are co-publications between 'star' bioscientists in leading research institutes in eight of the top ten Science Citation Index journals for recent periods: *Cell* (2002–04); *Science* (1998–2004); *Proceedings of the National Academy of Sciences* (2002–04); *Genes and Development* (2000–04); *Nature* (1998–2004); *Nature Biotechnology* (2000–04); *Nature Genetics* (1998–2004); *EMBOJ (European Molecular Biology Organization Journal)* (2000–04). In all, 9336 articles were checked. Abbreviations refer to research institutes.

Figure 4.1   Publishing collaborations

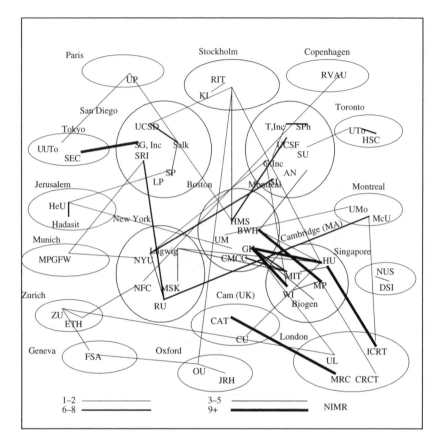

Source: Derived from American Patent Office database, 1998–2004.

Figure 4.2 Biotechnology co-patenting networks

internal knowledge capabilities are attenuated by the problems of dealing with new technologies outside their traditional field of research competence – in the case of pharmaceuticals in synthetic *chemistry* not bioscience. In ICT a similar story lay behind the rise of Intel and Microsoft at the expense of IBM.

Hence, to conclude this chapter, we have shown how knowledge moves from exploration to examination and exploitation stages and that it also changes in interaction with expertise along that chain as it shifts from pretacit (self-transcending) to tacit in interaction, with complicit knowledge intermediaries before it becomes explicit, codified and potentially commercializable as innovations. These interactions cannot simply be conducted distantly using the Internet. Rather, there is great value in many kinds of proximity, already discussed, the most important of which is geographical proximity when it comes to the transformation of raw knowledge into refined innovation. So powerful is the proximity impulse in the advanced technology sectors in focus in this book that they tilt the global economic geography of large firm location and more general interaction towards their clusters and away from former multinational redoubts like New Jersey, once tagged 'Medicine Chest of the World'. The past and recent investments by the likes of Novartis in Cambridge, Massachusetts, and in San Diego, AstraZeneca, Abbott, Amgen, and Pfizer, and Wyeth in Massachusetts too, to access bioscientific knowledge bears witness to this process in biopharmaceuticals. Comparable moves by the likes of Microsoft establishing outposts in Silicon Valley and Cambridge, UK, *inter alia* show this is by no means confined to biotechnology. These are the knowledge economy influences that justify a discourse of Globalization 1 evolving into Globalization 2 as opportunities from open science and open innovation are discovered asymmetrically embedded in global research and talent pools.

NOTES

1. In Crotty's (2001) biography of Nobel laureate David Baltimore, he notes how in 1976 he and Ptashne combined in testifying to the safety of genetic engineering against Mayor Vellucci and Cambridge City Council's moratorium suspending research on recombinant DNA. At one point Vellucci wrote to the president of the National Academy of Sciences (NAS) about two local newspaper reports that concerned him greatly. One claimed nearby sighting of a 'strange, orange-eyed creature' while the other spoke of a sighting of a 'hairy, nine foot creature'. He suggested the NAS should investigate any role of rDNA experiments in these sightings. No evidence of any such connection was ever forthcoming.
2. In 2003 US pharmaceuticals corporation Johnson & Johnson acquired global marketing rights for this product from Millennium (see Table 4.4).
3. University of Macerata economist, Ernesto Tavoletti, deserves major acknowledgement at this point for inviting me to join him in researching the Marche clothing and footwear industry during May 2005.
4. See, for example Becattini (2001) and Becattini and Dei Ottati (2006).
5. Dierickx and Cool (1989).
6. Malmberg and Maskell (1997).
7. Malmberg and Maskell (2002).
8. Cooke (2005a), where it is shown, schematically, that the *complicit* part of knowledge transformation in biotechnology involves at least 15 different 'communities of practice' before implicit knowledge of a therapeutic treatment may become an innovation utilized in patient health care.
9. Cooke (2005c).
10. Marshall (1920).
11. Conducted 2004–07 for UK, Netherlands, Norway, Sweden, Denmark, Finland under the European Science Foundation programme project: 'Technology, Talent and Tolerance in European Cities: A Comparative Analysis' funded by the respective research councils: UK ESRC grant no. RES-000-23-0467. Affiliate research teams from Germany, Switzerland and Spain are also involved on an informal basis.

12. Thus, as a case in point, the *OECD Science, Technology and Industry Outlook 2004* reports that the following are the R&D priorities of several OECD members: Canada – Biotechnology, ICT, Nanotechnologies; Japan – Life Sciences, ICT, Nanotechnology; S. Korea – Electronics and Biotechnology; Mexico – ICT, Biotechnology; New Zealand – Biotechnology, ICT and Creative Industries. Meanwhile in the USA, more 'pervasive' R&D priorities have been set such as 'Networking IT' involving large-scale networking; high confidence software systems; software design productivity; social, economy and workforce applications; and human–computer interaction. Other 'pervasive' priorities include: complex systems, climate, water and hydrogen, homeland security, and long-term nanotechnology applications, for example, in information processing.
13. Interviews were conducted in June 2005 with ITI heads as part of the ESRC project 'Collective Learning in Knowledge Economies: Milieu or Market?', Grant No. RES-000-23-0192.
14. Marx (1973).
15. 'British Columbia Premier's Health Strategy 1987–1991', www.healthypeople.gov/Implementation/Consortium/Annual_Meetings.
16. A further theoretical analysis of the relations between innovation capability and varieties of proximity is presented in Boschma (2005).
17. A seminal US article is Glaeser et al. (1992; see also for the USA, Florida (2002). For the Canada, see Polese (2002). For the UK, see Begg (2002). For Europe see Cheshire and Magrini (2002). For Israel, see Cooke and Schwartz (2003).
18. Marshall (1918: 603). Notice how 'dissatisfaction' with the status quo is the stimulant of restless innovative intent by both user and producer in this telling instance.
19. The difference between disruptive and radical innovation is as follows. Following Christensen (1997) and others, disruptive innovation is revolutionary regarding techno-logical profile, for example, candle loses out to light bulb. Yet context, that is room, house, remain largely unaltered and its built environment can persist for hundreds of years. However, in the innovation systems tradition radical innovation changes both technolog-ical profile and context or environment, for example, air travel versus all other historic forms of travel. That is, airports, air traffic control systems, and so on are radical, literally 'root and branch', innovations in historic terms. Equating the two, as tends to happen in some business school literature is less helpful in conceptual terms than distinguishing them analytically.
20. 'PWC Money Tree Survey, 2004', www.pwcmoneytree.com.
21. This firm representative made a presentation on this subject to the EU BioLink project meeting at Genopole, Evry, near Paris in January 2005, www.biolink.org.il.

REFERENCES

Audretsch, D. and Feldman, M. (1996), 'Knowledge spillovers and the geography of innovation and production', *American Economic Review*, **86**, 630–40.
Balasubramanyam, V. and Balasubramanyam, A. (2000), 'The software cluster in Bangalore', in J. Dunning (ed.), *Regions, Globalization, and the Knowledge Based Economy*, Oxford: Oxford University Press.
Becattini, G. (2001), *The Caterpillar & the Butterfly*, Florence: Felice le Monnier.
Becattini, G. and Dei Ottati, G. (2006), 'The performance of Italian industrial district and large enterprise areas in the nineties', *European Planning Studies*, **14**, 1139–62.
Begg, I. (ed.) (2002), *Urban Competitiveness*, Bristol: Policy Press.
Bellandi, M. and Di Tommaso, M. (2005), 'The case of specialised towns in Guangdong, China', *European Planning Studies*, **13**, 707–30.

Belussi, F. (2006), 'In search of a useful theory of spatial clustering: agglomeration vs. active clustering', in B. Asheim, P. Cooke and R. Martin (eds), *Clusters & Regional Development*, London: Routledge.

Boschma, R. (2005), 'Proximity and innovation: a critical assessment', *Regional Studies*, **39**, 61–74.

Bottazzi, G., Dosi, G. and Fagiolo, G. (2002), 'On the ubiquitous nature of agglomeration economies and their diverse determinants: some notes', in A. Curzio and M. Fortis (eds), *Complexity and Industrial Clusters*, Heidelberg: Physica-Verlag.

Breschi, S. and Lissoni, F. (2001), 'Localised knowledge spillovers versus innovative milieux: knowledge "tacitness" reconsidered', *Papers in Regional Science*, **80**, 255–73.

Burton-Jones, A. (1999), *Knowledge Capitalism*, Oxford: Oxford University Press.

Caniëls, M. and Romijn, H. (2003), 'Localised knowledge spillovers: the key to innovativeness industrial clusters?', paper presented to conference on Reinventing Regions in the Global Economy, Pisa, 14–16 April.

Chesbrough, H. (2003), *Open Innovation*, Boston, MA: Harvard Business School Press.

Cheshire, P. and Magrini, S. (2002), 'The distinctive determinants of European urban growth: does one size fit all?', *Research Papers in Spatial & Environmental Analysis*, **73**, Department of Geography, London School of Economics.

Christensen, C. (1997), *The Innovator's Dilemma*, Boston, MA: Harvard Business School Press.

Cooke, P. (2002), *Knowledge Economies*, London: Routledge.

Cooke, P. (2004), 'The accelerating evolution of biotechnology clusters', *European Planning Studies*, **12**, 915–20.

Cooke, P. (2005a), 'Rational drug design, the knowledge value chain and bioscience megacentres', *Cambridge Journal of Economics*, **29**, 325–42.

Cooke, P. (2005b), 'Regionally asymmetric knowledge capabilities and open innovation: exploring "Globalisation 2" – a new model of industry organisation', *Research Policy*, **34**.

Cooke, P. (2005c), 'Contrasts between biotechnology and ICT clustering: a UK study', presented at the Regional Studies Association conference on Regional Growth Agendas, Aalborg, 28–31 May.

Cooke, P. and Memedovic, O. (2003), *Strategies for Regional Innovation Systems*, Vienna: United Nations Industrial Development Organization.

Cooke, P. and Schwartz, D. (2003), 'Regional knowledge economy variations: an EU–Israel comparison', *Regional Industrial Research Report 45*, Cardiff: Centre for Advanced Studies.

Crotty, S. (2001), *Ahead of the Curve: David Baltimore's Life in Science*, Berkeley and Los Angeles, CA: University of California Press.

Department of Trade and Industry (DTI) (2005), *Comparative Statistics for the UK, European & US Biotechnology Sectors*, London: Department of Trade and Industry.

Dierickx, I. and Cool, K. (1989), 'Asset stock accumulation and sustainability of competitive advantage', *Management Science*, **35**, 1504–11.

Feldman, M. and Audretsch, D. (1999), 'Innovation in cities: science-based diversity, specialisation and localised competition', *European Economic Review*, **43**, 409–29.

Feldman, M., Francis, J. and Bercovitz, J. (2005), 'Creating a cluster while building a firm: entrepreneurs and the formation of industrial clusters', *Regional Studies*, **39**, 129–41.

Florida, R. (2002), *The Rise of the Creative Class*, New York: Basic Books.

Galison, P. (1997), *Image & Logic: A Material Culture of Microphysics*, London: University of Chicago Press.

Gereffi, G. (1999), 'International trade and industrial upgrading in the apparel commodity chain', *Journal of International Economics*, **48**, 37–70.

Glaeser, E., Kallall, H., Scheinkman, J. and Shleifer, A. (1992), 'Growth in cities', *Journal of Political Economy*, **100**, 1126–52.

Griliches, Z. (1992), 'The search for R&D spillovers', *Scandinavian Journal of Economics*, **94**, 29–47.

Haas, P. (1992), 'Introduction: epistemic communities and international policy coordination', *International Organisation*, **46**, 1–37.

Henderson, J., Dicken, P., Hess, M., Coe, N. and Yeung, H. (2002), 'Global production networks and the analysis of economic development', *Review of Political Economy*, **9**, 436–64.

Henderson, V. (2003), 'Marshall's scale economies', *Journal of Urban Economics*, **53**, 1–28.

Jacobs, J. (1969), *The Economy of Cities*, New York: Random House.

Jaffe, A., Trajtenberg, M. and Henderson, R. (1993), 'Geographic localisation of knowledge spillovers as evidenced by patent citations', *Quarterly Journal of Economics*, **108**, 577–90.

Klepper, S. (2002), 'The capabilities of new firms and the evolution of the US automobile industry', *Industrial & Corporate Change*, **11**, 645–66.

Krugman, P. (1995), *Development, Geography & Economic Theory*, Cambridge, MA: MIT Press.

Malmberg, A. and Maskell, P. (1997), 'Towards an explanation of regional specialisation and industry agglomeration', *European Planning Studies*, **5**, 25–42.

Malmberg, A. and Maskell, P. (2002), 'The elusive concept of localization economies: towards a knowledge-based theory of spatial clustering', *Environment and Planning A*, **34**, 429–49.

March, J. (1991), 'Exploration and exploitation in organisational learning', *Organisation Sciences*, **2**, 71–87.

Markusen, A. (1996), 'Sticky places in slippery space: a typology of industrial districts', *Economic Geography*, **72**, 293–313.

Marshall, A. (1918), *Industry and Trade*, London: Macmillan.

Marshall, A. (1920), *Principles of Economics*, London: Macmillan.

Marx, K. (1973), *Grundrisse*, Harmondsworth: Penguin.

Morgan, K. (2004), 'The exaggerated death of geography', *Journal of Economic Geography*, **4**, 3–21.

Organisation for Economic Co-operation and Development (OECD) (2004), *OECD Science, Technology and Industry Outlook 2004*, Paris: Organisation for Economic Co-operation and Development.

Owen-Smith, J. and Powell, W. (2004), 'Knowledge networks as channels and conduits: the effects spillovers in the Boston biotechnology community', *Organization Science*, **15**, 5–21.

Paniccia, I. (2006), 'Cutting through the chaos: towards a new typology of industrial districts and clusters', in B. Asheim, P. Cooke and R. Martin (eds), *Clusters & Regional Development*, London: Routledge.

Penrose, E. (1995), *The Theory of the Growth of the Firm*, 3rd edn, Oxford: Oxford University Press.

Polese, M. (2002), 'The periphery in the knowledge economy', www. inrs-ucs.uquebec.ca.

Porter, M. (1998), *On Competition*, Boston, MA: Harvard Business School Press.

Sapsed, J., Gann, D., Marshall, N. and Salter, A. (2005), 'From here to eternity? The practice of knowledge transfer in dispersed and co-located project organisations', *European Planning Studies*, **13**, 831–52.

Sarma, V. (2005), 'IT outsourcing: China versus India', *FS Market Insight*, 8 June.

Saxenian, A. (2000), 'Networks of immigrant entrepreneurs', in C. Lee, W. Miller, M. Hancock and H. Rowen (eds), *The Silicon Valley Edge*, Stanford, CA: Stanford University Press.

Zeller, C. (2004), 'North Atlantic innovative relations of Swiss pharmaceuticals and the proximities with regional biotech areas', *Economic Geography*, **80**, 83–111.

Zucker, L., Darby, M. and Armstrong, J. (1998), 'Geographically localised knowledge: spillovers or markets', *Economic Inquiry*, **36**, 65–86.

5. Varieties of business system and innovation

INTRODUCTION

We noted in the introductory chapter to this book that business systems literature exists and is complementary to our neo-Schumpeterian impulse in taking an institutional approach to economic evolution. It is particularly resonant with work conducted at the level of national innovation systems for the reason that it is national business systems that are the focus of study. Little sub-national research exists in this tradition, therefore its approach has only limited value at the *regional* level of analysis to the forefront here. Nevertheless, we shall return to this tradition later in the chapter when we examine its propositions concerning, particularly, distinctive entrepreneurship cultures as between liberal and coordinated market regimes. We already commented on the emphasis in national business systems research upon multinationals rather than SMEs, something shared to some extent with national innovation systems research. But there is somewhat more of an interest in the former, business systems approach, on aspects of talent-formation which also interest us. Indeed, there are lacunae in the national innovation systems approach regarding both talent and entrepreneurship, the latter surprising given Schumpeter's celebrated highlighting of this kind of actor in assisting the evolution of capitalism through facilitating 'creative destruction'. Although critical of the earlier predominance of the study of 'technology' in innovation studies, and correctly, seeking to present a more rounded picture, national innovation systems research has remained fairly relentlessly concerned with industries more than the inputs to industries except for its concern with 'science and technology' policies and the governance of that fairly narrow field. Hence, we wish to provide a somewhat more balanced conceptualization of innovation systems as embracing also talent and entrepreneurship in what follows.

As noted, much of the early work on innovation systems was conducted at the national level (Lundvall, 1992; Nelson, 1993; Edquist, 1997), responding mainly to the issue of whether the globalization process was undermining the ability of individual nations to influence their own technological sovereignty. However, more recently, researchers have tried to explore how innovative capabilities are sustained through regional communities of firms and supporting networks of institutions that share a common knowledge base and benefit from their shared access to a unique set of skills and resources. Innovative actions are increasingly highly dependent on localized or regionally based sources of knowledge and learning and, as production becomes more science-based, advantages such as developed research infrastructure, a highly qualified workforce and an entrepreneurial culture are becoming more important than natural resources. Heterogeneous and localized capabilities such as institutions, specialized resources and skills become invaluable assets to build firm-specific competences (Maskell and Malmberg, 1999). Regions are becoming increasingly laboratories of knowledge-creation and innovation (Wolfe, 2002; Cooke and Piccaluga, 2004) and the breeding ground for local policy networks. The local context, in which work and social life overlap, assumes a fundamental importance as the place where collective identity is produced and reproduced, mutual trust is reinforced and a flexible and effective network of economic and cognitive relations supporting knowledge creation and diffusion are embedded.

Within the same region different industrial agglomerations can coexist where disparate oligopolies and different sectoral specialization may be found. If several different types of economic activity are present then it is logical to expect different technological and sectoral systems of innovation to be found in the same region. But, as we argued in Chapter 4, there is also a regional 'sense of knowing' how institutionally such diversity is best integrated in relation to the institutional base. This gives impetus to possible socio-technical network differentiation within regions where innovation 'styles' differ. As we have hinted, biotechnology with its plethora of research-intensive entrepreneurship, needs different institutional resources from ICT with its stronger value-chain imperatives. Moreover, we must not forget the many regions which lack sectoral concentration and technological localization benefits. This may be due to low density, peripherality, lack of dynamic innovative firms and institutions, and the fact that they suffer debilitating knowledge asymmetries. The path dependency argument about the difficulty for embedded industrial complexes and trajectories to switch let alone restructure due to institutional 'lock-in' effects bears recollection here. To escape 'lock in', some peripheral regions focus on adapting key elements of their industrial age economy strength and the skills associated with that industry rather than 'leapfrogging'. However, this is by no means easy as the skills for opto-electronics, biotechnology or the media/multimedia industries are different from those of the industrial age. Nevertheless, one thing that has been learned is that the *skill content* of many 'information age' industries is less than many in the industrial age.

Information and communication technology utilization skills have even been identified as deskilling workers by providing generic not specific skills in 'flexible labour markets'.[1]

This chapter deals with the regional innovation system framework and the importance of entrepreneurship and talent as repositories of skills and knowledge crucial to regional innovation and development. Furthermore, it relates systemic regional innovation conceptually and institutionally to other kinds of innovation system, national, technological and sectoral in particular. Then it draws attention in a third section to the relationships with distinctive business systems, discussing three as per the literature, but finding two – entrepreneurial and associative – germane to the concerns of this book. Thereafter, the chapter proceeds by highlighting certain key dimensions of regional innovation systems to their 'economy culture', something foreshadowed in the introduction to the book. After exploring innovation, business regimes and talent formation systems in relation to the liberal market versus coordinated market paradigms, we begin to draw interim concluding propositions. These mostly concern the extent of convergence or divergence with respect to our selected industry sectors for Austria and the UK. These are then tested and elaborated in the empirical reportage arising from our primary comparative quantitative and qualitative research on contemporary and evolving modes of innovation. In doing these tasks attention is paid to the multi-level governance aspects of innovation and in particular how different varieties of capitalism may influence the development of a regional system of innovation. However, for fuller treatments of this the reader is directed to previous work by these authors.[2]

REGIONAL INNOVATION SYSTEMS: RELATIONS WITH NATIONAL SYSTEMS OF INNOVATION

It is often held that differences in economic performance between relatively more or less successful regions can be explained by looking at the mix of regional innovation policies and institutions that foster economic dynamism (Cooke and De Laurentis, 2002). The policies pursued by regional governments enhance the economy culture and identity of regions including the institutional capacity to attract, animate and construct competitive advantage. As already indicated in the introductory chapter and in Chapter 4, *collective entrepreneurship* by promotion of cooperative practices among actors may give regions distinctive trajectories in regional economic development. Cases in point where economic governance gave global identities to artificial regions are those of Emilia-Romagna and Baden-Württemberg respectively.[3] To become attractive for companies, territories can set up specific institutions to support their innovation strategies. Regions that have *constructed advantage* through supporting innovative enterprise, represent meaningful communities of economic interest, define genuine flows of economic activities and can take advantage of true linkages and synergies among economic actors. Regions have to seek competitive advantage from mobilizing all their assets including institutional and governmental ones where these exist, or press for them where they do not. As regions become more specialized and pull the institutional support structure along, so foreign direct investment (FDI) seeks out such centres of expertise by following domestic investment as part of a global location strategy. This is something predicted for countries like China and India, leaving Western regions pondering a future in which that stage of innovation system-building is now over (Cooke et al., 2000b).

What was learned in the 1990s is that assisting in the formation of network relationships among firms and the broader institutional setting supports firms' innovative activities but is not sufficiently powerful to embed them structurally when globalization conditions turn against such strategies. As Chapter 3 showed, regional innovation systems (RISs), while seen as a useful framework for studying economic and innovative performance, are also functional tools to enhance the innovation processes of firms (Asheim and Coenen, 2004). They do this by knitting together knowledge flows and the systems on which they rely, building trust and confidence in institutional reliability and, above all, they do it by generating institutional self-knowledge and a certain kind of collective dissatisfaction with the status quo. A regional innovation system comprises a set of institutions, both public and private, which produces pervasive and systemic effects that encourage firms within the region to adopt common norms, expectations, values, attitudes and practices, where a culture of innovation is nurtured and knowledge transfer processes are enhanced. A national system of innovation cannot adequately do this. Inevitably peak organisation leadership may be 'in the loop' but time economies alone, let alone distance decay effects, militate against thoroughgoing cognitive penetration. A dimension that seems almost completely missing in most national innovation system functionality is the capability to manage media communication about innovation in other than a dogmatic way. This may have devastating effects for furthering innovation strategies since labour supply and demand mismatches arise as market failures due to the absence of even 'virtual' knowledge spillovers. Contrariwise, regional innovation interaction among firms and other innovation organizations has been regarded as playing an important role in fostering regional innovation potential. Labour demand and supply are increasingly influenced by innovation, growth potential and linkage among firms within a defined location. We noted a certain

'portability' of ICT skills as a case in point where knowledge of labour
market opportunities filters through to appropriately qualified 'talent'. It
follows from this that the conditions of a system of innovation influence
the labour market dynamic and the ability of localities to generate, attract
and retain highly skilled workers who are essential for establishing and
growing innovative companies, as argued in Florida (2000).

Previous work has identified two sides of an innovation system: a supply
and a demand side (Braczyck et al., 1998). The former consists of the institutional sources of knowledge creation as well as the institutions responsible for training and the preparation of highly qualified labour power. The
demand side subsumes productive systems, firms and organizations that
develop and apply the scientific and technological output of the supply side
in the creation and marketing of innovative products and processes.
Bridging the gap between the two are a wide range of innovation support
organizations that play a role in the acquisition and diffusion of technological ideas, solutions and know-how throughout the innovation system.
These may include: skills agencies, technology centres, technology brokers,
business innovation centres, organizations in the higher education sector
and mechanisms of financing innovation such as venture capital systems.
One of the assumptions of the regional innovation systems approach is that
many innovative firms operate within regional networks, cooperating and
interacting not only with other firms such as suppliers, clients and competitors, but also with research and technology resource organizations,
innovation support agencies, venture capital funds, and local and regional
government bodies. Innovation, as we have seen, is a process that frequently
benefits from the proximity of organizations that can trigger this process.
Furthermore, regional authorities have an important role to play to
support innovation processes by offering services and other mechanisms
that augment the interlinkages between all these actors. A figure that summarizes the RIS assumption is provided in Figure 5.1.

We think this shows how the main connectivity vectors at regional level
are horizontal in contrast to those at national level, which are primarily vertical. Moreover although difficult to estimate it can be suggested that at least
half the key connectivity at national level is outwards to the global level.
This applies equally to larger firms and the activities of innovation ministers at numerous meetings on science and technology policy, technical
standards and designing international innovation programmes. These
include the Framework Programmes of the European Union, the broader
community in the European Research Co-operation Agency (EUREKA) or
bilateral collaborations. In some countries such ministries and their func-
tionaries are fully engaged in devising schemes to protect 'national champi-
ons' or to promote their defence industries to global military markets.

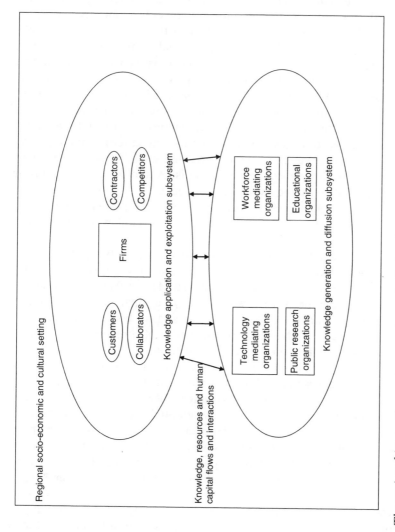

Figure 5.1   The regional innovation system: a schematic illustration

In most cases national states are engaged in science and technology strategy building.[4] It is difficult to avoid the inference that national innovation systems have lost salience considerably to the rise of, on the one hand, global innovation impulses – most notably from the USA – and, on the other, the more embedded approach feasible only at regional level but by no means generically present in all regions, probably still only a minority.

Where there is a rich innovation infrastructure, ranging from specialist research institutes, to universities, colleges and technology transfer agencies, and institutional learning is routine, firms have considerable opportunities to access or test knowledge, whether internally or externally generated to the region. A strong, regionalized innovation system is one with systemic linkages between external as well as internal sources of knowledge production (universities, research institutions, and other intermediary organizations and institutions providing government and private innovation services) and firms, both large and small. Most regions do not have these systemic innovation characteristics and in realistic terms an innovation system can be more or less systemic.

Following Cooke (1998) and Braczyk et al. (1998) different innovation systems can be measured upon (identified) the two following dimensions:

- The governance dimension, which comprises public policy, institutions and knowledge infrastructure; also known as soft infrastructure of enterprise innovation support. Here, reference is made to a networking propensity whereby key regional governance mechanisms, notably the regional administrative bodies, are interactive and inclusive with respect to other bodies of consequence to regional innovation. This may lead to an organizational setting in which the regional administration animates or facilitates associativeness among representative bodies inside or outside public governance.
- The business innovation dimension, namely, the industrial base characterized in terms of 'productive culture' and systemic innovation. This refers to the level of investment, especially in R&D; the type of firms and their degree of linkage and communication, in terms of networking, subcontracting, presence or absence of supply chains and degree of co-makership between customers and suppliers.

Following these two dimensions, Braczyk et al. suggested a taxonomy of regional innovation systems, and this is represented in Figure 5.2. Firms can range from possessing global to merely local reach. Three different types of RISs emerge: the *localist* one is characterized by the extent of the lack of domination by large indigenous firms and where the business innovation culture is one in which the research reach of firms is not very great,

	Grass roots	Network	Dirigiste	
Localist	Tuscany	Denmark	Tohoku (Japan)	Business innovation
Interactive	Catalonia	Baden-Wurttemberg	Gyeonggi (Korea)	
Globalized	Ontario	North Rhine-Westphalia	Wales	

Governance of enterprise/innovation support

Source: After Cooke et al. (2004).

Figure 5.2   Regional innovation systems: a taxonomy

although there may be local research organizations capable of combining with industry clusters within the region. A localist set up will probably have fewer major public innovation or R&D resources, but may have smaller privates one. Finally there will be a reasonably high degree of associationalism among entrepreneurs, and between them and local or regional policy-makers. An *interactive* RIS, on the other hand, is not particularly dominated by large or small firms but by a balance between them. The reach of this combination will vary between widespread accesses of regional research resources to foreign innovation sourcing as and when required. The mix of public and private research institutes and laboratories, in the interactive RIS, is balanced, reflecting the presence of larger firms with regional headquarters and a regional government keen to promote the innovation base of the economy. Concluding, in the third type of RIS, the *globalized* one, the innovation system is dominated by global corporations, often supported by clustered supply chains of rather dependent SMEs. The research reach is largely internal and highly privatistic rather than public, although a more public innovation structure aimed at helping SMEs may have developed.

Therefore, following Cooke et al. (2004), the governance dimension can generate three different RISs: grass roots, network and dirigiste. The RIS can be defined as *grass roots*, where the innovation system is generated and organized locally, at town or district level. Financial support and research competences are diffused locally, with a very low degree of supra-local or national coordination. Local development agencies and local institutional actors play a predominant role. A *network* RIS is more likely to occur when the institutional support encompasses local, regional, federal and supranational levels, funding is often guided by agreements among banks, government agencies and firms. The research competence is likely to be mixed, with both pure and applied, blue-skies and near-market activities geared to the needs of large and small firms. A *dirigiste* system is animated mainly from outside and above the region itself. Innovation often occurs as a product of central government policies. Funding is centrally determined, with decentralized units located in the region and with research competences often linked to the needs of larger, state-owned firms in or beyond the region. This offers the context within which we propose to develop the discussion that follows.

REGIONAL INNOVATION SYSTEMS AND THE LABOUR MARKET

The interface between labour supply and demand is increasingly based on regional distinctiveness, influenced by the specific innovative system in place, the presence or absence of universities, research and talent, the capacity of attracting new businesses, planning and housing policies, international migration, educational policies and the capability to generate, attract and retain a highly skilled labour force. Innovation activities require new ways of working, which pertain not simply to the relationships between firms, but also to interaction that reaches the public sphere of universities, research laboratories, technology transfer, training agencies and support organizations. These include project work, flexibility and outsourcing, including R&D outsourcing. Access to resources for innovation (skills and knowledge) have therefore become central to the competitive strategy of firms, the best of which have developed new flexible structures to better utilize and capture such advantages on a global scale. Knowledge exchange and learning are embedded within global, national, regional and local networks. The rapidly changing pattern of knowledge demand affects traditional labour market models and firms do not necessarily need to employ labour directly to gain access to the knowledge they require. Outsourcing, including to distant networks overseas is nowadays the key competitive strategy to in-house solutions. There is an increasing focus upon the need for reflexive practices in work, on the creation of communities of practitioners who can work reflexively in collaboration, and on collaborative knowledge-processing practice.

Institutions sensitive to the new context engage in partnership formation as a means of developing and promoting communities of practice. These have become the focus of the generation and management of knowledge within organizations. It is widely accepted that innovation networks are likely to develop within regions where there is a widespread policy interest in strategies supporting the development of networks of firms engaged in formal and informal vertical interactions and removing constraints on the development and functioning of such interactive processes. One of the assumptions of the regional innovation systems approach is that many innovative firms operate within regional networks, cooperating and interacting not only with other firms such as suppliers, clients and competitors, but also with research and technology resource organizations, innovation support agencies, venture capital funds, and local and regional government bodies. Innovation is a learning process that benefits from the proximity of organizations that can trigger this process.

Therefore the labour market dimension becomes an increasingly important element within regional innovation systems, and as suggested by Hommen and Doloreux (2003) it can be argued that the *embodiment* of knowledge in the regional workforce is one of the primary mechanisms through which processes of learning and knowledge transfer occur. Two features appear of particular importance in understanding the link among RISs and the labour market. On the one hand, according to Breschi and Lissoni (2001) a crucial mechanism through which knowledge flows across firms and regions is represented by the mobility of individual workers. As workers that embody relevant knowledge move locally, they help diffuse knowledge through a certain region and technological or sectoral innovation system. This type of externality was first identified by Marshall (1918) as external labour market economies, where a localized industry attracts and creates a pool of workers with similar skills, smoothing the effects of the business cycle both on unemployment and incomes through the effects of large numbers (Krugman, 1991).

In other words, the mobility of engineering, scientific and other talent across firms, between firms and academic institutions allows knowledge to diffuse locally. Thus the regional labour market may become an arena where a *talent pool* of technical knowledge and expertise is mobilized and a potential base where knowledge suppliers and users interact. Accordingly, the movement of people between labour markets, sectors and firms has important consequences for industrial functioning and innovation (Power and Lundmark, 2004). It follows that labour market policies and institutions affect the scope for the firm to appropriate rents generated through innovative activities and although product and labour market policies usually aim at objectives other than innovation, they may have important consequences for firms' innovative strategies (OECD, 2002). Alternatively, it can be argued that labour mobility is likely to create bonds and links between firms, workplaces and institutions, and thereby nurtures networking propensity (Power and Lundmark, 2004). The rise of the 'entrepreneurial university' (Smilor et al., 1993) and even promotion of the so-called 'triple helix' of interaction between industry, government and universities as a key feature of the knowledge economy (Etzkowitz and Leydesdorff, 1997) suggest how bonding processes may involve knowledge transfer through labour mobility as a key focus. This is also highlighted in the increasing interest in vocational educational networks (VET) (Cooke et al., 2002), consisting of educational partnerships, collaborative activities and cooperative ventures. These occur between universities and other educational providers and firms or other organizations that not only supply and maintain a skilled labour force but also enhance knowledge transfer within the economy through a variety of interactive learning and innovation processes (Rosenfeld, 2000).

This, in turn, obliges us to consider that labour mobility and networking propensity need to be supported by a local innovation culture, where the institutional and social context become increasingly central to fostering knowledge exchange. Regional authorities therefore have an important role to play to support this learning process by offering services and other mechanisms that augment the interlinkages between all regional actors. Not only has regional intervention become more important to economic success, but there has also been a qualitative shift in the form of local policy towards indigenous entrepreneurship and innovation, and to providing a more sophisticated environment for mobile capital so as to maximize local value-added (R&D and other high-status jobs, successful and therefore growing firms).

We must now turn our attention from talent and its role in constructing advantage for regional innovation systems and consider a key variant of talent – *entrepreneurship*. We will attempt this, in part, by reference to sectoral and technological systems literature. This is not least because the entrepreneurship literature places a great emphasis on the opportunities presented for entrepreneurship by the quality of the technological and sectoral offer in a given region. A study reported by Sternberg and Litzenberger (2004) is one of the first to spatialize the analysis of entrepreneurship, and in this instance Germany provides the analytical canvas. They note, in line with us, that entrepreneurial activities are perceived widely among policy-makers and regional scientists as providing an essential impetus for regional growth. Entrepreneurial attitudes and activities reveal an uneven spread over regional space. Where these concentrate geographically they may be a key stimulus to localized cluster formation. Sustainable clusters of this entrepreneurial type give rise to and draw resources from relational embeddedness in regional networks. When these have solidified institutionally, the resulting structural embeddedness may be said to constitute a regional innovation system (Granovetter, 1992).

In a critique of neoclassical economics' 'spaceless wonderland' of entrepreneurship, attention is drawn to the issue of spatial proximity we have found to be predominant in many aspects of regional innovation and growth, by indicating how proximity studies disclose the geographical influences exerted upon entrepreneurial start-up activity by local role models and the proximate 'me-too' effect. Moreover, entrepreneurship in proximity, whether or not leading to cluster networks forming, also creates markets for entrepreneurship support services in markets for management accountancy, advertising, training and e-business services, among others, as well as public support services, grant aid and so on. Thus Johannisson (2000) argues that entrepreneurship is a generic, social, collective phenomenon rather than an attribute solely of individuals in an argument that chimes well with our discussion in Chapter 4 on the subject of *collective entrepreneurship*. This further implies the presence of an entrepreneurial social infrastructure and associated actor networks as alluded to in Cooke and Wills (1999) after Flora et al. (1997). Although some important aspects of such infrastructure will be coterminous with the governance boundaries of the region to the extent services such as incubators or seed-funding are supplied at that level, others may be more global or local. Either of the last two might intersect, for example, with globalized technological systems and/or equally with localized sectoral specificities.

Thus as we will show, entrepreneurial start-up businesses in biotechnology clusters, as in Cambridge in the UK, habitually have other local firms from whom they buy inputs and others they may supply. But they also have close exploration and exploitation networks with firms and research institutes in Cambridge, Massachusetts, San Francisco and San Diego or Singapore with whom they are also part of a biotechnological system of innovation. Accordingly, they may have few direct network linkages with Cambridge ICT firms, though they may have strong links with some that engage in bioengineering, bioinformatics, or biosoftware. The ICT firms, by contrast, may belong to more than one technological system, for example designing software and circuitry for manufacture of advanced mobile telephony and Formula 1 motor sport control systems in the same factory, as shown in Cooke (2002). Alternatively, they may belong to subnetworks of similarly minded entrepreneurs engaged in global computer games design in which case they may have relatively few strong links with other ICT firms but extremely strong ties to common advice or financial sources such as business angel networks, copyright lawyers or venture capitalists. They may also be outsourcing targets for global corporations such as Sony, Nokia or Microsoft, not to mention producers of third and future generations of mobile telephony for whom 'content' is a key selling point.

Finally, talent availability both entrepreneurial and professional/technical is a sine qua non of the entrepreneurial social infrastructure. We have pointed to the global asymmetries in talent pool distribution and the attractiveness of these to many varieties of businesses seeking appropriately priced and qualified labour. Opportunities for corporate venturing, corporate spin-outs and corporate entrepreneurship are significant contributors to such talent pools. Thus merger and acquisition activity in pharmaceuticals often releases entrepreneurs or experienced managers who may benefit existing entrepreneurial businesses in the biotechnology sector, just as defence industry or telecommunications downswings and upswings often release bursts of entrepreneurship in proximity to the downsizing or 'rightsizing' incumbent corporations. On this, the case of Richardson, Texas, and such waves of entrepreneurship in telecommunications and military subcontracting was shown to be an exemplary case in Cooke (2002). Much the same is true in oil engineering locales such as Houston and Aberdeen, as suggested earlier. These are knowledge entrepreneurs with technical expertise in existing and evolving spheres of business in their own technological or sectoral innovation system able to sell such expertise to outsourcing firms and organizations on a worldwide basis. In Sternberg and Litzenberger's (2004) study they mobilize impressive quantities of data that show very clearly how the spatial distribution of industry concentrations is a major influence upon entrepreneurship which itself tends to concentrate in proximity to such areas of sectoral and technological concentration. Where these are also clusters rather than mere agglomerations the statistical associations are even stronger, demonstrating the power of collective entrepreneurship and entrepreneurial social networks in combination. They also report similar findings from other countries, notably the USA, suggesting that 'proximity knowledge' of this kind is relatively uncircumscribed by whether the economy in question is coordinated or liberal market in character.

FORMS OF ECONOMIC EVOLUTION: ENTREPRENEURIAL, ASSOCIATIVE AND DEVELOPMENTAL MODELS

This leads us neatly into a reconsideration of innovation from the national business systems perspective, which assists us in tying together numerous disparate conceptual and perspectival strands. The task is not as complex as might be expected since the master discourses of national business systems and varieties of capitalism are congruent and the intellectual approach of innovation systems perspectives is, in broad terms, equally institutional and evolutionary. This literature agrees that path-dependent trajectories of innovation propel often distinctive national or regional patterns of technological specialization and industrial development. We saw how national innovation systems literature addresses this but on a rather narrow range of institutions relating to science and technology policies and systems. The 'varieties of capitalism' and national business systems literatures (Hollingsworth and Boyer, 1997; Whitley, 1999; 2000; Hall and Soskice, 2001) pays attention to a much broader range of institutions, including financial markets, labour markets and the educational system. We have sought to highlight these also in our discussion of systems of innovation, particularly those of regional provenance. We think these institutional features to be of core importance to the development of particular forms of organization and related innovation trajectories within an economy. Building on this, we think knowledge production, entrepreneurship, talent and innovation strategies are strongly embedded in societal institutions. We have already anatomized key, competing alternative socioeconomic regimes, now we turn to models of development. Foremost are the following three developmental models, as distinct from varieties of capitalism, which distinguish the different institutional settings (see also Cooke and Morgan, 1998; Whitley, 2000):

- Entrepreneurial.
- Associative.
- Developmental.

In what follows, we outline the key institutional features of the three developmental models that are closely linked to the evolution of institutional and organizational capabilities sustaining innovation trajectories within different types of economies. These include: modes of economic coordination, labour market organization, education and training, capital and financial markets, R&D systems and their related constitutional set-up. What we describe below represents no more than ideal types. Many hybrid forms exist in practice, as noted by Whitley (2000). Industrialized countries in North America, Europe and East Asia take a position somewhere on a continuum between fully entrepreneurial and developmental state models of economic development. In general, however, one can say that industrialized countries are increasingly under pressure to adopt elements of the entrepreneurial model.

The Entrepreneurial Model of Development

The entrepreneurial model is associated with liberal market institutions that are thought likeliest to stimulate both disruptive and radical innovation. Economies organized around liberal market institutions such as the USA and the UK are believed to accommodate an entrepreneurial model of business organization and innovation styles and strategies. Coordination of these economies relies on market mechanisms, constrained, recently in the UK, by regulated competition in spheres that elsewhere are often state monopolies, though this is changing even in classic coordinated market settings such as Austria (see the introductory chapter). The labour market in this model is occupational in character offering a relatively high scope for job mobility between firms (Lam, 2000; 2002). Knowledge and skills are more individual-specific and portable than firm-specific and career-track determined. The efficient transfer and accumulation of knowledge arising from creative or scientific novelty is individualistic, hedged in by intellectual property rights, often assigned at least in part to the individual discoverer/inventor. Job mobility is based on opportunities to take risks and change jobs, especially where job opportunities are abundant elsewhere in the region (Florida, 2002).

The characteristic practice of limited-term, individualized employment contracts means this model enables firms to promote flexible employment strategies, rapidly changing their knowledge and talent base and competence structures as general change and more specific innovative imperatives demand. Accordingly radical innovation is stimulated (Casper et al., 1999; Hall and Soskice, 2001) at the risk of threats to social cohesion, especially during stock market downturns, and labour market polarization is more pronounced than in coordinated market settings. A somewhat polarizing educational system, which, in principle yields abundant highly qualified scientific personnel, nevertheless cannot keep pace with skills demand and substantial in-migration of professionally and technically skilled talent ensues, especially at the peak of the market. This can generate knowledge discontinuities, inadequate connectivity and increasing social distance between different talent categories hindering rapid diffusion and exploitation of knowledge in the absence of intermediaries. However, as we have shown, these are usually forthcoming at the behest of markets, but such assets are extremely spatially skewed and asymmetrically distributed. The R&D system is characterized by strong university-based R&D within some elite institutes and strong state-sponsored defence-oriented R&D. Entrepreneurship and intermediary market infrastructures to exploit such R&D results are often less than optimal, where for example the UK is more 'European' than the USA or Canada.

The corporate governance of this model is characterized by a high level of ownership coordination combined with a low level of alliance coordination (Whitley, 2000). Ownership is largely remote from the direction of the enterprises, there is little involvement of the workforce in the management of companies, and trade unions have a weak position. Even long-term connections between suppliers and customers may be terminated as cost or demand pressures develop. The model is dominated by large and liquid capital markets which, especially in boom times, promote risk and venture capital businesses for entrepreneurial, high-technology start-ups though, once again, asymmetrically in regional terms. The entrepreneurial model of development inhabits an institutional environment sceptical of long-term collaboration between economic agents. Rather, it facilitates an efficient reallocation of resources between firms and sectors, and a high level of flexibility and responsiveness according to market, and specifically rate of profit, signals. Temporary project and contract forms of organization thrive through outsourcing and a general history of subcontracting based on competitive tendering.

The Associative Model of Development

In contrast with the entrepreneurial model, the associative socio-economic model has more rule-bound, even legalistic regulation of economic activity. As we saw in discussing coordinated market regimes, the institutional framework for economic affairs rests upon consensual decision-making norms and high interdependence among economic actors. It may be regarded as a model of 'social partnership' between government, labour and industry, or even a model of 'public–private partnership', where the state negotiates with, and delegates important social and economic functions to, private associations. This is not common in entrepreneurial developmental models or their supporting liberal market regimes. Thus business and industry associations are key intermediaries regulating competition between business partners and facilitating collaboration in talent formation, business advice and representation of shared interests. This devolves into forms of self-management, which localize economic development and political stability (Cooke and Morgan, 1998).

Collaboration among economic institutions, egalitarian social norms and non-market patterns of coordination favour incremental innovation trajectories involving long-term cooperation between firms and cumulative organizational competence. Economic development through radical innovation and entrepreneurial start-ups in new fields needing organizational adjustments and rapid reallocation of resources across industries and technologies tend to be disfavoured (Soskice, 1997; Whitley, 2000).

The associative model is typically found in Germany, Austria, Switzerland, to some extent in the Netherlands and in distinctive, participatory forms in the Nordic countries. The associative model develops education and training systems that combine broad-based general education with deep vocational training for a wide spectrum of the workforce. This type of education system mixes formal academic knowledge and practical experience to develop organizational competence. Thus polarization of skills and labour markets is minimized, and decentralized work organization encourages interactive learning and the cultivation of tacit knowledge as key organizational capabilities. The dominant form of knowledge base for innovations is tacit: embedded in firm-specific product and process technologies combined with market and customer knowledge in the industry. It is conducive to successful high-quality incremental innovation strategies (Soskice, 1997).

The associative model also reinforces long-term cooperation among economic actors, emphasizing growth goals and the continuous exploitation of knowledge in established technologies. The business system usually combines a high level of ownership integration with strong interfirm linkages (Whitley, 2000). Risk-sharing is common between companies and there is cooperation among competitors in training and technical support assisted by industry associations and other intermediaries such as chambers of commerce and industry. The capital market is credit based, characterized by close links between banks and companies, including interlocking board-membership. Banks typically provide long-term investment or 'patient capital' for large investments with relatively low long-term risk. These are accordingly reluctant to finance risky, more entrepreneurial projects because of uncertainty regarding 'exit' options. The R&D system involves universities and public research establishments, often in cooperation. Cooperation with large and smaller enterprises also enables innovation gains from research findings to filter through the innovation chain rather than awaiting mainly entrepreneurial impulses from market signals. Close networking enables companies to combine their existing knowledge base with diverse sources of new knowledge to develop new product qualities for differentiated uses. Together, the R&D system and organizational competence support the development of complex product and process innovations, often at a leading scientific edge. In recent years the associative model has been challenged in several European countries by the interpretation of 'competition' and corresponding regimes of competition regulation provoked by the entrepreneurial developmental model. Despite this, the associative model remains persistent in many European countries. It is particularly notable that associative modes of governance play an increasingly important role in regions, transforming the regional level from being passive spaces subjected to corporate allocative decision-making into laboratories where regional institutions enhance trust and cooperation (Cooke and Morgan, 1998).

The Developmental State Model

On the face of it, this is less relevant to our deliberations in this book than the two other developmental models which, as was suggested, *dynamize* the hosting institutional regimes of liberal and coordinated markets respectively; except that, from a regional science perspective, the two regimes are too broad a brush to capture the asymmetries inherent in either of them. Thus within a liberal market economy it is easy to envisage a 'developmental state' economic strategy. Precisely this kind of analysis is performed in differentiating the economic governance strategies of the newly devolved regional administrations in the UK by Cooke and Clifton (2005). So we propose to examine this model to ensure elements that may be relevant at the regional level are not missed in reviewing conceptual frameworks that were devised to stereotype national entities. The developmental state model is widely practised in 'catch-up' contexts such as that of Asian economies, where a variant of the broader state-centric socio-economic model of development which can be found in France is practised (see Cooke and Morgan, 1998). Exemplars of the developmental state 'big-push' model of development are Japan and Korea but China, and to a lesser extent India, represent modern elaborations of Rosenstein-Rodin's (1943) famous stylization of this approach. Both Japan and South Korea, in seeking to catch up (and overtake) the West industrially and technologically, followed the route of the developmental or plan-rational state, rather than the market-rational or plan-ideological state. This was characterized by a strong, authoritarian, central government which deliberately and strategically supported large enterprises and industrial competitiveness (Nonaka and Reinmöller, 1998). Economic policy followed the sequence of import, import substitution, then export orientation. The emphasis of economic policy gradually shifted in time from trade policy to industry policy and, more recently, to innovation policy. There are many similarities between the state's role in Japanese and Korean socio-economic models of development, but in Japan it is more a case of 'state-guided' coordination, whereas Korea is the archetype of the centralized 'state-organized system'. Moreover, Japan was the pioneer while Korea followed, utilizing knowledge of Japanese mistakes to speed up the developmental process.

The education system is highly centralized and academic performance is a key to economic success. Unlike in the associative model, the public vocational system is weak, but this is compensated for by an effective firm-based training system which places a heavy emphasis on practical skills and on-the-job learning. A high value is also placed on practical skills, notably regarding engineers, not least because having borrowed ideas and imported technologies in order to develop, engineers played a crucial role in international technology transfer, and in translating theoretical knowledge into concrete operational details for production workers. These are important factors contributing to a strong organizational capacity for absorbing external technology, and the creation of knowledge through synthesis and combination. Large enterprises typically dominate labour markets, themselves characterized by long-term stable employment. Such long-term and stable relations between employers and employees encourage company investment in worker training in firm-specific skills inclined towards structures that create inflexibility in external labour markets.

The corporate governance in this developmental model is characterized by a high level of state-controlled ownership and coordination (Whitley, 2000). The state controls the capital market, banks are mostly state owned and provide favoured financial support for targeted large enterprises. Owing to stronger state involvement in banks, South Korea's large conglomerates (*chaebol*) are more strongly controlled by the state than in the case of Japan's dominating business groups (*keiretsu*). In addition, *chaebol* tend to be characterized by strong vertical integration processes and a centralized, hierarchical and kinship-based organization, whereas the *keiretsu* are less vertically integrated, centralized and hierarchical and have stronger networks with suppliers. In Japan, industrial districts are more important to endogenous knowledge creation and learning than in South Korea (Nonaka and Reinmöller, 1998). The innovation strategies of firms in developmental state economies have tended to focus on applied R&D projects to promote incremental product innovation. Science-based industries are relatively poorly developed, hence the relative weakness of such countries regarding biotechnology and Internet applications technologies. This is due to the weak position of universities in the national research system and a low level of public funding in basic research. Although the state does play an important role in helping firms to mobilize internal R&D resources, most R&D activities take place within private firms. The Japanese state funds far less R&D than any other major advanced economy. Moreover, formal linkages between university and industry in R&D collaboration are underdeveloped, leading to a lack of academic spin-offs and technology-oriented start-ups (Lam, 2002).

Thus we have reviewed the three predominant developmental models found in contemporary capitalism. Of importance are three features that bear upon the interests of this book. The first is the importance and attractiveness for innovation of the entrepreneurial model practised in North

American and European (mainly the UK and Ireland) liberal market regimes of market organization. This in turn places immense pressure on associative developmental models in coordinated market regimes to respond as swiftly to market signals because their coordinating mechanisms move much more slowly owing to the need to establish consensus prior to action. Second, this means many mechanisms favouring disruptive and even radical innovation can be relatively quickly mobilized and risk faced by specialist investment vehicles in the entrepreneurial model. The model is unstable, asymmetrically developed and prone to severe economic disequilibrium. However, it is capable of rapidly creating new industries, markets and employment opportunities – something that cannot be said of the coordinated market economies in the middle of the first decade of the twenty-first century. Finally, individual, person-specific skills reward flexibility and mobility of talent, including entrepreneurship talent in entrepreneurial developmental contexts, whereas firm-specific and rigidly prescribed skills substantially militate against this in coordinated market economies. In periods of lengthy economic stability and stable global trading patterns, coordinated market economies may outperform liberal market economies. However, once uncertainty, the rise of new competitor economies and the rise of new industries undermine such equilibrium conditions, liberal market economies are capable of adjusting and innovating more effectively, but with greater degrees of social polarization owing to the reward system entailed in stimulating entrepreneurship and innovation.

LINKING NATIONAL AND REGIONAL DIMENSIONS: CONVERGENCE OR DIVERGENCE?

We suggested in the concluding part of the previous section that it seems difficult for coordinated market economies to remain institutionally unchanging under contemporary global trading conditions, which favour entrepreneurship, talent and innovation in an industrial order where scientific, technological and creative *knowledge* are valued formally far more than ever before. This suggests convergence between regime categories is inevitable and to a noticeable extent if such economies are not to be overwhelmed by cost competition in their traditional trading sectors and left behind as incremental innovators, slow to move up the value chain owing to inadequate skills, financing systems, talent and innovative capabilities. It was shown conclusively in the introduction to this book that Austria, a paradigmatic case of an historically coordinated market economy, has sought to eschew precisely that fate by deregulating and introducing features of liberal market competitiveness into its institutional framework. The key question that this section of this chapter poses is, how does this operate on the ground in real time? For if a national system of governance is intent on a convergence path towards a new model, does that process proceed relatively evenly with regard to its regional spaces, or are new divergences set in motion at regional level? These questions are important because there is plenty of evidence that at the macro-level of the European Union precisely this paradoxical outcome occurred. That is, as national or member-state economic convergence proceeded, regional economic divergence was the result, something that is, as we have shown, exacerbated by knowledge economy spatial effects. The main one of these is increasing returns to proximity to large capital cities, mainly due to knowledge spillovers that give such locations quasi-monopoly knowledge-asymmetry advantages.

It goes without saying that the regional innovation systems approach is sensitive to such regional nuances. It also has as a key principle the idea of the region as an important level at which strategic innovation support is appropriate. Attention to the concepts of knowledge capabilities, interactive innovation, proximity, associational networking and clustering activities of public and private governance actors is coupled with the multi-level governance analysis of innovation systems. The literal meaning of the concept 'region'[5] recognizes the widespread existence of an important level of industry governance between the nation and the local. To varying degrees, *regional* governance as well as national governance are expressed both in private representative organizations such as branches of industry associations and chambers of commerce, and public organizations such as regional ministries with devolved powers concerning enterprise and innovation support, particularly for SMEs. Furthermore, there are few regions thus defined that do not possess increasingly important universities or polytechnics that can look outward to industry either for research commissions or as incubators for innovative start-up firms. This context acknowledges that, in line with the variation in regional innovation systems, there will also be different national components of each region that characterizes its capacity for transformation. Recall that our two comparator countries have very different post-war histories in terms of multi-level governance. Austria was immediately federalized after the Nazi interregnum. The UK was an almost completely unitary state with only Northern Ireland's devolved government occluding that unitariness. The UK has evolved its multi-level governance far more than Austria but it is still predominantly unitary with only perhaps Scotland now approaching something like federal powers.

An interesting question pursued in this book involves the extent to which, at regional economic governance level, there is convergence towards some set of common institutions and practices for supporting innovation in the knowledge economy. In the first, and so far only, longitudinal study of regional innovation systems, contained in Cooke et al. (2004), a definite convergence in innovation business and governance practices was visible. It is captured in Figure 5.3. The study tracked the noted regions and small countries from Europe, North America and Asia over the 1993–2003 period. While there are exceptions, such as Ontario and North Rhine-Westphalia that were in stasis having both experienced rightward political shifts, and Wales where a networking, integrative propensity had atrophied with the onset of the knowledge economy and offshoring by the inward investment firms that had driven innovation in the 1990s, elsewhere there was a common move towards networking in both dimensions. Could this be seen as convergence in disposition towards the imperatives of interactive innovation? Probably yes, since even the two liberal market regions had not regressed from a position in which they had governances and businesses that favoured network modes of interaction. In Wales political devolution meant that external agencies that had promoted it were taken into government, resulting in a regionally distinctive 'state-centric', possibly *developmental state* approach to promoting a liberal market and entrepreneurial development model. As noted in the account of such intra-member state spatial variation by Cooke and Clifton (2005), it is by no means unusual to observe varieties of business and economic governance practices between regions *within* countries. This has also been observed between regions in Germany (Esser, 1989), Italy (Garmise and Grote, 1990), and the USA (Osborne, 1990). In other words, returning to Figure 5.3, the rise of innovation up the economic governance and business practice agenda under knowledge economy conditions is mostly accompanied by the evolution of an 'economy culture' (Albert, 1993) of networking, interaction and connectivity both globally and regionally.

However, the research summarized in Figure 5.3 was not entirely focused on the kinds of high-technology sectors that are the focus in this book, where we would expect the networking propensity to be even more pronounced across regional and national regulatory regimes, albeit coordinated versus liberal conventions should also remain visible. Public-private risk investment versus private venture capital seems to continue as a fault line between the two regimes even at regional level. Research which asked regionally relevant questions of such 'varieties of economic governance' focusing on ICT and involving questions concerning entrepreneurship and talent formation is found in De Laurentis and Cooke (2005). Unlike the predominant emphasis in research on talent and entrepreneurship in high-technology sectors in different regional 'varieties of economic governance', which is elite orientated, this study[6] investigated the role of ethnic minorities in ICT, inquiring about training and entrepreneurship from within the

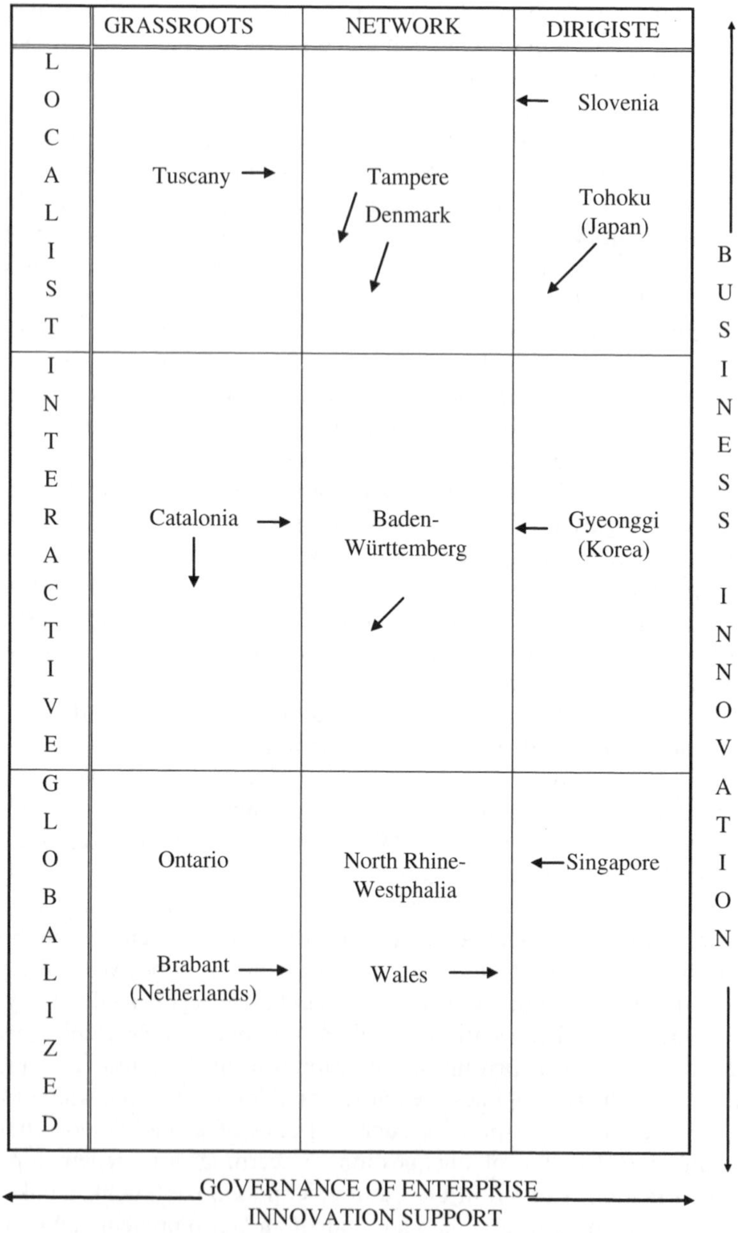

	GRASSROOTS	NETWORK	DIRIGISTE
L O C A L I S T	Tuscany →	Tampere / Denmark ↓	← Slovenia Tohoku (Japan) ↙
I N T E R A C T I V E	Catalonia → ↓	Baden-Württemberg ↗	← Gyeonggi (Korea)
G L O B A L I Z E D	Ontario Brabant → (Netherlands)	North Rhine-Westphalia Wales →	←Singapore

BUSINESS INNOVATION

GOVERNANCE OF ENTERPRISE
INNOVATION SUPPORT

Source: After Cooke (2004).

Figure 5.3 Regional innovation systems: typology and evolution

same perspectives – business systems, regulatory regimes, innovation systems – as this book adopts. The results were derived for six regions in three countries, in the UK, East Anglia, Scotland and Wales, in Sweden, Stockholm and East Gothia and in Italy, Lazio.

Before providing brief overviews that lead to an interim answer on the question of whether *innovation* imperatives override others noted above, causing a convergence at regional level towards networking and interactivity, recall that Sweden is firmly part of the coordinated market economic family. Sweden is usually identified with the Nordic variant, possessing a traditionally 'enabling' state that facilitates collaborative interfirm relations and cooperative labour management relations, with a high degree of labour cooperation, strong worker investment in skill acquisition, and cooperation and consensus in labour–management relations[7] (Culpepper, 2004). These differences in the national institutional frameworks of coordinated market economies, such as Austria, Sweden and elsewhere, notably Denmark, support different forms of economic activity, and their competitive advantage is mainly in diversified quality production. The UK, as we know, is supposedly different. Liberal market economies are most competitive in industries characterized by quite strongly science-based, technological, financial and creative, often radical, innovative activities.

As argued, the UK is usually identified among the liberal market economies, similar to the USA, Canada and Ireland; its institutional foundations are open and flexible markets, a state regulated to favour competition, residual welfare programmes and monetary stability. Despite its liberal market nature, the UK has strong regional development agencies especially, for over a quarter of a century, in Scotland and, until 2006, Wales. Nowadays nine English regions have had regional economic development agencies since 1999. These agencies promote liberal market ideals, but further, for example, entrepreneurship, innovation, clustering and talent formation through *associative* mechanisms. In this respect they are, in principle among the key change agents for economic convergence. Hence, indirectly, the UK government promotes networking and cooperation as an entrepreneurial business model utilizing the associative model *at regional level* as the key driver in innovation. This is new in that from 1979 to 1997 the UK government idealized the supremacy of liberal markets and thus eschewed regional devolution, government intervention in industry, and state ownership, indeed anything that smacked of 'coordinated market' type structures. So politics matter in institutional change, influencing modes of organization of the market, the education system, the labour market, the financial system and innovation processes, and contributing to the formation of divergent but also convergent 'business systems' (Whitley, 1999).

Varieties of Economic Governance: The Labour Market Dimension of Regional Innovation Systems

The interface between labour supply and demand is increasingly based on regional distinctiveness, influenced by the specific innovative system in place, the presence or absence of universities, the capacity for attracting new businesses, planning and housing policies, international migration, educational policies, and the capacity and the ability to generate, attract and retain a highly skilled labour force (Florida, 2002). Innovation activities require new ways of working, which pertain not simply to the relationships between firms, but also to interaction that reaches the public sphere of universities, research laboratories, technology transfer, training agencies and support organizations. These include project work, flexibility and knowledge outsourcing. Access to resources for innovation (skills and knowledge) have therefore become central to the competitive strategy of firms, which have developed new flexible structures to better utilize and capture such advantages on a global scale. Knowledge exchange and learning are embedded within global, national, regional and local networks. The rapidly changing pattern of knowledge demand undermines traditional labour market models and firms do not necessarily need to employ labour directly to gain access to the knowledge they require; institutions sensitive to this new context will engage in partnership formation as a means of developing and promoting communities of practice, which become the focus of the generation and management of knowledge within organizations. We have seen that innovation networks develop within regions where there is a widespread entrepreneurship and a policy interest in strategies supporting the development of networks. Firms are engaged in formal and informal, lateral and vertical interactions with only 'anti-trust' constraints on the development and functioning of such interactive processes.

One of the key assumptions of the regional innovation systems approach is that many innovative firms operate within regional networks, cooperating and interacting not only with other firms such as suppliers, clients and competitors, but also with research and technology resource organizations, innovation support agencies, venture capital funds, and local and regional government bodies. Innovation is a learning process that benefits from the proximity of organizations that can trigger this process. But do they do so in equivalent ways comparing liberal market and coordinated market regimes, particularly at regional level? The labour market dimension, therefore, increasingly becomes an important element within regional innovation systems and, as suggested by Hommen and Doloreux (2003), it can be argued that the *embodiment* of knowledge in the regional workforce is one of the primary mechanisms through which processes of learning and knowledge transfer occur. Two features appear of particular importance in understanding the link among RISs and the labour market. On the one hand, accordingly to Breschi and Lissoni (2001) a crucial mechanism through which knowledge flows across firms and regions is represented by the mobility of individual workers. As workers that embody relevant knowledge move locally, they help diffusing this knowledge through a certain region and industry. As discussed in Chapter 4, this type of externality was first identified by Marshall (1918) as labour market economies, where a localized industry attracts and creates a pool of workers with similar skills, smoothing the effects of the business cycle (both on unemployment and wage) through the effects of large numbers.

In other words, the mobility of engineers, scientists and other skilled workers among firms and between firms and academic institutions allows knowledge to diffuse locally. Indeed Best (2001) shows how the rapid development of economies like Singapore, South Korea and Taiwan can be pre-traced to their achievement rather rapidly of a tenfold increase in the proportion of science and engineering graduates per 10 000 population. China and India in the early years of the twenty-first century show comparable trend rates of increase in such graduate numbers. However, Japan, the Asian Tigers, the US and EU economies continue to register rates of 20 per cent and over, while China and India though increasing, had reached by 2004 only some 10 per cent (OECD, 2004). It is also clear that such labour markets are highly skewed in spatial terms. Therefore the regional labour market becomes a key arena where a pool of technical knowledge and expertise is mobilized and a potential base of knowledge suppliers and users interact. The movement of people between labour markets, sectors and firms has important consequences for industrial functioning and innovation (Power and Lundmark, 2004). It follows that labour market policies and institutions affect the scope for the firm to appropriate the rents generated through innovative activities and, although product and labour market policies usually aim at objectives other than innovation, they may have important consequences for firms' innovative strategies (OECD, 2002).

One of the key reasons for the relative success of regions such as Emilia-Romagna and Baden-Württemberg is the preponderance of distinctive but related varieties of engineering employment. These often exist in distinctive labour market pools or districts. This has an effect of minimizing labour market mobility somewhat, but it facilitates rapid knowledge flow and transfer between 'clusters', giving to the regions concerned the air of 'engineering platforms' of considerable power and strength.

Regional Innovation Systems: What Role in Promoting Integration?

The mobility of labour within more 'liberal' regional innovation systems is also regarded as a key mechanism through which knowledge flows and regions are increasingly becoming the arena where interaction and networking take place. The channel of interaction between different actors develops within specific social networks. Social networks affect individuals' abilities to attain their job, to function effectively on the job, and to gain promotion. According to Granovetter (1992), networks of social contacts provide access to valuable labour market information that is unavailable through more formal means. Hence, social contacts can act as an entry ticket into the labour market and can have a snowballing effect where early labour market success, mediated through social networks, can lead to more and more contacts being acquired enhancing career opportunities. Conversely, the homogeneity of most social networks means that using social networks in the labour market tends to produce and reproduce occupational segregation. Nonetheless, recent research undertaken in Silicon Valley (Saxenian, 1999; Saxenian and Hsu, 2001) shows the Valley's immigrant engineers rely on local social and professional networks to mobilize the information, know-how, skill and capital needed to start technology firms. In so doing, they have enhanced their own entrepreneurial opportunities as well as the dynamism of the regional economy overlapping with sector-specific networks, becoming part of the mainstream to grow. This research highlights that the most successful immigrant entrepreneurs in Silicon Valley today are those who have drawn on ethnic resources while simultaneously integrating into mainstream technology and business networks.

Within the regional innovation system framework, social networks can be regarded as an essential tool for explaining the logic of institutional arrangements between firms and their environment, the patterns of innovative activities and labour market dynamics within a specific locality. The guiding idea behind this analysis is that social networks are an important and often successful means that link people interpersonally, providing membership to local networks, allowing collective learning among firms and enhancing capabilities to link also governance bodies, including government programmes and policies. It can be argued therefore, that there is an important link between the regional innovation system framework and social integration as the challenge that involves building social and professional linkages as well as local networks that go beyond *self-contained* ethnic communities or homogenous networks. The governance of such an innovation system can therefore play an important role in building bridges between both mainstream and ethnic or female professional networks as well as between different ethnic and women's associations. Regional and local authorities establish forums that facilitate interaction between these traditionally separate communities and help them to articulate their shared problems as well as jointly to develop solutions. Through this process, policy-makers can learn more about measures they might undertake to improve the support infrastructure for innovation, facilitating relationships with local business communities.

Entrepreneurship: Ethnic Minority and Women Entrepreneurs

Ethnic groups and women can gain access to jobs in the ICT sectors not only by getting the necessary training (and participation in the relevant professional networks) but also by establishing their own firms. This section briefly reviews the literature on ethnic and female entrepreneurship to highlight the key determinants that drive entrepreneurial behaviour among these two groups. There is a growing evidence of the rapid expansion of 'ethnic entrepreneurship'. The trend towards a multicultural society, reflected particularly in urban areas, has created the seedbed conditions to develop new entrepreneurial activities that find their origin in the specific sociocultural habits of ethnic segments within the population. The concept mainly refers to SME activities undertaken by entrepreneurs with a specific sociocultural or ethnic background (Masurel et al., 2002).

Theories of ethnic entrepreneurship focus on how the ethnicity or ethnic minority status of an individual has an effect on his or her decision to enter business and highlight several types of explanations for business entry in the case of an individual belonging to an ethnic minority group. These can be summarized as follows:

- the role of discrimination in the labour market that pushes members of ethnic minorities into self-employment and small business
- the distinctive cultural features and values of different ethnic minority groups that explain their relative representation in business and emphasise their self-sufficiency and predisposition towards self-employment
- the importance, in many ethnic cultures, of strong family structures that facilitate access to resources such as family capital, family labour as well as 'free' information and advice
- the importance of business opportunities and constraints. The greater the resources and opportunities available to the ethnic minority

migrant, the higher the probability that he or she will venture into business.[8]

It emerges that the social network in which the ethnic community is embedded plays a key role as a driving force for starting a new business among ethnic population groups, regulating customer relationships, labour and capital conditions and business motivation (Masurel et al., 2002; Menzies et al., 2003). However, the literature highlights how this network may also hinder the success of the ethnic business, limiting the opportunities of entering more promising larger market segments and accessing adequate professional personnel and information sources.

A considerable volume of research has focused on gender differences in labour market status and participation, and female entrepreneurship has been indicated as a major force for innovation and growth. Much research about women entrepreneurs has sought to shed some light on the motivations that drive women to set up their own business. Although the literature has emphasized that the entrepreneurial motivations between women and men are often very similar, with independence and the need for self-achievement always being ranked first, some distinctions can be highlighted and only a small part of entrepreneurial motivations are acknowledged as gender based. Many authors distinguish among 'pull' and 'push' factors in order to explain different motivations for women to start a business. The former refers to elements of necessity such as insufficient family income, dissatisfaction with a salaried job, difficulty in finding work and a need for a flexible work schedule because of family responsibilities. The latter, pull factors, relates to independence, self-fulfilment, entrepreneurial drive and desire for wealth, social status and power (Orhan and Scott, 2001). Gender-based differences relate often to push factors such as the need for a flexible schedule, which reflects the family caring role that is still expected from women, the problem of the 'glass ceiling', which encourages women to start their own business operations because of an inability to obtain proper recognition by their employers, and the need to accommodate work and child-rearing roles simultaneously (Cromie, 1987; Orhan and Scott, 2001). Recent studies have also focused on gender-based differences in entrepreneurship and women's social orientations, arguing that women develop an interconnected system of relationships that include family, community and business (Brush, 1992). Carter et al. (2001) show the literature has also identified gender differences in the way networks are created and used, enabling improved access to finance and the development of strong relationships with financial backers. Differences emerge in both the establishment and management of networks (that is, the process of net-working) and in the contents of social networks (that is, what networks are used for), suggesting the women are more likely to have networks com-posed entirely of other women.

COMPARING REGIONAL INNOVATION SYSTEMS

Given the two dimensions of governance of innovation support and busi-ness innovation in the analysis of each regional system, the ensuing com-parative analysis is based on two important issues. On the one hand, we compare six regional innovation systems and appraise the type of govern-ance systems and the posture of firms in the regional economy both towards each other and in relation to other actors. On the other hand, this comparative exercise is also informative, offering an opportunity to analyse the dynamics that have occurred in the regional innovation systems exam-ined. First, some clear distinctions between the six regions emerge. The RIS of Wales is one characterized by a dirigiste governance system and global-ized business innovation, where initiation has been typically a product of central government policies with research infrastructure playing a marginal role in the emergence of its ICT clusters. Women entrepreneurs in ICT are rare but are found in media, including animation and multimedia. In terms of integration of ethnic minorities in the ICT sector, this primarily occurs through entrepreneurship in ICT retailing and the recruitment of 'to-and-fro' software engineers by large software producers, foremost among which is British Telecom which employs some 200 in its telecoms software facility in Cardiff. Wales like the rest of the advanced world has insufficient talent or training capacity to meet its software engineering needs and is forced to recruit, principally from India.

The Scottish innovation system can, to a lesser extent, also be seen as a dirigiste system where the government plays a key role in supporting innovation within the economy. However, Scotland is also character-ized by interactivity, where a combination of indigenous and foreign firms and public and private research infrastructure are interacting to promote and develop Scotland's position as a science-based economy. Entrepreneurship involving women and ethnic minorities is comparable to Wales and the labour market is similarly stretched to meet higher professional and technical demands. However, Scottish ministers perceive this as a weakness and have embarked on a policy of stimulating a more cosmopolitan labour market by encouraging foreign students to remain in Scotland as entrepreneurs and employees, with more impact in biosciences than ICT.

The East Anglian innovation system captures most of the features of the localist and interactive RIS, in which innovation and research competence are mixed and structured in a networked innovation architecture (network RIS), with a good representation between associationism in business, labour and entrepreneurial culture. Cambridge is also interesting for its indigenous public innovation institutions and its cluster building and strong networking oriented policies. Regarding ethnic and female entrepreneurship, Cambridge itself has a sizeable Chinese and some other Asian technical and professional cohort both employed and entrepreneurial in ICT. Cambridge and East Anglia more generally is rather minimally the subject of innovation governance, not least because liberal market processes function extremely well and both disruptive and radical ICT technologies emerge. A major cluster of advanced ICT exists near the British Telecom R&D laboratory (BT Exact Technologies) in East Anglia, at Adastral Park, Martlesham, near Ipswich. This facility recruits many Indian (elite) technically and professionally skilled workers through BT Mahinda, its Indian affiliate. Employment rather than entrepreneurship predominates and BT has in its 3000 workforce here only 9 per cent female employment and 3 per cent from ethnic minorities. Labour market shortages are endemic in this region to the extent that 'fast-tracking' specifically for ethnic elite ICT engineers has been introduced in the UK, favouring such ICT clusters as Cambridge and Martlesham.

The Swedish cases of Mjärdevi (Linköping, East Gothia) and Kista (Stockholm) draw attention in different ways to a more general pattern in Sweden of coordinated market regional innovation. In Sweden, the role of the central state is not simply limited to social and educational policies, but also extends to economic development policy, including development initiatives at the local or at the regional levels. This pattern is more pronounced in outlying regions, such as East Gothia. Thus, Saab's location in Linköping, as well as the creation of Linköping University, was decided by the Swedish central government. More recently, numerous local agencies that support ICT industry in places like Mjärdevi – for example the Technology Bridging Foundation – have been established as local branches of national organizations supported by central government funding. Ethnic and female entrepreneurship and employment in ICT here are both scarce and concentrated as female employees in Ericsson's R&D laboratory. Thirty per cent (370) of Ericsson's Mjärdevi workforce is female, of which 80 per cent occupy professional positions, echoing Ericsson's non-discriminatory policy and practice. Ericsson's workforce is 12 per cent (148) ethnic minority, mainly Middle Eastern (especially Iran) in origin. There is little enterprise support policy for ethnic and female entrepreneurs here and fewer practising ICT entrepreneurs from such backgrounds. However, a European Social Fund-backed programme in support of generic entre-preneurship for women and ethnic minorities was planned to start in early 2005. East Gothia's regional innovation system is becoming more of a 'network' RIS, where institutional support has encompassed local, regional, national and supra-national levels, funding is guided by agree-ments among banks, government agencies and firms, and research compe-tences are usually mixed, reflecting an even balance between basic and applied research, and between large- and small-firm needs.

In the Stockholm case, on the other hand, the ICT cluster, employing some 30 000 before the technology downturn, is an illustrative example of a globalized RIS (IBM, Sun Microsystems, Oracle, Ericsson) with a net-worked innovation architecture, and firms that have displayed rapid and high rates of growth influencing local supply industries. The development of Kista as a local growth pole has benefited to a great extent from urban-ization economies associated with location in a large capital city. Hence, the central state and lower levels of government have played a less prominent leading role than in Linköping/Mjärdevi. They are nevertheless involved in a supporting role, especially in the current development of Kista Science Park, where they figure as important actors. However, it is also true that changes have taken place in the past few years. The system of innovation in the Stockholm region has undergone several changes. The ICT downturn forced firms to move out from Kista and to change their specialization strategy. The outcome has been one of increased entrepreneurship, inter-activity and collective learning. Skilled workers moved into temporary organizations, creating project teams, and this support mechanism has pro-vided a framework for a formalized innovation measure, the Kista Innovation and Growth initiative. Many global firms on Kista such as IBM (25 per cent) and Ericsson (35 per cent) have sizeable ethnic minority work-forces. These are spread among all occupational levels, but 14 per cent of Kista's ethnic minority workers are in unqualified occupations. However, some ethnic minorities previously employed in larger firms have generated 'corporate spin-out' businesses. Female employment in these two firms is 31 per cent and 28 per cent respectively, and in Ericsson's workforce 17.5 per cent are women with PhDs and 18.5 per cent are qualified engineers. This shows the labour market demand is for less elite ethnic employees because talent formation policy in the Stockholm region is clearly non-discriminatory and women take up professional vacancies. This contrasts with the UK regions where talent is in short supply, perhaps because voca-tional training in ICT has been less inclusive.

Finally, the Lazio system of innovation can be seen as both dirigiste and localist because it is home to a number of ICT oligopolies associated with Rome's capital city function. Accordingly, a few large firms and organizations like national public research institutes predominate and display little institutional interaction or knowledge transfer. Other firms in ICT tend to be small to medium sized and entrepreneurial, concentrated in software and telecom services, supplying both local and national outsourcing markets. Italy had for a long time one of Europe's most homogeneous populations but this is changing. The demand for low-wage ethnic minority workers, which is where many concentrate in other sectors, is low in ICT; ethnic minority entrepreneurship is also negligible. Female entrepreneurs have a strong presence in an ICT incubator in Cinecittà where 30 per cent are women. More generally, as in Wales, women entrepreneurs are found in the film and other media sectors more than in most others in ICT. The largest employment sector for women in ICT in Lazio is in the call centre sector where 62 per cent of the workforce is women. Another important sector is telecom equipment manufacturing where 22 per cent are women. In neither case are these occupying other than a few high-skilled occupations. This analysis, incorporating elements of often overlooked entrepreneurship, ethnicity, gender and labour market elements in typologizing regional innovation systems shows interesting intra-country variation in which Wales and Scotland hardly conform to the UK's supposedly liberal market traditions, whereas East Anglia or Eastern England clearly does, suggesting liberal markets are strongly Anglo-Saxon, rather than Anglo-Celtic constructs, as is often stated as shorthand in the relevant literature. Accordingly, there appear to be 'coordinated market' asymmetries within the UK's unitary state. East Gothia and Stockholm are far more similar than the UK exemplars, while Lazio, with no national comparator, nevertheless looks very different from Emilia-Romagna, Marche or Tuscany. The remaining issue in this analysis concerns the dynamics of these regional trajectories given our firm belief that ICT (and biotechnology) innovation conventions force firms and economic governances towards a networking and interactive predisposition. Thus the key question is: are these innovation systems like those shown in Table 5.1, actually converging towards a common point. The answer is given in Figure 5.4, which shows in bold terms that the majority of regional innovation systems are converging, as were the ones in Figure 5.3 towards a networking and interactive character. Of course, some are more clearly doing this than others and Wales, as we saw before, seems to becoming more state-centric or dirigiste. Also, noticeable in Table 5.1 is that some systems have a kind of shadow occupancy of more than one innovation predisposition as they have evolved. But the strong hypothesis is of convergence towards the centre, at least at regional level. The question this raises is to what extent should a middle category be invented to reflect the fact that not all economic regimes occupy binary space? The case is now strongly made that, for example, Denmark has

Table 5.1   Economic governance and innovation dispositions of regional innovation systems

Economy regime	*Liberal market* e.g. UK	*(Social market)* e.g. Denmark	*Coordinated market* e.g. Austria
Development model	*Entrepreneurial* e.g. USA	*Associative* e.g. Sweden	*Developmental state* e.g. Singapore
RIS innovation (ICT)	*Localist* East Anglia Lazio	*Interactive* East Gothia Stockholm (Scotland)	*Globalized* Wales Scotland (Stockholm)
RIS governance (ICT)	*Grass roots* East Anglia	*Network* East Gothia Stockholm	*Dirigiste* Wales Lazio Scotland

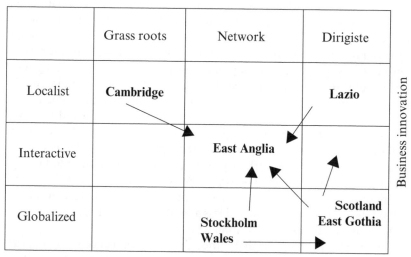

Governance of enterprise/innovation support

Figure 5.4   Regional ICT innovation system trajectories

become different to some degree from its Nordic counterparts and from its own history, with a highly deregulated market sector that is liberal market in all but name. Correspondingly, however, Denmark retains the key institutions of its traditionally generous welfare state arrangements. Hence in

Table 5.1 this was inserted as an 'economy regime' under the shadow category of 'social market', something we shall also be alert to exploring in relation to our empirical findings on Austria presented in Chapter 8.

CONCLUSIONS

This chapter aimed at highlighting variations among regional economies in a context of varying institutional and developmental models. First, we could argue that networking regimes were established around the issues of growth in the ICT and related science sector. A constellation of forces came together to promote growth through strategic use of resources and decision-making. Some of the cases under investigation, such as East Anglia, Stockholm and East Gothia, showed how strong university involvement and an active business sector matched local, regional and national government efforts. Wales and Scotland, on the other hand, relied more on public support to promote systemic innovation. Although the coalitions varied from case to case, all cases show how processes of learning and knowledge transfer occur. Knowledge became embodied within the regional system and the mobility of workers within different types of institutions allowed knowledge to flow. Universities became key actors in this process of mobility and transfer, moving from a strict role as provider of skills to now becoming more engaged in supporting innovation. They are also leading local policies towards a more entrepreneurial approach providing a more sophisticated environment for mobile capital to maximize local value-added. Other regional institutions like the local innovation culture and entrepreneurship context also became important in fostering knowledge. Regional authorities thus play an important role in providing an adequate infrastructure to nurture a networking propensity and augment the interlinkages between regional and global actors. However, as the case of Wales shows, an innovation system heavily based on public knowledge generation and exploitation can become rigid and less adaptable to changes, thus hindering systemic interaction and innovation within the region.

The chapter also argued that although the region is becoming increasingly the level at which strategic innovation support is appropriate, there will also be different national components that influence a region's capacity for transformation. The 'varieties of capitalism' framework was suggested in order to understand some of the differences that the regions, particularly in the UK and Sweden, have experienced in their innovation trajectories. On the one hand, it can be argued that within the two Swedish case studies, the national state played a key role in the formation and the success of their regional innovation systems. As argued, Sweden as a coordinated market economy has shown a natural bent to coordination which facilitated networking and collaborative partnership among key actors. On the other hand, in parts of the UK the state has not played a key role in promoting interactive innovation and the market has been the dominant force, as the cases of Cambridge and Martlesham in East Anglia showed. However, despite the liberal market nature of the UK, Wales (until 2006) and Scotland over a quarter of century developed strong regional agencies, aimed at developing the regional economy and they were the main vehicles for managing policies in support of innovation and coping with the demands of the knowledge economy. Both were quite hierarchical, technocratic bodies, but Scotland's agency has embraced knowledge economy principles closely while Wales was more wedded to a 'public enterprise' mode of top-down stimulation of entrepreneurship and innovation, especially since foreign direct investment is no longer as prominent as it was as a source of new knowledge. In 2006 the Welsh Development Agency was closed.

NOTES

1. This was a key conclusion of research in the UK ESRC Learning Society research programme. See, for example, Cooke et al. (2000a).
2. In particular, Cooke et al. (2000b) and Cooke (2002).
3. On this, see in particular, Cooke and Morgan (1998).
4. Recall note 12 in Chapter 4 where a remarkable similarity was to be seen among the science and technology strategies of many OECD member countries. Organizations such as the OECD and the EU are the forums at which such 'me-too' strategizing inputs are formed. Biotechnology and ICT sectoral strategies are favoured nearly throughout the OECD. Notable in its exceptionalism is the USA, for whom the sectoral innovation approach seems almost passé compared with an approach that proposes more complex innovation platforms seeking to tackle pervasive socio-technical problems.
5. Region (Latin *regio* from *regere* meaning 'to govern') can be defined as a meso-level political unit set between the national or the federal and local levels of government that might have some cultural or historical homogeneity but which at least has some statutory powers to intervene and support economic development, particularly innovation.
6. RISESI: The Regional Impact of the Information Society on Employment and Integration. Funded by the Information Society and Technology (IST) Action under EU Framework Programme 5.
7. However, Sweden, like Austria, has in recent years moved to a lower coordination, with more decentralized wage-bargaining to sectoral level. This change was largely driven by employers, who were responding to the increasing attraction of firm-level bargaining owing to technological change in the production process (Culpepper, 2004).
8. The ethnic resources referred to are the access to relatively cheaper labour and capital from within the ethnic minority community trading experience, as well as cultural characteristics such as family structure and the trading ethic of certain groups. Opportunities include the demand for ethnic goods and labour-intensive services, the existence of unattractive or neglected markets such as those in inner city areas, and industries that face unstable market conditions (Aldrich and Waldinger, 1990).

REFERENCES

Albert, M. (1993), *Capitalism Against Capitalism*, London: Whurr.

Aldrich, H. and Waldinger, R. (1990), 'Ethnicity and entrepreneurship', *Annual Review of Sociology*, **16**, 111–35.

Asheim, B. and Coenen, L. (2004), 'The role of regional innovation systems in a globalising economy: comparing knowledge bases and institutional frameworks of Nordic clusters', paper presented at DRUID Summer Conference, Industrial Dynamics, Innovation and Development, Elsinore, June.

Best, C. (2001), *The New Competitive Advantage*, Oxford: Oxford University Press.

Braczyk, H., Cooke, P. and Heidenreich, M. (eds) (1998), *Regional Innovation Systems*, London: UCL Press.

Breschi, S. and Lissoni, F. (2001), 'Localised knowledge spillovers versus innovative milieux: knowledge "tacitness" reconsidered', *Papers in Regional Science*, **80**, 255–73.

Brush, C, (1992), 'Research on women business owners: past trends, a new perspective and future directions', *Entrepreneurship: Theory and Practice*, Summer, 5–30.

Carter, S., Anderson, S. and Shaw, E. (2001), 'Women's business ownership: a review of the academic, popular and Internet literature', report to UK Small Business Service.

Cooke, P. (1998), 'Introduction: origins of the concept', in H.J. Braczyk, P. Cooke and M. Heidenreich (eds), *Regional Innovation Systems*, London: UCL Press.

Cooke, P. (2002), *Knowledge Economies*, London: Routledge.

Cooke, P. (2004), 'The regional innovation system in Wales: evolution or eclipse?', in P. Cooke, M. Heidenreich and H.J. Braczyk (eds), *Regional Innovation Systems*, 2nd edn, London: Routledge.

Cooke, P. and Clifton, N. (2005), 'Visionary, precautionary and constrained "varieties of devolution" in the economic governance of the devolved UK territories', *Regional Studies*, **39**, 437–51.

Cooke, P. and De Laurentis, C. (2002), 'The index of knowledge economies in the European Union: performance rankings of cities and regions', *Regional Industrial Research Report No. 41*, Cardiff: Centre for Advanced Studies.

Cooke, P. and Morgan, K. (1998), *The Associational Economy*, Oxford: Oxford University Press.

Cooke, P. and Piccaluga, A. (eds) (2004), *Regional Economies as Knowledge Laboratories*, Cheltenham, UK and Northampton, MA, USA: Edward Elgar.

Cooke, P. and Wills, D. (1999), 'Small firms, social capital and the enhancement of business performance through innovation programmes', *Small Business Economics*, **13**, 219–34.

Cooke, P., Heidenreich, M. and Braczyk, H. (eds) (2004), *Regional Innovation Systems*, 2nd edn, London: Routledge.

Cooke, P., Boekholt, P. and Tödtling, F. (2000b), *The Governance of Innovation in Europe, Regional Perspectives on Global Competitiveness*, London: Pinter.

Cooke, P., Cockerill, A., Scott, P., Fitz, J. and Davies, B. (2000a), 'Working and learning in Britain and Germany: findings of a regional study', in F. Coffield (ed.), *Differing Visions of a Learning Society*, Bristol: Policy Press.

Cromie, S. (1987), 'Motivations of aspiring male and female entrepreneurs', *Journal of Occupational Behaviour*, **8**, 251–61.

Culpepper, P. (2004), 'Focal points and power plays in institutional change: an empirical assessment', *Urban Studies*, **41**(5/6), 1025–44.

De Laurentis, C. and Cooke, P. (2005), *Regional Innovation Systems and Social Inclusion: From Promoting Innovation to Sustaining Integration*, Report to the EU FP5 IST Project 'The Regional Impact of the Information Society on Employment and Integration' (RISESI), European Commission, Brussels.

Edquist, C. (1997), *Systems of Innovation*, Cheltenham, UK and Northampton, MA, USA: Edward Elgar.

Esser, J. (1989), 'Does industrial policy matter? *Land* governments in research and technology policy in federal Germany', in C. Crouch and D. Marquand (eds), *The New Centralism: Britain Out of Step?* London: Macmillan.

Etzkowitz, H. and Leydesdorff, L. (eds) (1997), *Universities and the Global Knowledge Economy: A Triple Helix of University–Industry–Government Relations*, London: Pinter.

Flora, J., Sharp, J. and Flora, C. (1997), 'Entrepreneurial social infrastructure and locally initiated economic development in the non-metropolitan United States', *Sociological Quarterly*, **38**, 623–45.

Florida, R. (2002), *The Rise of the Creative Class*, New York: Basic Books.

Garmise, S. and Grote, J. (1990), 'Economic performance and social embeddedness. Emilia-Romagna in an interregional perspective', in R. Leonardi and R. Nanetti (eds), *The Regions & European Integration: The Case of Emilia-Romagna*, London: Pinter.

Granovetter, M. (1992), 'Problems of explanation in economic sociology', in N. Nohria and R. Eccles (eds), *Networks & Organisations: Structure, Form & Action*, Boston, MA: Harvard Business School Press.

Hall, P. and Soskice, D. (eds) (2001), *Varieties of Capitalism: The Institutional Foundations of Comparative Advantage*, Oxford: Oxford University Press.

Hollingsworth, R. and Boyer, R. (eds) (1997), *Contemporary Capitalism: The Embeddedness of Institutions*, Cambridge: Cambridge University Press.

Hommen, L. and Doloreux, D. (2003), 'Is the regional innovation system concept at the end of its life cycle?', paper presented to the conference 'Innovation in Europe: Dynamics, Institutions and Values', September, Roskilde University, Denmark.

Johannisson, B. (2000), 'Modernising the industrial district – rejuvenation or managerial colonisation', in M. Taylor and E. Vatne (eds), *The Networked Firm in a Global World: Small Firms in New Environments*, Aldershot: Avebury.

Krugman, P. (1991), *Geography & Trade*, Cambridge, MA: MIT Press.

Lam, A. (2000), 'Tacit knowledge, organizational learning and societal institutions: an integrated framework', *Organization Studies*, **21**, 487–513.

Lam, A. (2002), 'Alternative societal models of learning and innovation in the knowledge economy', *International Social Science Journal*, **171**, 67–82.

Lundvall, B. (ed.) (1992), *National Systems of Innovation*, London: Pinter.

Marshall, A. (1918), *Industry and Trade*, London: Macmillan.

Maskell, P. and Malmberg, A. (1999), 'Localised learning and industrial competitiveness', *Cambridge Journal of Economics*, **23**, 167–85.

Masurel, E., Nijkamp, P. and Vindigui, G. (2002), 'Motivations and performance conditions for ethnic entrepreneurship', *Growth and Change*, **33**, 238–60.

Menzies, T., Brenner, G. and Filion, L. (2003), 'Social capital, networks and ethnic minority entrepreneurs', in H. Etemad and R. Wright (eds), *Globalization and Entrepreneurship*, Cheltenham, UK and Northampton, MA, USA: Edward Elgar.

Nelson, R. (ed.) (1993), *National Innovation Systems*, Oxford: Oxford University Press.

Nonaka, I. and Reinmöller, P. (1998), 'The legacy of earning: toward endogenous knowledge creation for Asian economic development', in H. Albach, M. Dierkes, A. Antal and K. Vaillant (eds), *Organisationslernen – institutionelle und kulturelle Dimensionen*, Berlin: edition sigma (WZB-Jahrbuch 1998), pp. 401–32.

Organisation for Economic Co-operation and Development (OECD) (2002), *Labour Market Institutions, Product Market Regulation, and Innovation: Cross Country Evidence*, Paris: OECD.

Organisation for Economic Co-operation and Development (OECD) (2004), *Science, Technology & Industry Outlook*, Paris: OECD.

Orhan, M. and Scott, D. (2001), 'Why women enter into entrepreneurship – an explanatory model', *Women in Management Review*, **16**, 232–43.

Osborne, D. (1990), *Laboratories of Democracy*, Boston, MA: Harvard Business School Press.

Power, D. and Lundmark, M. (2004), 'Working through knowledge pools: labour market dynamics, the transference of knowledge and ideas, and industrial clusters', *Urban Studies*, **41**, 1025–44.

Rosenfeld, S. (ed.) (2000), *Learning Now*, Washington, DC: Community College Press.

Rosenstein-Rodin, P. (1943), 'Problems of industrialisation of Eastern and Southeastern Europe', *Economic Journal*, **53**, 202–11.

Saxenian, A. (1999), *Silicon Valley's New Immigrant Entrepreneurs*, San Francisco, CA: Public Policy Institute of California.

Saxenian, A. and Hsu, J. (2001), 'The Silicon Valley–Hsinchu connection: technical communities and industrial upgrading', *Industrial & Corporate Change*, **10**, 893–920.

Smilor, R., Dietrich, G. and Gibson, D. (1993), 'The entrepreneurial university: the role of higher education in the United States in technology commercialisation and economic development', *International Social Science Journal*, **45**, 1–11.

Soskice, D. (1997), 'German technology policy, innovation and national institutional frameworks', *Industry and Innovation*, **4**, 75–96.

Sternberg, R. and Litzenberger, T. (2004), 'Regional clusters in Germany – their geography and relevance for entrepreneurial activities', *European Planning Studies*, **12**, 767–91.

Whitley, R. (1999), *Divergent Capitalisms: The Social Structuring and Change of Business Systems*, Oxford: Oxford University Press.

Whitley, R. (2000), 'The institutional structuring of innovation strategies: business systems, firm types and patterns of technical change in different market economies', *Organization Studies*, **21**, 855–86.

Wolfe, D. (2002), 'Social capital and cluster development in learning regions', in A. Holbrook and D. Wolfe (eds), *Knowledge, Clusters and Regional Innovation*, Montreal: McGill-Queen's University Press.

PART II

Empirical Findings

6. Introduction to key research results

In this chapter we first describe and compare the main features of the two industries that are an important focus of the empirical research of Chapters 7 and 8, ICT and biotechnology. We briefly point out their development and key activities, characterize their knowledge base and innovation processes, and deal with the character of their knowledge flows and innovation links. We will see that, although both can be considered as 'knowledge-intensive sectors', there are striking differences in this respect. In the second section of this chapter we point out the main research questions as well as the methodology and the steps of our empirical research.

BIOTECHNOLOGY AND ICT IN A COMPARATIVE PERSPECTIVE

Two key knowledge-based industries are at the centre of the following empirical analyses for the UK (Chapter 7) and Austria (Chapter 8): ICT and biotechnology. These two sectors have been selected for several reasons:

- ICT and biotechnology both rely on knowledge as a key input and innovation is at the core of competitive strategies. They can be regarded, therefore, as representative of the knowledge economy. However, there are quite different types of innovation and related processes involved, revealing the multifaceted character of the knowledge economy.
- They rely on codified, tacit and complicit forms of knowledge by exploiting a widely distributed knowledge base (Chapter 3; Smith, 2002). Both local and global knowledge sources and partners have to be accessed and used. A key challenge in order to stay competitive is in fact the successful integration of these various forms and sources of knowledge in the innovation process.
- They both represent 'platform' or general purpose technologies (Helpman, 1998) having impacts across many industries. Considerable effects can be seen not only in the ICT and biotechnology sectors themselves, but also in many other sectors such as ICT users and sectors affected by biotechnology.

- The two sectors are growing strongly and, thus, are becoming important motors for employment generation and wealth creation in the emerging knowledge economy.

Besides these similarities there are also some striking differences between these two sectors, as can be seen from the following sections.

Information and Communication Technologies: Short Cycles, Few Patents

The ICT industry has its roots in the radio and telephone industry as well as the electrical appliances industry of the early twentieth century. Owing to its strategic and military importance in the two world wars it experienced public protection, regulation and attracted public investments. After the Second World War the sector was boosted by the development of the electronics, computer and software industries. The telecom system later shifted from analogue to digital technology, a revolutionary change which has triggered many subsequent innovations in the sector. Technological change has been so fast that the 1990s saw a speculative 'New Economy' boom, driven largely by Internet and dot.com companies. This resulted in a spectacular bursting of the bubble in 2001 followed by a modest recovery since then.

The ICT sector is rather heterogeneous. It comprises the production of ICT-related goods such as office machinery, computers and peripherals, telecommunication equipment (including telephones, cables and switches), radio, television and electronic instruments. Then it includes services such as telecommunication services, software development, consultancy, data processing and database activities. This heterogeneity is also reflected in firm sizes and firm types. At one end of the spectrum we find large and global companies in areas where scale or network economies are important (for example, semiconductors, processors, telecom providers). At the other end there are an increasing number of small and local companies, in particular in ICT-related services such as software development and consulting, which often are part of local clusters.

In the past decades we have seen considerable changes in the organization of the sector as was demonstrated for the computer industry. In this sector the dominant firms in the 1960s and 1970s were vertically integrated corporations that controlled all aspects of hardware and software production (the 'IBM' or national champion model). The rise of an alternative industrial model (the Silicon Valley model) spurred the introduction of the personal computer and initiated a radical shift to a more fragmented industrial structure organized around networks of increasingly specialized producers (Saxenian, 2002: 184). Today independent enterprises produce all components that were once internalized within a single large corporation, from microprocessors and other components to computers, operating systems and application software. Within each of these segments again there is further specialization of production and a deepening social division of labour. Saxenian describes this change towards a decentralized system as one which looks like a shift towards market relations. But in fact the decentralized system goes beyond the classic auction market mediated by price signals since it depends on the coordination provided by social structures and institutions, particularly varieties of networks.

The fragmentation of industrial structure and the decentralized production system have led to the development of local industry clusters in particular segments of the ICT industry (Swann et al., 1998). Some scholars have argued that these clusters are providing advantages in terms of production and transaction costs (Scott, 1988; Swann et al., 1998). Others have pointed out that it is rather advantages in terms of knowledge links and spillovers which give rise to local clustering in ICT (Saxenian, 1994; Cooke, 2002; van Winden et al., 2004; see also Chapter 4).

There are also global geographic shifts of ICT industries and clusters. In the early years of decentralized production, clusters were concentrated in favourable locations and innovative milieux of advanced countries such as Silicon Valley, Route 128, Texas in the USA, Cambridge, the Western Crescent and the South East region in the UK, Munich and Köln in Germany, Paris and Grenoble in France. More recently ICT clusters have sprung up in Taiwan (Hsinchu Science Park), India (Bangalore and Hyderabad), South Korea, Hong Kong and China. Often these industry agglomerations were initiated by reverse migrants from the USA or Europe who were transferring their knowledge from their places of education to the new locations (Saxenian, 2005). At the same time the return migrants kept their links to the countries or regions where they were educated, thus linking established clusters with the new ones. Or, as Saxenian states, 'the social structure of a technical community appears essential to the organisation of production at the global as well as the local level' (2002: 85).

Not just the production and firm structure, but also the knowledge and innovation process in ICT differs from biotechnology. Innovations are to a higher extent stimulated by and related to customer needs, new components or materials from suppliers, and new solutions from software developers. New knowledge and innovations come less often from science and research than in biotechnology. The ICT sector, thus, combines synthetic, analytic and symbolic knowledge bases. Moreover, product cycles are much shorter than in biotechnology. In some segments they are two to three years, on others as short as six months (Saxenian, 2002). As a consequence, patenting is far less common and important than in biotechnology. Knowledge is often protected through speedy R&D and commercialization, and the use of tacit knowledge of employees (embrained knowledge) and of knowledge 'embedded' in networks and in communities of practice (Brown and Duguid, 1991; Amin and Cohendet, 2004).

Biotechnology: Science Driven and Long Development Cycles

Biotechnology understood as 'use of biological systems and living organisms to make or modify specific products or processes' has existed since human beings used fermentation to make bread, cheese, beer and wine. Modern biotechnology refers to the understanding and application of genetic information of animal and plant species, and it is of more recent origin. Genetic engineering modifies the functioning of genes in the same species or moves genes across species resulting in genetically modified organisms (GMOs; Christensen et al., 2002).

Modern biotechnology encompasses several fields, the most important being bio-pharmaceuticals and customized drug development, various forms of medical applications (diagnostics and gene therapy), biomaterials (tissues, organs), agro-food (genetic modification of existing plant and animal species), biofuels (ethanol), bioremediation (enzymes) and computing (Bergeron and Chan, 2004). Key features of biotechnology are the strong and driving role of science, its intricate knowledge links to universities and the strong interrelationships with 'big pharma' or agro-food and energy corporations, as well as the prominent role of venture capital.

Strong science links
Although biotechnological principles such as fermentation have been applied since prehistory, 'modern' biotechnology was brought forward by scientific discoveries such as the following (Bergeron and Chan, 2004):

- 1863: Gregor Mendel (Austrian) discovered laws of genetic inheritance.
- 1869: Friedrich Miescher (Swiss) isolates DNA.
- 1953: Watson and Crick (USA/UK) discover helical structure of DNA; together with discovery of RNA in the late 1950s this led to the cracking of the genetic code mid-1960s.
- 1976 onwards: foundation of first biotechnology companies in California and Massachusetts.
- 1980: first patents on genetically engineered life and on gene cloning granted by the United States Patent and Trademark Office (USPTO).

- 1988: development of transgenic mice and start of USA/UK-led Human Genome Project.
- 1997: cloning of mammals with birth of Dolly the sheep (UK).
- Late 1990s: Craig Venter's (USA) Celera Genomic enters race of gene sequencing with aid of proprietary computer methods.

The mapping of the human genome, that is, the identification of the about 30 000 genes that encode the hereditary characteristics of the human being, has been described as a quantum leap in biology (Christensen et al., 2002). Despite occasional backlashes such as the 2005 'scientific scandal' around genome research in South Korea, biotechnology continues to be a science-driven industry where scientific discoveries push the innovation and firm formation process. 'Star scientists' continue to play an important role in setting up spin-off companies and usually there are narrow and ongoing relationships with university research.

A key challenge for the biotechnology industry results from the explosion of data that came along with the human genome project and related research. This has led to a boom in new research results and to a fragmentation and broad distribution of relevant knowledge among a huge number of organizations, ranging from universities, to start-up companies to global pharmaceutical giants. As Ernst & Young (2001: 5) have stated, nowadays 'no single organisation, however big, has all the answers. The only way to survive is through integration with others in the industry'.

Dense knowledge links among dedicated biotechnology companies, universities and pharmaceutical companies can be observed in various forms, such as:

- spin-offs, where the founders often keep their intense relationship with their original university or science organization
- star scientists from well-known research organizations who take on consulting or advisory roles in start-up companies
- cooperations/alliances in the fields of research and development
- market relationships in the form of, for example, contract research or licensing agreements
- mergers and acquistions (M&A) and financial relationships with the intention to gain direct or indirect control over companies and their technology.

These links give rise to knowledge interactions at various spatial levels, reaching from local to global (Powell et al., 2005; McKelvey, 2004; Gertler and Levitte, 2005). There are still many open questions as to the nature of local and global links (formal/informal, static/dynamic: see Chapter 3) and the kind of knowledge exchanged. These aspects are investigated in Chapters 7 for the UK and 8 for Austria.

Strong relationships with the pharmaceutical industry
There are close relationships between biotechnology and the pharmaceutical industry. 'Big pharma' is – besides venture capital – an important financial source as well as a key customer for biotechnology start-ups and companies (Malerba and Orsenigo, 2002). The biotechnology industry holds the following promises for the pharmaceutical industry and its corporations:

- It may help them to decrease time to market.
- It supports the development of custom drugs (such as patient specific drugs) through rational drug design (use of computer methods).
- It helps pharmaceutical firms in filling up the dwindling product pipeline. This is urgent since the number of drugs on patent is shrinking (about half of sales would be affected by running out of patents by 2010 – Ernst & Young, 2001).

There are also promises and applications of biotechnology in medicine in fields such as the following:

- cancer therapy
- cloning (duplicate tissues and therapeutic cells)
- enhanced diagnosis by examining genes
- gene therapy and genetic engineering
- xenotransplantation (organs and tissues).

Key challenges for biotechnology and pharmaceutical companies lie in the very long drug development process including the phases of drug discovery, screening, lead development, pre-clinical trials, regulatory approval, clinical trials (phases 1–3). This process can take up to 15 years and implies a high failure rate (see Bergeron and Chan, 2004: 12). For pharmaceutical companies a major challenge also lies in the fact that customized designer drugs run counter to the current business model of big pharmaceutical companies, which is to push a select few blockbuster drugs as fast and hard as possible.

Owing to such difficulties, and some failures of biotechnology drugs in clinical trials, biotechnology stocks suffered a marked decline from a peak in the year 2000. However, the industry has survived the bursting of the 'new economy bubble' better than many dot.com firms, partly because established pharmaceutical companies continued to invest in biotechnology in an attempt to fill their pipelines with new drugs.

Despite the aspiration in healthcare biotechnology to become an independent industry it is still strongly interrelated with the pharmaceutical industry. Pharmaceutical companies are key customers, R&D partners, capital providers and sometimes controlling units. In 2001 the pharmaceutical industry expected to see some 50 per cent of sales coming from product and technology acquisitions – in-licensing, partnering, co-development, co-marketing and/or M&A activities (Ernst & Young, 2001).

Comparing Biotechnology and ICT

The main characteristics of the two sectors and the major differences are summarized in Table 6.1. Biotechnology, and in particular the therapeutic and drug development segments, relies to a high extent on academic and university based analytical knowledge, whereas in ICT and its different sub-sectors elements of all three types of knowledge base (synthetic, symbolic and analytic: see Chapter 3) can be identified. As several studies have shown (Zucker et al., 1998; Cooke, 2004), knowledge in biotechnology is more often documented and of a codified nature, whereas ICT relies to a higher extent on a mix of explicit, implicit and complicit knowledge. On the other hand, ICT as technology is an important tool for the codification of knowledge.

The innovation process also differs between biotechnology and ICT. In biotechnology, innovations often result from scientific and basic research, and they more frequently have a radical character. Research, development, examination and testing typically take many years (Bergeron and Chan, 2004) and, owing to the uncertainties and risks involved, the commercialization of knowledge and innovations often takes the organizational form of spin-offs and start-up companies. In ICT, as a contrast, we find much more customer-driven innovation, often of an incremental character and based on applied R&D. Life cycles of products are much shorter, in computers and information technology, often as low as two to three years.

Owing to these differences in the knowledge and innovation process the knowledge interactions and innovation networks also differ. In biotechnology, science and university links are key, and they are often formal and contract based (Zucker et al., 1998; Powell et al., 2005). Regarding their geography we find a coexistence of local and global relationships (Cooke, 2004; McKelvey, 2004; Gertler and Levitte, 2005), where in particular for the young and catching up clusters, global links are of key importance. In comparison, for ICT, interfirm links and networks are more relevant.

Table 6.1   Characteristics of biotechnology and ICT

	Biotechnology	ICT
Knowledge base	Mainly analytical with synthetic elements	Combination of synthetic, symbolic and analytical
	Strong role of codified and complicit knowledge	Combination of codified, tacit and complicit knowledge is relevant
Innovation process	Radical innovation	Often incremental from time to time radical shifts
	Strong role of science and basic research	More applied R&D customer-driven innovation
	Long duration of research, development, examination and testing (up to 10+ years)	Short product life cycles (two to three years)
Networking	Strong links to universities through spin-offs and R&D cooperations	Interfirm networks far more important
	Relationships are often formal or contract based	Both formal and informal knowledge links
	Both local and global links: global links are key for young clusters	Both local and global links: local networks are combined with 'virtual communities'
Regulation	Highly important for most fields	Important for industries such as telecom
	Strong regulations for new drugs and medical applications but also in agro-food biotechnology	Efforts to liberalize markets
Finance and capital	High financial requirements for R&D and testing	High financial requirements in telecom, networks and scale intensive hardware production
	Need of different kinds and several rounds of seed and venture capital	Lower financial requirements in services, software, content and Internet business
Firm structure	Strong role of new firms and university spin-offs	Combination of established large firms (hardware and telecom) and young and small firms (software, content protection services, etc.)
	Some established biotech companies (mainly US) Big pharma is 'monitoring' partnering, taking over, etc.	

Here we also find both local and global networks, but in ICT young clusters are said to be more often characterized by local links and networks, whereas in more mature ICT clusters we often find global knowledge links (Swann et al., 1998). In addition, Cairncross (1997) and Amin and Cohendet (2004) have pointed out that local networks might be increasingly substituted by 'virtual communities' in the knowledge process.

Furthermore, for both sectors regulatory bodies and rules are setting important framework conditions. In biotechnology most fields are subject to strong regulation. This is in particular the case for bio-pharmaceutical drug development (clinical trials I to III, patent protection, and so on), but also for medical and agro-food applications. Within ICT in particular telecom has been subject to national regulation in the past. National telecom companies were monopolies normally in public ownership and/or protected from competition. Since the 1980s we find a strong move towards deregulation, the enhanced role of 'regulators' and privatization, which started in the 'liberal' market economies (USA, UK) but has also reached coordinated market economies such as Northern European countries as well as Germany and Austria.

We also find differences between biotechnology and ICT in terms of firm structure and finance. In biotechnology, research-oriented start-up firms and university spin-offs take over a key role ('dedicated biotechnology firms': DBFs). They require, owing to the long time for research, development and testing, different kinds and rounds of seed and venture capital for the start-up stage and subsequent growth stages. Many DBFs do not earn revenues for lengthy periods, since they are still at the stage of research and development, or in specific phases of clinical trials. A smaller number of biotechnology companies are in the phase of earning profits, and some of them (such as the US Genentech) have reached a very high-value market capitalization. Big pharma is another group of players: they are closely monitoring the field, looking out for interesting ventures in order to fill up their dwindling drug pipeline or to complement their product or technology portfolios. They often license new technologies, enter into R&D partnerships or take over promising young firms.

Information and communication technology is quite heterogeneous in terms of sub-segments, firm groups and financial requirements. In the segment of hardware and telecom we often find high capital requirements for investments and very large firms. In segments such as software development, IT consulting, customer-specific IT services and development we usually find lower capital requirements and entry barriers and, as a consequence, a large number of competing small companies.

RESEARCH FOCUS AND METHODOLOGY

The empirical part of this volume analyses important aspect of locational preferences, emergence and growth of firms in knowledge-based industries with a special emphasis on information and communication technologies and biotechnology. Comparisons have been undertaken between those sectors as well as between the UK as a representative of a liberal or entrepreneurial business system and Austria as a coordinated or associative type of system. Within the UK five regions have been dealt with (South East, Wales, Scotland, London and Eastern). For Austria secondary data on knowledge-based sectors including ICT have been analysed for all political districts of the country, whereas biotechnology has been studied for three selected regions where concentrations of biotechnology activities can be found. These include Vienna as the strongest biotechnology agglomeration in Austria, and the regions of Styria and Tyrol as more embryonic biotechnology locations.

Research Questions

Important aims of the research are:

- To identify locational preferences and geographical concentrations of knowledge-based sectors and to analyse recent changes of these spatial patterns.
- To describe industrial, locational and organizational characteristics of knowledge-based clusters and ascertain the extent to which clustering is beneficial to firm performance.
- A particular emphasis is laid on the intensities and kinds of knowledge interactions, differentiating between formal and informal as well as static and dynamic effects. We have distinguished between the static categories of (traded) market links and (untraded) spillovers as well as the dynamic types of (formal) networks and (informal) milieu effects (see Chapter 3).
- This classification allows us to test assumptions about 'free-riding' or 'untraded dependencies' advantages that are supposed to derive from clustering. We thus explore the extent to which local knowledge spillovers are available as literature predicts in cluster settings and which channels they take. We also investigate the extent to which knowledge transfer contains more market transactions and formal networks than suspected in the dominant literature that stresses local knowledge spillovers and milieux.
- Finally, the research also compares support policies for enhancing knowledge transfer at the regional level as well as interregionally and

internationally. Of particular interest here is the question of to what extent those policies differ between liberal (UK) and coordinated economies (Austria) as well as between different types of regional innovation systems.

Methodology

Methodologically, the research consisted of four phases:

1. *Assessment of national secondary data sources*. This includes in the UK an analysis of local authority and district-level employment and workplace data from the Annual Business Inquire (ABI) data set.[1] In Austria data from the census of companies and employment for the years 1991 and 2001 could be used. This allowed us to indentify areas of strong and weak geographical concentration of the targeted knowledge-based industries, and to investigate respective changes. This data analysis furthermore helped to sharpen the geographical and sub-sectoral focus of the subsequent firm samples.

2. *Postal firm survey*. Postal questionnaires were mailed out to firms in specific knowledge-based industries, in particular ICT and biotech. In the UK about 7195 ICT firms and 156 biotech firms were drawn from the targeted industries and locations. In Austria the questionnaire was sent to 2228 firms of a broader subset of knowledge-based sectors, including ICT. The questionnaire focused on the importance and role of clustering, and surveyed the nature and scope of firms' innovation practices, including the formal and informal dimensions of interfirm and firm–laboratory partnership and network linkages for knowledge-transfer and collective learning.

3. *In-depth interviews with firms*. Firms of particular interest and relevance to key theoretical and empirical propositions of the research were selected and interviewed based on a partly open interview guideline to gain stronger qualitative insight and explanation for clustering and the mechanisms of knowledge interactions.

4. *In-depth interviews with policy-actor informants*. Interviews were conducted also with key policy-actors and support agents on the role of policy organizations for cluster development, innovation support and the stimulation of knowledge interactions and networks. The results of the research were shared and interpreted in workshops with policy-actors.

Defining the Sectors

The following research focused on spatial aspects of the knowledge economy investigating the location pattern, innovation activities and knowledge inter-actions of respective sectors. In this research process, different sectoral aggregates were used and investigated. A part of the analysis in the Austrian study is addressed to 'knowledge-based sectors' in a wider definition; other parts of the research are focused on the two specific sectors of ICT and biotechnology.

Knowledge-Based Industries

Knowledge-based industries were defined according to the OECD classification (OECD, 2001) differentiating between three groups of knowledge-based industries:

- *high-technology industries (HT)*, comprising the NACE[2] sectors Pharmaceuticals (244), Office, Accounting and Computing Machinery (30), Radio, TV and Communication Equipment (32), Medical, Precision and Optical Instruments (33), and Aircraft and Spacecraft (353)
- *medium-high-technology industries (MHT)*, comprising the NACE sectors Chemicals excluding Pharmaceuticals (24 except 244), Machinery and Equipment n.e.c. (29), Electrical Machinery and Apparatus n.e.c. (31), Motor Vehicles, Trailers and Semi-Trailers (34), and Railroad Equipment and Transport Equipment n.e.c. [not elsewhere classified] (352, 354, 355)
- *technology-related knowledge-intensive business services (TKIBS)*, comprising the NACE sectors Computer and Related Activities (72), Research and Development (73), Architectural and Engineering Activities and related Consultancy (742), and Technical Testing and Analysis (743).

ICT Industry

The ICT industry is conventionally referred to as a group of activities including mainly IT hardware, electronics components and systems, telecommunications and IT services sectors (including software). However, estimating the level of employment in the ICT industry is not always straightforward since some definitions of ICT do not accord with the classification system used for employment statistics.

For the purpose of this research ICT has been defined following the OECD ICT sector definition approved in 1998 (DSTI/ICCP/AH/M(98)1/

REV1; in OECD, 2002). This definition includes a combination of manufacturing and services industries that capture, transmit and display data and information electronically.

A few modifications have been made, however. The research team decided to exclude from the analysis the Wholesale sector, where the industrial class 5150 was seen to be too broad and covered much more than ICT activities. Similar issues were raised considering the class ISIC 3130 – Manufacture of insulated wire and cable – owing to its inclusion of transmission cable for electric power. However, because of the perceived growing importance of optic fibre cables as part of this broader industry, the team agreed to include this industry and where possible only firms producing specific products have been included in the sample. Table 6.2 shows the sectors that have been included for the purpose of the research.

Types of Spatial Concentration of Knowledge-Based Sectors and ICT

In order to find out to what extent knowledge-based sectors, biotechnology and ICT were concentrated or dispersed in geographical space, location quotients for employment and companies were calculated using the national secondary data sources mentioned above.

However, examining employment concentrations alone is not always sufficient for identifying sources of regional competitive advantage. Employment location quotients (LQe) do not provide an indication of the number of firms involved – whether the sector is aggregated, disaggregated or somewhere in-between. For instance, a sector's large number of employees may be concentrated in a single large firm. Location quotients for numbers of firms (LQd) are important because each establishment represents a locus of activity, a set of business and employment opportunities, and a discrete decision to begin operations or to locate an operation in that specific locale for some specific reason. Thus, for the purpose of the research, it is useful to compare location quotients based on employment with location quotients for the same industry that are based on the number of establishments. This can reveal where a particular industry group in the region is more disaggregated than it is in the country.

In an attempt to identify specialization in the local economy, location quotients were calculated for various knowledge-based, biotechnology and ICT sectors. This allowed us to identify where there was a concentration of employment and firms in a particular ICT or biotechnology sector and whether these could be regarded as clusters, districts or as dominating firms (see Table 6.3).

Table 6.2   ICT definition: sectors selected from Standard Industrial Classification (SIC 2003)

Computing and peripherals	Telecom service and manufacturing	Software and private R&D	Electronics and opto-electronics
30.01 Manufacture of office machinery and computers	32.201 Manufacture of television and radio transmitters (32.20)	72.10 Hardware consultancy	31.62 Manufacture of electrical equipment not elsewhere classified (excl. engines and vehicles)
30.02 Manufacture of computers and other information processing equipment	32.202 Radio and electronic capital goods (32.20)	72.20 Software consultancy and supply	33.10 Manufacture of medical and surgical equipment and orthopaedic appliances
	33.20 Manufacture of instruments and appliances for measuring, checking, testing, navigating and other purposes, except industrial process control equipment	72.30 Data processing	33.402 Manufacture of optical precision instruments
	64.2 Telecommunications	72.40 Database activities	33.403 Manufacture of photographic and cinematographic equipment
	32.30 Television and radio receivers, sound or video recording or reproducing apparatus and associated goods	73.10 Research and experimental development on natural sciences and engineering	3130 Insulated wire cable (optic fibre cable; optical networks)
	32.10 Manufacture of electronic valves and tubes and other electronic equipments		

Table 6.3   Categories of spatial firm concentration

Types of regional concentration	LQd < 1.25	LQd > 1.25
LQe > 1.25	Dominating firm	Cluster
LQe < 1.25	No concentration	District

Table 6.4   Survey methodology

Firm survey	UK	Austria
Media	Postal questionnaire	Postal questionnaire Internet questionnaire
Data collection	Spring 2003	Spring 2003
Stratification	By economic activity, firm size and district	No stratification
Sectors covered	ICT as defined above Biotechnology	ICT and other knowledge-based sectors (HT, MHT, TKIBS)
Geographical area	115 districts	All districts of Austria
Total sample covered:	7351	2228
Respondents	297	189
Response rate	4%	9%

Firm Survey

Table 6.4 summarizes the main features of the firm surveys carried out and the primary data collection method. Although the basic approach and the underlying questionnaire were the same in the UK and Austria, there were some differences in coverage and stratification as can be seen from Table 6.4. These differences in approach were mainly due to size differences between the two countries.

In both countries a broadly identical questionnaire was used covering the following areas:

- basic characteristics of the firm and its performance (employment and turnover and respective changes within a five-year period)
- intensity and type of innovation activities (R&D, types of innovations, patenting)

- types and location of knowledge sources used
- mechanisms and channels of knowledge transfer (traded/non-traded, formal/informal)
- cooperation partners and their geography
- importance of the region for knowledge exchange
- role of policy support.

These variables and indicators allowed us to analyse the nature and scope of firms' innovation activities and their knowledge links, including the formal and informal dimension of interfirm and firm–laboratory partnership and network linkages for knowledge transfer and innovation. Also relationships to firm performance could be investigated as well as the role of policy support for innovation and knowledge exchange.

NOTES

1. The ABI is a business survey which collects both employment and financial information. This survey has replaced the Annual Employment Survey as the source of information on employee jobs.
2. Nomenclature générale des Activités économiques dans les Communautés Européennes (General Industrial Classification of Economic Activities in the European Communities).

REFERENCES

Amin, A. and Cohendet, P. (2004), *Architectures of Knowledge*, Oxford: Oxford University Press.
Bergeron, B. and Chan, P. (2004), *Biotech Industry – a Global, Economic, and Financing Overview*, Hoboken, NJ: John Wiley.
Brown, J.S. and Duguid, P. (1991), 'Organizational learning and communities of practice: towards a unified view of working, learning and innovation', *Organization Science*, **2**(1), 40–57.
Cairncross, F. (1997), *The Death of Distance: How the Communications Revolution Will Change Our Lives*, Boston, MA: Harvard Business School Press.
Cooke, P. (2002), *Knowledge Economies: Clusters, Learning and Cooperative Advantage*, London: Routledge.
Cooke, P. (2004), 'Regional knowledge capabilities, embeddedness of firms and industry organisation: bioscience megacentres and economic geography', *European Planning Studies*, **12**, 625–41.
Christensen, R., Davis, J., Muent, G., Ochoa, P. and Schmidt, W. (2002), 'Biotechnology: an overview', EIB sector papers.
Ernst & Young (2001), *Integration: Ernst & Young's Eight Annual European Life Sciences Report 2001*, London: Becket House.
Gertler, M. and Levitte, Y. (2005), 'Local nodes in global networks: the geography of knowledge flows in biotechnology innovation', *Industry and Innovation*, **12**(4), 487–507.

Helpman, E. (ed.) (1998), *General Purpose Technologies and Economic Growth*, Cambridge, MA: MIT Press.

Malerba, F. and Orsenigo, L. (2002), 'Innovation and market structure in the dynamics of the pharmaceutical industry and biotechnology: towards a history-friendly model', *Industrial and Corporate Change*, **11**, 667–703.

McKelvey, M. (2004), 'What about innovation collaboration in biotech firms? Revisiting occurrence and spatial distribution', Biotech Business Working Paper No. 02-2004, Copenhagen Business School.

Organisation for Economic Co-operation and Development (OECD) (2001), *OECD Science, Technology and Industry Scoreboard: Towards a Knowledge-Based Economy*, Paris: OECD, available at: www1.oecd.org/publications/e-book/92-2001-04-1-2987/.

Organisation for Economic Co-operation and Development (OECD) (2002), Reviewing the ICT sector definition: Issues for Discussion, DSTI/ICCP/IIS(2002)2, available at www.oecd.org/dataoecd/3/8/20627293.pdf.

Powell, W., White, D., Koput, K. and Owen-Smith, J. (2005), 'Network dynamics and field evolution: the growth of interorganizational collaboration in the life sciences', *American Journal of Sociology*, **110**, 1132–205.

Saxenian, A. (1994), *Regional Advantage: Culture and Competition in Silicon Valley and Route 128*, Cambridge, MA: Harvard University Press.

Saxenian, A. (2002), 'Transnational communities and the evolution of global production networks: the cases of Taiwan, China and India', *Industry and Innovation*, **9**(3), 183–202.

Saxenian, A. (2005), 'From brain drain to brain circulation: transnational communities and regional upgrading in India and China', *Studies in Comparative International Development*, **40**(2), 35–61.

Scott, A. (1988), *New Industrial Spaces*, London: Pion.

Smith, K. (2002), 'What is the "knowledge economy"? Knowledge intensive industries and distributed knowledge bases', paper presented at the DRUID Summer Conference on The Learning Economy – Firms, Regions and Nation Specific Institutions, Aalborg, June.

Swann, P., Prevezer, M. and Stout, D. (eds) (1998), *The Dynamics of Industrial Clustering*, Oxford: Oxford University Press.

Van Winden, W., van der Meer, A. and van den Berg, L. (2004), 'The development of ICT clusters in European cities: towards a typology', *International Journal of Technology Management*, **28**(3/4/5/6), 356–87.

Zucker, L., Darby, M. and Brewer, M. (1998), 'Intellectual human capital and the birth of the U.S. biotechnology enterprises', *American Economic Review*, **88**, 290–306.

7. UK ICT and biotechnology performance: the significance of collaboration and clustering

INTRODUCTION

As highlighted in previous chapters of this book, the UK has moved more towards the knowledge economy than other European countries. According to the OECD Science and Technology Outlook (2004), the UK is well positioned in both knowledge-intensive services, such as telecommunications, finance, insurance and business services, and high-technology industry, such as pharmaceuticals, aircraft, ICT equipment and precision instruments. Knowledge-intensive services account for approximately 23 per cent of UK value-added, putting the UK behind only Switzerland, Luxembourg and the USA. High-technology industries account for almost 35 per cent of UK manufacturing exports, with only Ireland, South Korea, Switzerland and the USA having a larger share of these industries in total exports. The UK also accounted for over 4 per cent of worldwide value-added in manufacturing in 2002, making it the sixth-largest manufacturing nation in the world, behind the USA, Japan, China, Germany and France (OECD, 2004). Subsequently, the UK surpassed France to rank fifth. Nonetheless, the UK's most obvious knowledge-based strengths are in ICT and pharmaceutical biotechnology, industries with a major presence in both the manufacturing and service categories of the OECD's knowledge-based industries classification (OECD, 1999).

As already mentioned in the previous chapter, the ICT industry is conventionally referred to as a group of activities including IT hardware, electronics components and systems, telecommunications and IT services. The manufacturing side of ICT – which comprises IT hardware, electronic components and systems – is characterized by much industrial 'related variety' and the main areas of activity within the supply chain are raw material suppliers, component/sub-component manufacture, design, production, distribution and logistics, R&D and quality control. Most firms are micro-sized (fewer than 10 employees) with the exception of electronic component and process control equipment manufacturers where there is a higher concentration of small to medium firms (fewer than 250 employees). These employ about half the UK industry while the few large UK-based companies employ the other 50 per cent of the workforce (DTI, 2004a). Only a small number of firms have over 1000 employees and most of these are non-UK owned (DTI, 2004a; 2005). In 2002, the electronics manufacturing sectors contributed more than £21 billion (or 2 per cent) to UK GDP (DTI, 2004a).

The UK's software and computer services industries were among the fastest growing sectors of the UK economy from 1995 to 2005. These industries are characterized by extremely rapid technological change and firms in the sectors include companies who:

- produce generic software for the business and domestic market
- produce bespoke software for commercial and public sector clients
- install systems software for commercial and public sector clients
- provide consultancy services for business and the public sector leading to software solutions.

The software industry now accounts for around 3 per cent of the UK's gross domestic product, and its impact is even greater as a source of innovation. Thus the UK software industry also stands out as a leader in R&D investment. It contains 117 of the UK's top 750 R&D spending companies, contributing 5.2 per cent of total UK R&D investment (DTI, 2004b). This is coupled with the fact that the UK is consolidating as a world player in R&D, also attracting a substantial number of international companies. The largest proportion of US and Japanese R&D facilities in Europe are based in the UK and more global companies are establishing their own international R&D centres benefiting from new product research, design and development. Many other multinational corporations have formed strategic partnerships with universities or contract research teams in the UK.

With respect to biotechnology, the UK is Europe's leading location for biotechnology and ranks second in the world, some way behind the USA. The biotechnology sector in the UK typically refers to firms in the core business of large molecule drug discovery and high-throughput screening of small molecule drug candidates derived from synthetic chemistry. That is the therapeutic core, but diagnostics and platform technologies like the manufacture of diagnostic kits, development of industrial applications for gene-based technologies, providing support services such as contract R&D or supplying antibodies, cell lines, enzymes and reagents are then included in medical biotechnology. Some firms double-up in agro-food biotechnology, or are involved in some other way in the value chain, such as clinical research organizations (CROs) that conduct drug testing and trialling, or pharmaceutical development companies with no in-house R&D facilities. There is interdependence between large pharmaceutical companies ('big pharma') which need smaller specialist drug discovery R&D biotechnology firms to carry out the basic laboratory science to originate new biotechnology drug candidates for licensing. In the academic literature, these are referred to as DBFs (dedicated biotechnology firms) that often lack the resources to develop their drug candidates beyond pre-clinical and early-stage clinical trials. Approximately three-quarters of biotechnology firms are either small (11–50 employees) or medium (51–250 employees). The remainder can be divided roughly equally into micro-sized firms (fewer than 10 employees) and large firms (more than 250 employees). The largest firms are foreign-owned pharmaceuticals companies that have gained a foothold in the biotech sector through the acquisition of DBFs.

Pharmaceutical biotechnology is the UK's fastest growing industry, both because it is so relatively new and because it holds the promise of enormous returns on investment in the future. Biotechnology holds the key to most of the next generation of blockbuster drugs, without which big pharmaceutical companies would see a huge downturn in profits as the patents expire on their existing products. Furthermore, biotechnology has the potential to provide new retroviral drugs and antibiotics, as well as therapies for many inherited conditions, cancers and degenerative diseases. Because pharmaceutical biotechnology is a relatively new industry at an early stage in its development, its contribution to GDP is still very small, at a fraction of a percentage of the total; however pharmaceuticals as a whole accounts for about 1 per cent of UK GDP (OST, 2002).

United Kingdom universities and DBFs (see Figure 7.8) commonly enter into collaboration with overseas partners in addition to domestic ones for product innovation, distribution, licensing deals and supply contracts. As such UK biotechnology operates in a global marketplace. Many international DBFs have manufacturing or research operations in the UK, and many are located here as a result of acquisitions of mergers. For example, Acambis has divisions in Cambridge, UK, and Cambridge, Massachusetts. Amgen is headquartered in Los Angeles but has a base in Cambridge Science Park. DakoCytomation came about through the merger of a Danish firm and an American firm. Genzyme, headquartered in Cambridge, Massachusetts, has made numerous acquisitions worldwide and has UK plants in therapeutics and biosurgery at Oxford, diagnostics in Maidstone, Kent, bulk pharmaceuticals (enzymes) in Haverhill, Suffolk, with R&D and biomedical and regulatory affairs at Cambridge.

SPATIAL PATTERNS AND CHANGE IN ICT

United Kingdom ICT[1] accounted for 986 000 employees in 2002 (3.9 per cent of the total UK employment and approximately 105 000 data units[2] (4.8 per cent of the total for UK industry). The significance of the sector varies markedly across the UK regions. At regional level, the UK ICT industry is highly concentrated in the 'Greater South East', as shown in Table 7.1; South East, London and Eastern regions accounted for about 60 per cent of total ICT employment in the UK. Location quotients, which allow us to standardize for differences in regional scale and offer a relative comparison of the significance of ICT in each region were calculated on employment data for 2002. This confirmed a regional hierarchy led by these three core regions, with the North West, Wales, and Yorkshire and the Humber at the bottom (with location quotients of 0.72, 0.70 and 0.61 respectively).

Nonetheless, employment location quotients do not provide an indication of the number of firms (or more accurately 'units') involved and whether the sector is aggregated, disaggregated, or somewhere in-between (for instance, a sector's large number of employees may be concentrated in a single large firm). When comparing location quotients calculated from numbers of data units, a similar picture to the one described above emerges, with the South East, Eastern and London at the top of the league, with location quotients of 1.47, 1.35 and 1.25, respectively. Interestingly, the location quotient for Scotland drops significantly, showing that a significant portion of employees is concentrated in a few larger firms. Wales and the North East with Scotland and Yorkshire are at the bottom with location quotients (LQd), respectively, 0.54 for Wales and the North East, 0.63 for Scotland and 0.66 Yorkshire and the Humber.

Between 1995 and 2002, the UK experienced an increase in total ICT employment and data units of 20 per cent and 51 per cent respectively. Figures 7.1 and 7.2 show how ICT changed between 1995 and 2002 by activities and regions. It is noteworthy that the highest increase occurred in data processing and database activities in both data units and employments. On the other hand, employment in ICT manufacturing decreased in all regions, with Wales, London and the South West losing respectively 44 per cent, 37 per cent and 26 per cent of their workforce.

In an attempt to identify specialization in the local economy, location quotients have been calculated for each ICT activity. This allows us to identify clearly where there is a concentration of employment and firms in a particular ICT activity. Splitting the industry into manufacturing, software, telecoms, data and R&D, location quotients present two distinct geographies, with a concentration of manufacturing in the peripheral regions,

Table 7.1 UK ICT employment by region 2002

	ICT employees	ICT % regional empl.	ICT % all UK	LQe^	ICT data units	% on total industry regional*	% UK ICT data units	LQd**
East Midlands	46 202	2.66	4.69	0.69	5 838	3.93	5.60	0.82
Eastern	110 430	4.90	11.21	1.26	14 240	6.44	13.65	1.35
London	153 869	3.92	15.62	1.01	22 051	6.00	21.14	1.25
North East	37 714	3.79	3.83	0.98	1 704	2.58	1.63	0.54
North West	83 603	2.80	8.43	0.72	8 218	3.64	7.88	0.76
Scotland	82 179	3.65	8.34	0.94	4 972	3.00	4.77	0.63
South East	237 968	6.53	24.15	1.68	24 871	7.03	23.84	1.47
South West	76 723	3.68	7.79	0.95	7 665	3.92	7.35	0.82
Wales	29 651	2.73	3.01	0.70	2 281	2.59	2.19	0.54
West Midlands	76 805	3.33	7.80	0.86	7 329	3.97	7.03	0.83
Yorkshire and the Humber	50 660	2.37	5.14	0.61	5 151	3.14	4.94	0.66
GB	985 804	3.88	100.0	1	104 320	4.78	100.0	1

Notes:
^LQe = (region ICT/all regional employment)/(national ICT/all national employment).
* ICT region/all region data units.
** LQd = (region ICT/all region data units)/(national ICT/all national data units).

Source: ABI, Nomis website.

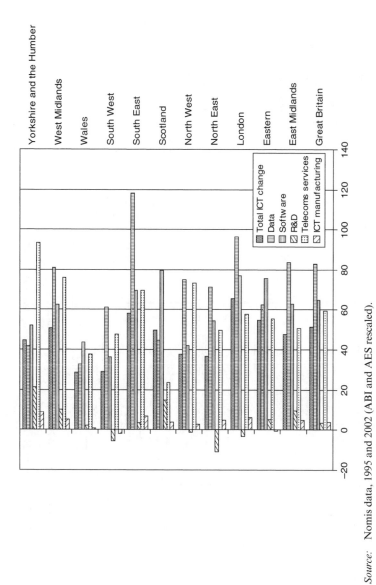

Source: Nomis data, 1995 and 2002 (ABI and AES rescaled).

Figure 7.1   Changes in ICT activities by data unit, 1995–2002

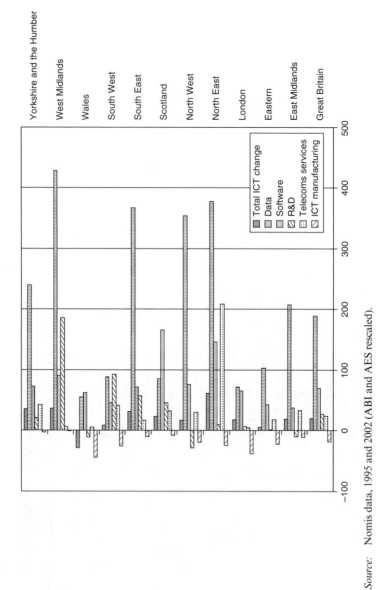

Source: Nomis data, 1995 and 2002 (ABI and AES rescaled).

Figure 7.2   Changes in ICT activities by employees, 1995–2002

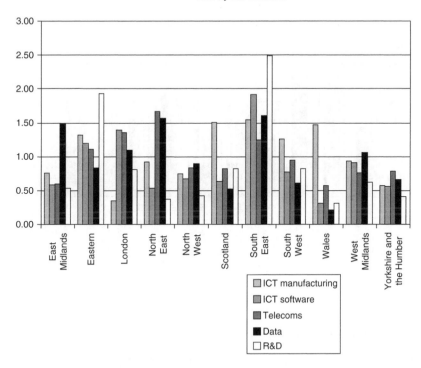

Source: Nomis website, ABI, 2002 data.

Figure 7.3 Location quotients by employees for UK regions and ICT activities

while the software elements of the industry are strongly concentrated around the South East, as shown in Figures 7.1, 7.2 and 7.3. Interestingly, Scotland is also becoming a more **R&D** specialized region as its employment figures and data units grew 46 per cent and 15 per cent respectively between 1995 and 2002, as shown in Figure 7.2.

Information and communication technology has important concentrations of employment in the Thames Valley and Cambridgeshire, while Scotland has high employment concentration in electronics with computer equipment (over 2.0 LQs) and Wales has a high concentration in opto-electronics (LQe 2.9) and was, in 2002, the leading region in the production of television/radio transmitters (LQe 5.0). Similarly, when analysing UK ICT location quotients calculated by number of establishments, the picture that emerges is one of high concentration of R&D units in the Eastern region, South East and Scotland, and major clusters

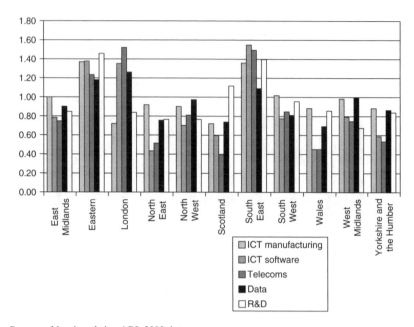

Source: Nomis website, ABI, 2002 data.

Figure 7.4   Location quotients by data units for UK regions and ICT activities

for software and computing services companies in 'Silicon Fen' (Cambridgeshire) and the Thames Valley/M4 Corridor. Nowadays some ICT is found in most British regions.

From the range displayed in Figures 7.1, 7.2 and 7.3 the regions selected for in-depth study were the South East, London, Eastern, Scotland and Wales, the first three being the leading employment and firms concentrations for the sub-sectors of computing and peripherals, telecom services and manufacturing, and software; the second two, Scotland and Wales, although lower-ranking in scale of ICT activity, nevertheless being good candidates for electronics, computer equipment and opto-electronics as shown in Figures 7.4 and 7.5.

As with the regional pattern, location quotients can also be used to give a better indication of the relative strength or weakness of each local authority[3] in ICT. This is important because smaller areas that are relatively strong in ICT would not be detected in consideration of absolute employment and data units. Table 7.2 shows that, at local

Table 7.2   Top ten ICT location quotients, by local authority 2002

Rank	Local authority	LQd	Rank	Local authority	LQe
1	Bracknell Forest	3.22	1	Bracknell Forest	4.53
2	Wokingham	3.16	2	Inverclyde	3.88
3	Reading	2.16	3	Reading	3.64
4	Milton Keynes	2.15	4	West Berkshire	3.50
5	Windsor/Maidenhead	2.13	5	Wokingham	3.43
6	Kingston/Thames	2.11	6	Windsor/Maidenhead	2.95
7	Hertfordshire	2.06	7	West Lothian	2.73
8	West Berkshire	2.04	8	Slough	2.71
9	Slough	2.00	9	Hackney	2.52
10	Merton	1.94	10	Oxfordshire	2.20

authority level, an even more uneven pattern emerges, when we consider data on employees and data units for the chosen regions at local authority level. The data in Table 7.2 show a high concentration of ICT firms occurring in the M4 corridor. The whole data set (121 local authorities) show lower concentrations in remote areas of Scotland and the Welsh Valleys.

Since one of the purposes of this study is the identification of industrial locations and organizational characteristics of knowledge-based clusters, and to differentiate these from other forms of industry concentration (agglomeration) and weak concentration (dispersed firms), the first step in identifying firms to include in the sample was to select areas of strong and weak geographical concentration in ICT within the targeted regions. Recalling Table 6.3 in Chapter 6, we can define candidate clusters as those agglomerations that present a location quotient by employees over 1.25 and a location quotient by data units over 1.25, as shown in Figure 7.5. Figure 7.5 displays the different types of regional concentration in the five UK regions selected. Of the selected local authorities, 20 could be identified as containing potential clusters, 11 are classified as having a 'dominating firm', 14 are 'district' (city concentration) and 71 are locations with no ICT concentration.

A similar exercise to the one conducted for ICT could not be performed for the biotechnology industry, mainly due to the characteristic associated with biotechnology as an enabling technology that encompasses two main industry groups, namely, pharmaceuticals and medical devices. Secondary resources were utilized to pinpoint areas of higher concentration of biotechnology firms. Evidence gathered for the DTI's cluster mapping exercise (DTI, 2001) shows comparatively strong concentrations for biotechnology

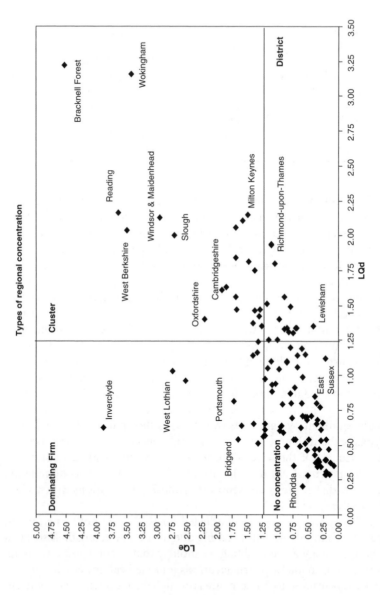

Figure 7.5 Types of regional concentration in five UK regions

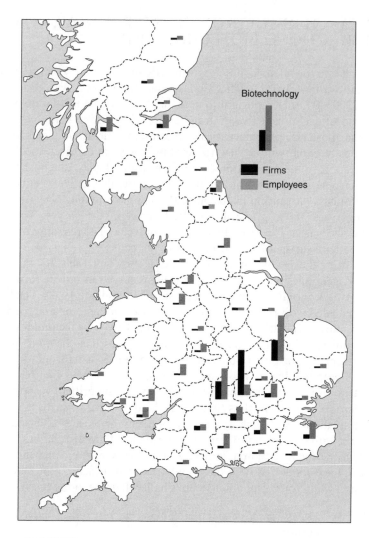

Source: DTI (2001).

Figure 7.6   UK biotechnology cluster map

employment in Cambridgeshire, Oxfordshire and south of London, with
lesser concentrations in Scotland and Wales, as shown in Figure 7.6. This is
also consistent with Shohet's (1998) location map of UK biotechnology
firms in 1993.

RESEARCH RESULTS ON CLUSTER INNOVATION INTERACTIONS: BIOTECHNOLOGY

Research into the microeconomics of the UK biotechnology sector's structure, innovation characteristics and spatial distribution was conducted during 2003–04. It is important to note that the firms sampled were in biopharmaceuticals thus active in drug development and in genomics. As it happens, most UK pharmaceutical biotechnology firms are also active in genomics research with a view to developing genomics-based medicines. With respect to the production of platform technologies, diagnostics and devices, some genomics biotechnology firms are also active in such production lines, warranting inclusion in the sample, though their genomics activities were those primarily inquired about in the survey. Thus the sample size was, at 156 firms, less than the universe. The response rate to a questionnaire survey of such firms was 16 per cent.

Regarding the age of sampled firms, this is shown in Table 7.3. It reveals that UK genomics bio-pharmaceuticals firms surged in rate of formation during two economic boom periods in the 1980s and late 1990s. The latter date coincided with the publication of the Human Genome and, perhaps more importantly, with the stock market boom driven by technology business equities, including biotechnology. This is revealed as the more important driver since the post-boom rate of new genomics firm formation in the UK has been far slower. The respondents' size by number of employment and turnover is shown in Table 7.4. As shown, the majority of respondents are micro to small firms (less than 50) with 16 per cent of firms following in the medium size category. The respondent firms are also small in turnover, with 76 per cent of firms registering less than £1 million.

When the microeconomic indicators for survey respondents shown in Table 7.4 are compared with the population of healthcare biotechnology firms as a whole, our respondents were smaller than the industry average in

Table 7.3   Date of establishment of UK biotechnology firms conducting genomics

Date	Frequency	%
<1980	4	3
1980–1989	62	39
1990–1995	17	11
1996–2000	56	36
2001<	17	11

Source: CASS ICT and Biotechnology Collective Learning Survey.

Table 7.4   Respondents by size of turnover and employment, 2003

Number of employees N = 25	Turnover size N = 25
Less than 10 (10) 40%	Less than £1 million (19) 76%
Between 10 and 50 (10) 40%	Between £1 million and £20 million (5) 20%
Between 51 and 250 (4) 16%	More than £20 million (1) 4%
Over 250 (1) 4%	

Source: CASS ICT and Biotechnology Collective Learning Survey.

Table 7.5   Economic and innovation indicators for UK genomics biotechnology firms, 2003

Indicator	Quantity
Mean employment	97
Mean turnover	$28 million
Mean exports/turnover	24%
Mean R&D expenditure/turnover	21%
Firms with new patents	47%
Mean patents per firm	4.5

Source: CASS ICT and Biotechnology Collective Learning Survey.

terms of turnover and employment size. Nevertheless, the indicators in Table 7.5 show that the responding firms have a healthy turnover, fall squarely in the small firm size category, are moderately high exporters of their output, and spend at 21 per cent more than, for example, the high pharmaceuticals average of 18 per cent of turnover on R&D. Foremost, they are shown to be significant patent holders and, importantly, nearly half had registered new patents during the year prior to the survey in 2004.

Finally, of key interest in assessing the extent the results indicate that there is clustering activity associated with bio-pharmaceuticals firm interaction patterns, Table 7.6 is strongly demonstrative of substantial clustering, looser collaboration and more focused cooperation activities. Moreover cooperation is especially pronounced with regard to *innovation* activities and interactions. However this is by no means limited to interactions within the home region cluster. United Kingdom bio-pharmaceuticals firms interact most with global partners, next with UK partners and only then with cluster and EU partners. Thus while they are hypothesized and prove to share geographical coexistence with similar firms and knowledge sources, notably research laboratories or centres of research excellence, local links

*Table 7.6   Business and innovation partnering by UK genomics
              biotechnology firms*

Indicator	Percentage
Collaboration with firms/institutes	77
Clustering spatially with collaborators	78
Cooperating specifically on innovation	70
Cooperating on innovation in home region	18
Cooperating on innovation in UK	23
Cooperating on innovation in EU	18
Cooperating on innovation globally	28

Source: CASS ICT and Biotechnology Collective Learning Survey.

are the 'looser' collaborative kind (78 per cent) while focused, cooperative actions concerning innovation attract partners from far and wide as well as locally.

This is highly suggestive of the extent to which the 77/78 per cent of the sample that cluster spatially with collaborators who are a mix of firms and research institutes access knowledge spillovers in so doing. This compares with only 18 per cent of firm interactions involving specifically cooperating with local partners for innovation activities. A rather substantial 59–60 per cent of pharmaceutical biotechnology firms in the UK may be hypothesized to cluster for broad collaborative interactions. However, when it comes to more specific innovation cooperations, this geographical proximity indicator declines and non-proximate *relational* and/or *cognitive* proximity increases significantly. Clearly respondents could cooperate with more than one geographically or relationally proximate partner, indeed it can be surmised that 17 per cent of firm cooperations are in this category.

In making further judgements regarding this, Table 7.6 reveals that, with specific regard to innovation, as distinct from other interactions such as research, joint patenting, purchasing or supplying, and other more informal collaboration, the act of *commercializing* new knowledge in the form of a product or service new to the firm or new to the market, firms behave distinctively. As noted, firms innovate in partnership with other actors in their region, mainly their cluster, to an equivalent amount that they innovate in partnership with actors in the EU. However, they innovate more than either of those categories with partners in the UK more generally, and finally their innovation partner is most likely to be outside Europe, actually the USA or to a lesser extent Asia in most cases. Here, survey data show the global disaggregation to be pronounced for *innovation*, that is, near-market commercialization of new knowledge.

Table 7.7   Scaling for proximity by UK genomics biotechnology firms

Proximity factor	Range	Mean	Standard deviation
University research	1–4	3.53	0.697
Qualified workforce	2–4	3.26	0.806
Business environment	2–4	3.16	0.898
Genomics services	2–4	3.00	1.000
Regional agency/grants	1–4	3.00	1.000
Other public research	2–4	2.74	1.147
Collaborators/competitors	1–4	2.68	1.157
Suppliers	1–4	2.63	0.955
Technology transfer	1–4	2.47	1.124
Private research	1–4	2.26	1.098
Customers	1–4	2.26	1.046

Source: CASS ICT and Biotechnology Collective Learning Survey.

Accordingly, we learn that geographical proximity is relatively *unimport-ant* for what are primarily genomics firms transforming new knowledge into new products or services or conducting cooperative innovation. So why do genomics biotechnology firms in the UK cluster so frequently? The answer lies in Table 7.7, which reports on the scaling importance ranked from 1 to 4 accorded by firms to proximity for accessing a variety of services. It is clear from this result that the most important reason given by UK genomics firms for locating in proximity is access to university research, followed by specific services expertise required in genomics based biotechnology. Of slightly less value than these factors are three of equivalence: the proximate business environment covering such things as legal, financial and other business-related services; presence of an appropriately skilled workforce, not wholly university-trained since technical and ICT skills are also important, and regional development agencies with grant-giving capacity.

Hence *exploration* or basic research knowledge and finance may be said to be the key business enhancements arising from proximity. Of these, university research scales highest. Since, in the UK, all key genomics universities pre-date genomics and most pre-date biotechnology, it is reasonable to infer that for a science-driven industry such as genomics, the presence of university research is responsible for the 'spatial knowledge domain' and 'spatial knowledge capabilities' that assist formation of the regional cluster. This is significantly more important than proximity to other kinds of R&D or business transaction relations. Regarding R&D specifically, Figure 7.7 shows the *regional* university ranks high but not

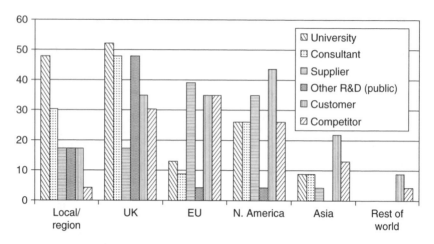

Source: CASS ICT and Biotechnology Collective Learning Survey.

*Figure 7.7   Economic geography of R&D collaborators of UK genomics
            biotechnology firms*

highest, which is university research links elsewhere in the UK, and in third
position R&D links in North America. 'Collaborator' is defined broadly, to
include market and non-market exchange. Perhaps surprising are the low
R&D links from UK genomics cluster firms to collaborators in the EU but
relatively higher numbers of R&D collaborators who are customers for the
research, suppliers of inputs for the research, or competitors of the R&D
practising firm. Most of the latter collaborations arise through EU
Framework Programmes for Science, Technology and Innovation in which
a few universities mobilize larger networks of SMEs. An interesting sidelight on this is that genomics firms either have few direct competitors in
their region, or they do not collaborate with them significantly if they are
present.

Contrariwise, they collaborate substantially with competitors elsewhere.
The fact that 37 per cent of firms have collaborations with competitors in
the UK may suggest the former interpretation is more likely. If so, it casts
an interesting sidelight upon the 'spatial knowledge capabilities' thesis.
That is, it indicates genomics firms have no desire to conduct R&D with
local competitors because they already know its likely content due to 'open
science' and localized knowledge spillovers among firms competing in
highly specific local niches (Owen-Smith and Powell, 2004; Caniëls and
Romijn, 2006).

Figure 7.8 shows how this operates in a cross-section of the UK's
Cambridge cluster. Both proximate specialist firms and university research

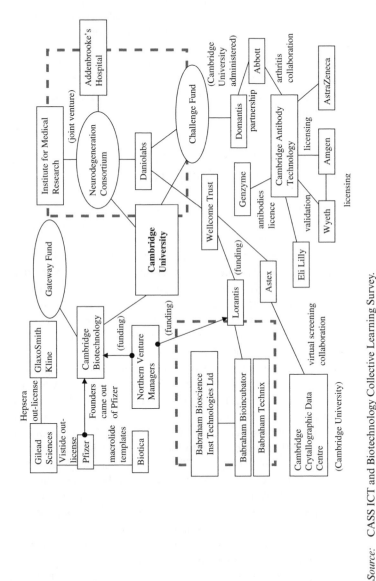

Source: CASS ICT and Biotechnology Collective Learning Survey.

Figure 7.8 Sample of proximate and distant networking in the Cambridge genomics cluster

are networked in joint genomics projects. This applies also to micro-firms in, for example, the Babraham Bioincubator. But noticeable also are the strong partnership links with distant 'big pharma', normally the ultimate customer partners for genomics drugs being researched in world-leading firms like Cambridge Antibody Technologies (in 2006 acquired by 2004 customer AstraZeneca). It is noticeable that regional hospitals, venture capitalists, and bioimaging suppliers also contribute to Cambridge's specific knowledge domains in neurology and inflammation. Moreover not all firms are university spin-outs; one is a former Pfizer employee start-up. Thus absence of distant spillovers means firms form collaborator relations with 'distant networks' (Fontes, 2006) to augment R&D knowledge for themselves. These occur broadly equally in the EU and North America, as well as more extensively in the home country.

RESEARCH RESULTS ON INDUSTRY AND CLUSTER INNOVATION INTERACTIONS:[4] ICT

Precisely the same methodology was elaborated for the study of reasons firms cluster or do not cluster in ICT as for genomics biotechnology. The key difference is that the universe and sample size of firms surveyed was vastly bigger for the former postal questionnaire survey than the latter. Basically sectors were grouped according to manufacturing, software, tele-coms, data and R&D. Table 7.8 summarizes the sample composition for the postal survey by region. Reference to Table 7.9 gives a comparison for ICT with data on longevity of firms in biotechnology (from Table 7.3). It shows, unsurprisingly, that the earliest establishment date of ICT firms is often before that for biotechnology, but that a similar bunching of firm forma-tion among survey respondents in ICT occurred in the 1980s and, particu-larly, in the 1990s. Similarly, founding activity after 2000, when the technology stock markets nosedived worldwide, is comparably meagre. More surprisingly, for the 1990s is the exact similarity at 47 per cent in the percentage of all ICT and biotechnology formed during that decade. Thus for both, very differently composed, activities just under half the respon-dent firm populations arose during the 1990s. Moreover, proportionately more biotechnology firms were founded than ICTs in the 1980s. This sug-gests that arguments that the biotechnology firm formation cycle is different from other sectors because it is science rather than finance driven may be wide of the mark, at least by comparison with ICT.

Of course, science plays an important role in ICT too, but lead-times from discovery to innovation are generally much shorter, as are timescales from firm formation to flotation on stock markets. Moreover, testing and

Table 7.8   Sample composition and respondents[1] by regions

	ICT data units in the region*	%	Sample	%	Respondents	%	Response rate (%)
Wales	2 281	3.3	346	4.8	42	15.4	12.2
Scotland	4 972	7.3	456	6.3	19	7.0	4.2
Eastern	14 240	20.8	1 580	22.0	56	20.6	3.5
London	22 051	32.2	1 297	18.0	28	10.3	2.2
South East	24 871	36.3	3 516	48.9	127	46.7	3.6
Total	68 415	100.0	7 195	100.0	272	100.0	3.8

Notes:
1 Representativeness of the sample in terms of geographic location and sectoral groups was guaranteed by stratifying the sample by location and ICT activity. It is worth noting that the figures also indicate that the London location is somewhat under-represented in the sample. This is due to the fact that only some London postcodes were randomly selected, including just a few high and low concentration of ICT areas. Secondary data were available for all the 7195 firms in the sample with respect to SIC codes, location, number of employees and turnover; this enabled us to compare respondents and non-respondents on these variables. First, the figures indicate that the survey overrepresents Welsh firms as their proportion is 5 per cent and 15 per cent in the sample and in the respondents group, respectively. Second, there is an overrepresentation of companies in the ICT manufacturing group. Nonetheless, when comparing the size of the firm and turnover there was no bias evident for these variables. In our sample 85.8 per cent of firms had fewer than 50 employees, 9.8 per cent had employees numbers between 50 and 199; 2.6 per cent between 200 and 499 and 1.8 per cent more than 500 employees. These figures are reflected in the respondents' firm size, which are 85.5 per cent, 9.9 per cent, 2.4 per cent and 1.9 per cent respectively. Comparing the turnover figures it is worth noting that in our sample there were fewer companies with turnover between £100 000 and £500 000 (20 per cent in sample and 32 per cent in the respondents) and a higher number of companies with turnover less than £100 000 (26 per cent in the sample and 13 per cent in the respondents). Significance tests (Paired samples 't' test) were used to compare the means in term of turnover and employees number of the sample with that of the respondents to test for sample differences. In both cases, the significance values were high and the confidence interval for the mean difference did not contain zero, indicating that there are no significant differences between the means for the sample and the respondents group.
* NOMIS data, 2002.

Source: CASS ICT and Biotechnology Collective Learning Survey.

trialling while regulated for safety reasons worldwide is less complex and drawn out because much of it is computerized, as with structural calcula-tion of finite elements algorithms.

Although in ICT, a proportion of firms are larger organizations, the UK ICT sector is dominated by small and micro-sized firms (fewer than 50 employees) which account for 80–90 per cent of total UK units. In the sample, respondent firms split 4.3 per cent large firms (over 200 employees), 9.9 per cent of medium size (between 50 and 199) and 85.8 per cent small

Table 7.9   Date of establishment of UK survey respondent ICT and biotechnology firms

Date of establishment	ICT frequency	ICT percentage	(UK biotechs percentage)
<1947	8	3	0
1948–1970	16	6	0
1971–1980	24	9	3
1981–1990	69	26	39
1991–1999	124	47	47
2000–2003	23	9	11

Source: CASS ICT and Biotechnology Collective Learning Survey.

Table 7.10   Economic and innovation indicators for UK ICT and biotechnology firms, 2003

Indicator	ICT	Biotechnology
Mean employment	89*	97*
Mean turnover	$26 million*	$28 million*
Mean exports/turnover	40%	24%
Mean R&D expenditure/turnover	16%	21%
Firms with new patents	11%	47%
Mean patents per firm	8.0	4.5

Note: * Data drawn from biotech and ICT population, from which the samples were derived.

Source: CASS ICT and Biotechnology Collective Learning Survey.

firms (less than 50). Regarding the corporate headquarters of respondents, the majority as we have seen being SMEs, 37 per cent were UK domiciled, 33 per cent North American, 23 per cent European and 7 per cent Asian.

With respect to key performance indicators of UK ICT respondent firms, the data in Table 7.10 provide this and facilitate comparison with the UK biotechnology firms.

Foremost, the two populations of firms are relatively similar in terms of mean SME size and turnover, with biotechnology firms being of slightly bigger size both in terms of turnover and employment. Both sectors spend a significant share of turnover on R&D with biotechnology, true to its high science-driven character, outstripping ICT although the ICT ratio is also rather high. As indicators of intellectual property regimes in the two

*Table 7.11   Business and innovation partnering by UK ICT and
                biotechnology firms*

Indicator	Percentage	
	ICT	Biotechnology
Collaboration with firms/institutes	80	77
Clustering spatially with collaborators	27	78
Cooperating specifically on innovation	73	70
Cooperating on innovation in home region	27	18
Cooperating on innovation in UK	48	23
Cooperating on innovation in EU	20	18
Cooperating on innovation globally	30	28

Source: CASS ICT and Biotechnology Collective Learning Survey.

sectors, our results conform to the conventional wisdom in the literature, namely that ICT tends to be a low patent-holding sector, while, again, biotechnology's science-driven nature and heavier R&D means nearly half had new patents within the year before filing the questionnaire. However, the mean number of patents held per firm tells a different story, which is that the average ICT firm holds nearly twice as many patents over the longer term as the average of the respondent biotechnology firms. This is probably due to two factors, the greater size by exports of ICT firms and their greater longevity. But it suggests strongly a different regime between sectors where ICT firms may exploit patents over a lengthy period of time, while biotechnology firms exploit theirs swiftly and, because of the complex nature of discovery in this sector, in patent 'families'.

Moving on, Table 7.11 begins comparative analysis of central importance to understanding the distinctive roles of proximity and collaboration, the heart of the matter of determining whether and why clustering among firms occurs as well as more distant networking. Recall that a strong finding of the UK biotechnology research was that firms cluster for *research* but utilize *distant networks* (often with large pharmaceuticals firms) for *innovation*. The comparison in Table 7.11 is in some ways remarkably similar for these distinctive sectors, and in other ways remarkably different. First, the similarities: strikingly *collaboration* in general is remarkably and similarly high for both at 80 per cent for UK ICT and 77 per cent for biotechnology. Comparably, specifically cooperating on innovation is almost identically high among respondents – 73 per cent of ICT firms doing cooperative innovation, 70 per cent of those in biotechnology. Collaboration is generic, cooperation is specific.

Cooperating on innovation in the home region is comparably low, as it is in the EU and with global cooperations at between 18 per cent and 30 per cent respectively. We summarize this as two highly cooperative sectors for innovation with varying amounts of this locally and globally. However, cooperating nationally is much higher in UK ICT than biotechnology. Notably clustering with collaborators is also much lower for ICT than biotechnology. Thus, on key *proximity* indicators for ICT we have a picture of a large majority of firms cooperating on innovation with other domestic but not particularly local firms and organisations. Contrariwise, biotechnology firms are nearly equivalently cooperating on innovation non-domestically.

Moving to the issue of proximity in the ICT firm rationale for interaction, the picture is different between ICT and biotechnology in this regard. It will be noticed, as Table 7.12 summarizes, that for biotechnology, proximity to universities for accessing research is the first imperative, whereas recall proximity (Table 7.11) was relatively less important for innovation. This frequently took place in partnership with distant network actors, with US or other transnational corporations. For ICT, the proximity rankings for business interaction are given in Table 7.12. These data show that, unlike biotechnology, universities are ranked only medium as 'proximity partners'. Most strikingly, 'customers' ranked lowest[5] in biotechnology, ranked highest for ICT, and other public research, such as that conducted in non-university laboratories is ranked very low by ICT but of medium influence in terms of proximity drivers by biotechnology firms.

Thus a picture is relatively easily formed of ICT and biotechnology having distinctive rationales for proximate interaction in research and innovation. Whereas biotechnology firms cluster around universities and, to a lesser extent, other public laboratories for research knowledge and related interactions, meanwhile interacting distantly with customers, many of which are pharmaceuticals transnationals, ICT firms prefer to cluster close to customer firms, keeping research at a distance. This is an original finding for both industries and tells us much about the nature of and differences between them. First, both collaborate intensively but ICT more nationally than either locally or globally as in the case of biotechnology. Second, ICT is more market than science focused in its proximity practices, a sign that innovation is more important and swifter than in biotechnology. Third, and of policy relevance, a region is well advised to have localized ICT multinational customers to help promote its nascent ICT cluster, while for biotechnology this is relatively unimportant and proximity to an accomplished medical or other biosciences research capability is of greater importance for cluster-building.

Table 7.12 also seeks to ascertain differences between collaborators and non-collaborators within ICT.[6] It can be argued that although the ranked

Table 7.12   Scaling for proximity by UK genomics biotechnology and ICT firms

Proximity factor	ICT mean			Biotech mean*
	All respondents	Collaborators	Non-Collaborators	
University research	2.38 (5)	2.57 (6)	2.21 (6)	3.53 (1)
Services	2.57 (3)	2.78 (4)	2.40 (4)	3.00 (4)
Business environment	2.48 (4)	2.67 (5)	2.33 (5)	3.16 (3)
Qualified workforce	2.83 (1)	2.88 (1)	2.79 (2)	3.26 (2)
Regional agency/grants	2.06 (7)	2.24 (8)	1.92 (7)	3.00 (4)
Other public research	1.92 (9)	2.07 (10)	1.76 (9)	2.74 (5)
Collaborators/ competitors	2.36 (6)	2.54 (7)	2.21 (6)	2.68 (6)
Suppliers	2.72 (2)	2.81 (3)	2.64 (3)	2.63 (7)
Private research	1.94 (8)	2.14 (9)	1.79 (8)	2.26 (9)
Technology transfer	1.70 (10)	1.81 (11)	1.61 (10)	2.47 (8)
Customers	2.83 (1)	2.85 (2)	2.82 (1)	2.26 (9)

Note: * Data on biotech refers to collaborating firms.

Source: CASS ICT and Biotechnology Collective Learning Survey.

order remains almost the same, with the exception of collaborators vis-à-vis the importance of qualified workforce as main proximity driver, collaborators tend to value proximity with customers of relatively more importance (2.85 compared to 2.82 for non-collaborators). Finally, we can to a considerable extent compare the economic geography of R&D collaborations by ICT and biotechnology firms. Figure 7.9 summarizes the position for UK ICT. Recall that the main lineaments of such collaborative economic geography for biotechnology were as follows. First, UK biotechnology's favoured R&D collaborator was UK universities, followed mostly by UK 'other (public) R&D', consultants and customers. Competitors and suppliers in the UK were as popular as the best scoring collaborator in the host region. This was the regional university, followed by regional consultancy, then supplier, public R&D, while regional customers and competitors were negligible R&D collaborators. Indeed customers anywhere

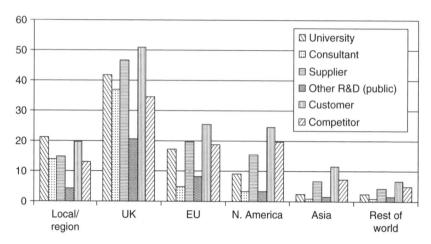

Source: CASS ICT Collective Learning Survey.

Figure 7.9   Economic geography of R&D collaborators of UK ICT firms

globally were of more importance (Figure 7.7). For ICT, the picture of
R&D collaboration is significantly more national in orientation as Figure
7.9 shows. Here, it is clear that most UK ICT collaboration in R&D occurs
nationally, with the host region some way behind, but relatively more
engaged for universities and consultants. 'Suppliers' and 'competitors' are
relatively important to R&D collaboration in both the EU and North
America, as indeed are customers. Europe is also the location where much
collaboration with other public R&D laboratories takes place.

Accordingly, a picture forms of UK ICT firms much engaged in transatlantic supply chains bolstered by UK and regional R&D collaborations
with a wide range of support actors, including universities. Hence, R&D is
less a factor in proximate location for UK ICT firms, especially compared
to the proximity force of innovation and market partners. Research and
development collaborations with UK regional, national, European or
North America is of minor significance except among competitors.
Nevertheless, such R&D collaboration is more important than that from
elsewhere overseas, including Asia and Rest of the world, currently a nexus
of R&D collaboration of minor significance.

Thus, in terms of the thesis advanced at the outset of this book that clusters gather for different reasons these results are clearly supportive. ICT and
biotechnology clustering in the UK, driven as it is by different imperatives
– research for biotechnology, innovation for ICT – also sees firms intimately
involved in interacting cooperatively with customer firms with whom
they engage for purposes of conducting 'open innovation' and/or 'R&D

outsourcing' kinds of collaboration. Further, these firms value proximity in this regard: to repeat, with national and regional consultants, customers and universities for ICT firms, and with national and regional universities, but more transatlantic customers and suppliers, for biotechnology firms. Hence, a further elaboration is a greater valuation by the latter of *relational* or *functional* proximity than *geographical* for innovation through distant networks.

MEASURING THE EFFECTS OF PROXIMITY AND COLLABORATION ON FIRM PERFORMANCE

Of key importance to this book is an answer to the question whether firm performance is affected by business 'clustering'.[7] The answer to this question, unanswered in the literature up to now, ascertains if UK firms in the biotechnology and ICT industries perform better in collaborative proximity (a synonym for 'clusters') than not. In particular, it is interesting and important to separate collaborator performance from general performance and key indicator data for ICT are presented in Table 7.13, which compares key performance indicators for firms that collaborate, do not collaborate, and their location. The analysis that follows is performed principally for ICT firms since scarcely any biotechnology firms sampled existed outside clusters. Thus their performance is intimately related to that locational characteristic.

Analysis was undertaken with respect to measures of firm performance, namely, employment change, turnover change, research and development expenditure change, and innovation, all between 2000 and 2003. In this respect, levels of innovation were measured by asking firms about the number of products/service and changes to products/services in the past three years from survey time, the number of patents announced, R&D activities and the firm's capacity to introduce new products/services compared to competitors. Table 7.13 shows that on most indicators of economic performance, collaborating UK ICT firms' mean performance is generally better than the mean scores in the respondent group as a whole, consisting of both collaborators and non-collaborators. Thus collaborators have superior performance regarding market share, capacity to introduce new products and services, higher R&D as a share of turnover in 2003, more employees per firm and greater turnover, showing a slight increase in both figures between 2000 and 2003. Thus collaboration clearly pays in most dimensions of measurable firm performance. Hence we continue to believe, on the basis of our evidence, that collaboration provides a competitive advantage. Accordingly, it is not unreasonable to propose that ICT

Table 7.13   Collaborator and non-collaborator performance of UK ICT firms

Selected performance indicators	All respondents	Collaborators	Non-collaborators	Collaborators		Non-collaborators	
				Cluster N = 55	Non-cluster N = 40	Cluster N = 71	Non-cluster N = 44
Mean employment	105	180	40	57	53	19	41
Employment increase 2000–03	36%	40%	34%	45%	41%	39%	39%
New products/ services 2000–03	80%	88%	73%	89%	87%	74%	69%
Patents 2000–03 +/–	0%	2%	–2%	6%	4%	–3%	0%
Turnover 2003	£9 m.	£13 m.	£6 m.	£5 m.	£5 m.	£2 m.	£5 m.
Turnover increase 2000–03	61%	69%	55%	70%	76%	49%	70%
Mean R&D expenditure/ turnover	16%	17%	14%	19%	13%	10%	11%
R&D expenditure increase 2000–03	31%	32%	30%	39%	16%	22%	33%

Source: CASS ICT Collective Learning Survey.

Table 7.14   Collaboration and performance of UK ICT firms

	Non-collaborators and collaborators	Effect of external collaboration	Product innovation in collaboration
Capacity to introduce new products/services	0.059	0.213**	0.024
High or low market share	0.273**	0.181	0.112
Patent announcements	0.158*	0.100	0.075
Turnover change	0.172*	0.236*	0.199*
Employment change	0.013	0.182*	0.145
R&D change	0.038	0.107	0.086
Improvements of companies' best products	0.102	−0.096	0.323**
New products for company but not new for market	0.029	0.117	0.142
New products new for company and new for market	0.230**	0.152*	0.203**
n. new products	0.102	0.011	0.157*
n. changes	0.081	−0.003	0.096
External coll. effect on company performance	0.538**	1.000	0.257**
Non-collaborators and collaborators	1.000	0.538**	

Notes:
** Correlation is significant at the 0.01 level (2-tailed).
* Correlation is significant at the 0.05 level (2-tailed).

Source: CASS ICT Collective Learning Survey.

firms engaging in collaborative activity with others are more capable on the R&D and patenting input side of the innovation relation and they benefit on the output side with greater market share. This is also confirmed in Table 7.14 which shows how collaboration is significantly and positively associated with key performance and innovation indicators.

The key question remains to what extent are ICT collaborators – not forgetting non-collaborators – found consciously locating in clusters. Recall, most biotechnology firms surveyed are located in clusters. However, the numbers are far greater in ICT, it has more variety and substantial numbers of co-locating, isolated, collaborative and non-collaborative firms accordingly. The answer provided in Table 7.13 illustrates that geographical (cluster) proximity for UK ICT firms is important, as shown by the number of non-collaborating firms that consciously decide to co-locate in a cluster

(56 per cent). Analysing data for the biotechnology firms, we hypothesized the existence of 'knowledge spillovers' as a locational attractant, particularly for non-collaborating firms. This seems to be consistent with what we found in the ICT sector, where substantial numbers of non-collaborators are found in clusters, indicating that there is a 'knowledge spillovers' attraction effect even for those who envisage non-collaborative interactions with their neighbouring firms. These may be assumed to be those seeking to exploit knowledge that is 'in the air'.

To further test the effects on ICT performance in multiple dimensions, questions were posed of collaborator and non-collaborator ICT firms and answers correlated with two categories of outcome concerning external collaboration and product innovation. Table 7.14 shows that the strongest effects, registered as correlations with high significance and positive directionality, occurred with new product introduction, turnover change and market share. Regarding the effects of collaboration in product innovation per se, it was improvements to best products, innovating wholly new products and the overall effect on company performance that had the highest significance.

Returning to Table 7.13 and focusing on the collaborator side, it shows that for some indicators of economic performance collaborators in clusters perform better than collaborators in areas with less concentration of firms. Collaborators in clusters tend to have superior performance regarding innovation *input* indicators like higher R&D as a share of turnover in 2003 (19 per cent compared to just 13 per cent), a higher number of firms recording an increase in R&D expenditure between 2000 and 2003, more firms announcing patents and number of patents announced in both 2000 and 2003. But clearly notable also is the fact that non-collaborators not in clusters have significantly higher *output* performance indicators like turnover and turnover increase than those in clusters. Additionally, collaborators not in clusters perform better on turnover than those in clusters. To test the cost of such apparent cluster *diseconomies*, further research revealed cluster land rents to be up to three times the equivalent non-cluster land rents only 30 kilometres away. When added to probably significantly higher wage costs this shows why cluster firm financial results are worse than those of unclustered ICTs. Moreover, it provides a proxy – particularly for non-collaborators – of the price of entry to the cluster 'quasi-monopoly' and its knowledge spillovers. This, for cluster isolates is the price of 'eavesdropping'.

Clustered collaborators tend to be bigger than non-clustered collaborators and have a higher number of firms that increased their employment size between 2000 and 2003. However, while mean turnover of the two sub-groups is similar (£5 million in 2003), 76 per cent of non-clustered collaborators increased their turnover between 2000 and 2003, highlighting that collaborators in clusters tend to be superior on the R&D and patenting input side. Turning the discussion back to the non-collaborators, we noted it was perhaps surprising that a significant number of non-collaborators consciously locate in clusters. However, in line with the generally negative financial *out-turn* numbers discussed above, the data reveal that non-collaborators in clusters only perform better than their counterparts in non-clustered areas on a few innovation outputs (employment increase and new products/services). Contrariwise, non-collaborators in non-clustered areas spend more on R&D, and have a higher proportion of firms that increased their R&D expenditure and turnover growth in the 2000–2003 period. This reinforces the point that non-collaborating cluster incumbents 'free ride' on knowledge spillovers.

It can be argued that clustering can provide competitive advantage to non-collaborators as the non-collaborators that co-locate in geographical proximity are smaller in size (19 average employees compare to 41 for non-clustered firms and £2 million average turnover compared to £5 million for the isolated non-collaborators), and this is somewhat a confirmation that economic spillovers are available as literature predicts in cluster settings. This is consistent with the 'knowledge spillovers' attraction effect since small firms would be more likely to need to seek out such spillovers than more self-contained larger ones. However, as shown, this may help innovation but may also, as we have seen, have some performance-hindering characteristics.

To further investigate the matter of proximity and performance, questions were asked of firms regarding whether and what advantage, if any, spatial proximity gave to firms' cognitive or knowledge capabilities. Furthermore, as a cross-check, firms were asked the extent innovation activities were conducted in a local cluster, the definition 'closely proximate collaboration' being provided to specify such a context. The results from asking these two questions reveal that collaborators favour spatial proximity to a greater extent than the respondent group as a whole and that of non-collaborators. Respondents answering these questions were low in number, with even those stressing clustering cooperation for innovation a minority compared to those conducting such activities with intra-firm cooperation (larger firms), and intra- or even extra-UK innovation cooperation. Spatial proximity is thought beneficial by clustered firms that collaborate as it facilitates informal communication (27 per cent), facilitates knowledge exchange (24 per cent), reduces uncertainty (17 per cent) and interaction costs (15 per cent). This is, of course, totally consistent with Table 7.13 regarding clustering in ICT. Interestingly, the reasons that motivate proximate location differ among collaborators and non-collaborators

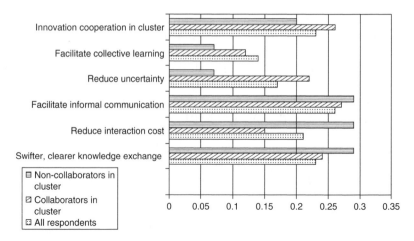

Source: CASS ICT Collective Learning Survey.

Figure 7.10   Proximity, cognitive and innovation advantages for UK ICT Wrms

in clustered location. One in three non-collaborating firms agreed that spatial proximity reduces interaction costs, speeds the knowledge exchange process, and facilitates informal communication as shown in Figure 7.10.

Exploring this more deeply in face-to-face interviews, UK ICT firms were discovered to access their knowledge sources through a variety of market links, spillovers and milieu effects in the innovation process. However, there are significant differences in the frequency and importance of channels uti- lized: research contracts and reading of scientific literature and patents are most commonly used in research firms, conferences and fairs mainly by ser- vices industries, while manufacturing ICTs tend to rely more on consulting. All firms, however, consider informal contacts important for knowledge exchange. Interestingly, firms predominately use informal contacts as pre- ferred knowledge channels and less so market links and formal networks, confirming that social interactions have economic value in transmitting knowledge and ideas. These social interactions differ in spatial scale: at local level spontaneous and informal meetings predominate; knowledge exchange with more global collaborators occurs in more structured personal meetings. It is also noteworthy that within their sector, 'virtual communities' (Cairncross, 1997; Amin and Cohendet, 2004) are becoming increasingly important in the knowledge transfer process. Contrariwise, employment of highly knowledgeable skills, although regarded as important, is used less fre- quently owing to the small size of the respondent firms and the opportuni- ties for knowledge spillovers. United Kingdom biotechnology firms balance up loose, often informal, channels within their cluster and virtual community accessed at the many trade fairs typical of the sector, with tight, often contractual channels to distant customers frequently located in a different continent.

Hence we find convincing evidence from the biotechnology, but especially the ICT, results from surveyed firms for the following. First, firms that collaborate perform better on all performance indicators than firms that do not. Collaboration thus gives to firms in these industries an added competitive advantage. Second, collaborative firms in clusters perform better than collaborators not in clusters in just some financial outcome metrics, mainly related to innovation inputs. Finally, within the cluster we find an unexpectedly large portion of even non-collaborating (59 per cent) biotechnology and (56 per cent) ICT firms. This arises from their conscious decision to access knowledge spillovers from the interaction effects and knowledge 'free-riding' opportunities available to firms within earshot of other incumbents with whom they have no intention of collaborating. The possibility that recruitment of talent is an element of knowledge spillover advantages being sought by such firms must also be taken into account. However, recall the significantly smaller employment size of these firms, which suggests that this may not be their primary interest. The constructed advantage of the knowledgeable cluster thus derives from its local linkages and conveys degrees of competitive advantage directly and indirectly to its collaborators and non-collaborators alike.

KNOWLEDGE NETWORKS IN KEY UK CLUSTERS

As indicated, knowledge networks displaying formal and informal links as well as more formal cooperations among companies and other organizations exist in both the biotechnology and ICT sectors. Evidence from the research conducted shows that networks can be proximate and/or distant, and that, in both instances, collaboration plays a fundamental role, influencing companies' innovative potential and performance. One of the key propositions this book tests is whether innovative firms operate within regional networks, cooperating and interacting not only with other firms such as suppliers, clients and competitors, but also with research and technology resource organizations, innovation-supporting agencies, venture capital funds and local and regional government bodies. Our results show that such network relationships produce pervasive and systemic effects that encourage firms to adopt common norms, expectations, values, attitudes and practices, where a culture of innovation is enforced and a learning process is enhanced. This, when combining good governance and the creative attractions to recruit and retain talent, constitutes a definition of constructed advantage.

Before turning our analysis towards the comparison of the different policy regime and institutional settings that emerged in the five devolved administrations of Scotland, Wales, the Eastern region, the South East and London, we may recall what we argued in Chapter 5 of this book. Within liberal market economy institutional settings at the UK level, the policy regime and its institutional stance may differ, indeed evidence exists to show those of the UK's devolved territories differ quite markedly from those that have less devolution and from each other.[8] Economies organized around liberal market rules and conventions, such as that of the UK, accommodate an entrepreneurial model of business organization and innovation styles and strategies that are thought to stimulate both disruptive and radical innovation. However, this entrepreneurial model does not capture the asymmetries that are often found at regional level and a more interventionist model has sometimes perforce been introduced. This model is more reliant on 'a developmental state' economic strategy, which focuses on innovation measures and instruments to promote incremental product innovation. Possibly the state may also play a role in helping firms to mobilize internal R&D resources; in other words, this model envisages a state-centric approach to promote a liberal market and entrepreneurial development. Two UK territories, Scotland and Wales, adhere to some extent to the *developmental state* model, while the other three, the South East, Eastern and London, are more of an entrepreneurial nature.

Interviews were conducted about this and related matters with some 70 policy intermediaries in the five target regions. Four themes were identified to contrast the institutional stance of the five target regions, namely:

1. *Networking interaction* – this hypothesized that in the two high-technology sectors investigated networking propensity is increasingly pronounced across regional and national regulatory regimes.
2. *Financial governance* – hypothesizing that this varies from a more public-private risk investment versus a strong reliance on liberal market mechanisms and private venture capital.
3. *University–business relationships* – hypothesizing a rise in interactions between firms and the local research base, including universities and other research organizations, ranging from licensing, spin-out, and research programme to joint venture.
4. *Geographical proximity and clustering* – hypothesizing the importance that geographical proximity and clustering policies play in shaping innovation strategies at regional and local level.

We shall discuss these in turn for the two categories of *entrepreneurial* and *developmentalist* regions in our broadly liberal market national model.

Networking Interaction

In Cambridge and Oxford, the local level is more important than the regional and, although interactions do happen with regional, national and international institutions, local firm-related actors are the drivers in networking interactions (such as the Cambridge Network and St John's Innovation for Cambridge and Oxford Trust and OxIT (Oxfordshire Information Technology) in Oxford). In London, the London Technology Network plays a comparable role, albeit scale gives it an almost regional quality too. Cooperation and coordination between regional and national policy levels is perceived to have a problem in that, being wealthy, they are the poorest funded of the regions. Government is perceived putting too much money into the poorer regions and not enough into the more successful regions. Until relatively recently such locales perceived themselves as having been too unsophisticated at an international level. A report written for the Greater Cambridge Partnership, also the sub-regional economic partnership (SREP), produced the case for employing a full-time person on a local basis who would handle international relations working with UK Trade International and others, trying to support at a local level and on the regional level much more of an international trade policy. This was possibly a vehicle for targeting China, and perhaps not just China but specifically Shanghai, to promote collaboration between businesses. Hitherto performance of such a function was considered only average and undifferentiated from cities like Birmingham or Newcastle. There is nowadays a perceived need for a more regional, or at least sub-regional, focus on building up international relations and alliances with those areas who will best benefit businesses that exist in those areas.

For biotechnology, a valuable resource in Oxford was the BioTechNet bioincubator managed under the Oxford Trust. Until its demise in 2005 it performed a valuable function housing spin-out firms emerging from Oxford University's Isis Knowledge Transfer Office (KTO), and linking to clinical research organizations (CROs), local IPR expertise, and regional public and private venture capital funds. Oxford BioTechNet exemplified a facility meeting a key need in the area but which did not ultimately find favour for ongoing funding from SEEDA (the South East England Development Agency). This signifies the difficulties even in accomplished innovation settings in finessing the networking propensity even where regional development agencies in both South East and Eastern England, SEEDA and EEDA (Eastern England Development Agency), respectively, value collaboration and see partnerships as critical to the their functioning. Both regional development agencies have appointed science advisory councils (Science and Industry Council, in the Eastern region, and the South East Science and Advisory Council) to complement and develop a science strategy, to advise the board, to promote knowledge transfer, to interact with other regions and to promote the region at international level.

In the EEDA region networking has long been formalized through the Cambridge Network Limited and there is perceived to be much collaboration and cross-fertilization between businesses. The Cambridge Network has some 1300 members, involving a smaller basic needs business-to-business club for start-up companies that has about 600 members. Hence relationships between local companies are thought to be good, they cooperate where it is sensible to cooperate but intellectual property is very important and requires protection, a matter of shared understanding. However, what was once seen as a problem of reticence, whereby Cambridge entrepreneurs failed to communicate unless first introduced, had been transformed. Reticence to talk to anybody had gradually been broken down and there was a much more collaborative culture here than there is in many other parts of the UK. Accordingly, there is a widespread perception of the rise of an associative, networking culture. This is also the case in Oxfordshire but it is recognized sometimes to be quite disjointed, multiple networks are known to be pursuing distinctive network agendas so Oxford Trust, for example, plays a brokering role, providing a focus for coordination. This means fostering good relationships with the county councils, the district councils, with the likes of Oxford Economic Partnership, the Oxford University spin-out intermediary *Isis*, universities, and companies. It means that organizations improve *absorptive capacity* for local knowledge of consequence to business. It is also widely perceived that when a consensus has formed, the business community operates more effectively because of collaboration and efficient communication. Existence of a critical mass of advanced technology firms helps in that professional services, like patent agents, the banks and the lawyers, and so on, each have a high-tech arm with all fighting for the business of the high-tech companies.

London Development Agency (LDA) chiefs regard ICT as a priority sector, yet it is not one with which they have worked actively of late. Now there is a fuller recognition of London's problems of scale and fragmentation regarding the knowledge value chain and a heightened consciousness of the desirability of stimulating and engaging in more collaborative interaction. For example, London possesses 42 higher education institutions (HEIs), nine research and technology organizations, a wide range of large business, 640 000 small businesses and a vast array of enterprise-related organizations and institutions. A key role for the LDA is perceived to be bringing them together where there is fragmentation and lack of coordination, something that is included in the LDA innovation strategy. There are three strategic priorities, the second of which is specifically about the knowledge base, about London's higher education institutions, and harnessing London's world-class knowledge base to benefit London's businesses. Thus the issue in these buoyant regions where 'entrepreneurial' competitiveness has been little influenced by formalized network-building in the past, is for institutional and organizational networking conventions and structures to be constructed for *regional* advantage in a globalized knowledge-based innovation economy.

The picture given above differs substantially for the devolved territories of Scotland and Wales. This is not least because of the relative absence of 'entrepreneurial competitiveness' and the lengthy existence of compensating, partnership-minded development agencies that sought, since their foundation in 1976, to offset the effects of market failure in advanced technology industries. Scottish Enterprise (formerly the Scottish Development Agency) and until its closure in 2006 the Welsh Development Agency (WDA) became key drivers of innovation within those countries. In Wales several innovation initiatives like the Technium Network were led by the WDA, but a widely perceived problem was an absence of genuine innovation expertise. Politics also bedevilled the engagement of academic institutions, with a knowledge exploitation fund (KEF) being earmarked bizarrely for further education (community) colleges rather than universities on grounds of the former's traditional 'Cinderella' status in this regard. By 2006 the WDA was shown to be more worried about the need for better networking *within* the WDA as an organization than outside. Either way, its fate was sealed when the First Minister of the Welsh Assembly Government (WAG) announced on Bastille Day (14 July) its powers were being taken inside the newly named Department for Enterprise, Innovation and Networks from April 2006.

Ironically one of the most valued networks in Wales owed no provenance to the WDA. The Cardiff University Innovation Network has proved a successful attempt by the university to link with small companies in the region and for those companies to link effectively with each other. Sixty or seventy businesses regularly meet at a series of ten annual events focused on particular issues as determined by an industrial adviser who directs the activities of the innovation network. Such techniques as innovation matchmaking, marketing and inter-university partnering are used in addition to network outreach and mentoring meetings for innovative businesses seeking cooperation opportunities. Separately, an Electronics Forum was also created in 1998 to develop a unique and consistent voice for the ICT sectors in Wales, run by the industry for the industry. However, the forum is perceived to be still in its early years and has not yet developed into a robust and stable network of the type found in Cambridge and Oxford.

Broader than Wales's Electronics Forum, Scotland evolved a technology forum also to enhance network-construction opportunities. The majority of its members are relatively small companies and many employ less then 50 people. Hence the theory that collaboration is a virtue under such circumstances is perceived as correct for innovation. However, business culture is perceived as individualistic and many small companies collaborate unreflectively, if at all, through specialist contacts locally and globally. Of formal collaborations between smaller and larger companies evolving firmer cooperative relations, there are many, albeit mainly subcontracting with all the power asymmetries thereby implied. In terms of evidence of substantial inter-company small-firm collaboration and with the research base in Scotland, the perception is of a relative absence of this form of interaction. By contrast, university research is seen as impressive in its degree of international collaboration outside Scotland. However, it is less evident in the business community and in relation to knowledge transfer, something officially recognized as a key policy issue Scotland must address. One key weakness in both territories is a history of overreliance on supply-side measures and scarcely any on demand-side improvement (Roper et al., 2006). Hence, in relation to the hypothesis posed for this theme, we conclude networking to be a sine qua non of business culture in the leading ICT and biotechnology cluster regions, to be emergent among the policy community in London and in retreat somewhat in Wales where an even more state-centric developmental model than Scotland's has been imposed since 2006.

Financial Governance

As an illustration of steps made to foster investment in innovative firms by LDA, the London Technology Fund is a £30 million (€45 million) seed equity fund for technology-based businesses, its first investment having been made in mid-2005. A second initiative is the Gateway to Investment programme that assists technology-based businesses to raise finance from elsewhere. A 'Proof of Concept' fund, modelled on Scotland's exemplar is also envisaged. For Cambridge, the area is one of the most sophisticated outside the USA for the provision of innovation services and venture capital. A belief expressed by a leading incubator manager is that 20 per cent (actually 8 per cent is a more accurate statistic) of Europe's venture capital is invested in Cambridge companies. However, there is a perceived problem in the quality of management, especially financial management of technology businesses, while another concerns specifically finance and fund-raising for start-up businesses. Only 20 per cent of start-ups receive seed funding, then for those successful enough later to attract venture capital there are two undesirably swift outcomes: an initial public offering (IPO) to the stock market or a trade sale. Of these, one in seven is an IPO, which means that the company achieving IPO is sustainable. If it is a trade sale, however, very often US investors buy it and, typically, within some three years the company is no longer in Cambridge, or in Europe but in the USA. In London, this is a key strategy of the London Technology Fund, which deliberately develops companies (such as software firm London Bridge) to the point of exit, where the aim is actually to sell it to a US company for a cash return. The contrast in aspirations for successful spin-outs as between a kind of techno-nationalism in Cambridge and a culture based upon global trading by London financiers is noteworthy.

In Scotland access to business funding is healthy, especially through business angel funding, which is co-funded by the Scottish Co-Investment Fund. While small amounts of capital are not easy to raise, they are nevertheless capable of being raised. However, second- and third-round funding are perceived to be where problems are found. Numerous US and indigenous venture capital firms have had a presence in Scotland but subsequently withdrew. Hence expansion not start-up investment capital characterizes Scotland's financial problem. In Wales, facing a similar dearth, Finance Wales was created essentially as a development bank for Wales with a variety of equity and loans funds, including a Creative IP fund devoted mainly to investment in film production. A case in point is funding opto-electronics firms, a technology cluster speciality in Wales that faces difficulties in early-stage funding, since often ideas that come out of universities are possibly two years away from the market and it is difficult keeping a company going in the interim until the product has matured and venture capital can be found. Hence the barriers to knowledge transfer are seen to be the first two years of a company's life. Even Finance Wales investments are stage 2 funding in the main, though its loan funding is not confined to innovative ICT or biotechnology businesses. Scotland's most significant innovation system improvement since 2003 has been the intermediary technology institutes (ITI) for ICT, Life Sciences and Energy. These are the intended vehicles for integrating Scotland's innovation system. Three specific kinds of disconnection which the ITIs are addressing are identifiable. First, despite the 'Proof of Concept' fund and the 'Co-investment Fund', such activities result in small, often slow-growing new businesses. Second, universities and other institutions of knowledge generation do not connect well with key financial institutions or those involved in innovation governance. Finally, Scottish Enterprise is the main economic intervention vehicle but modern innovation support requires flexibility,

coherence and a clear commercial ethos. Intermediary technology institutes are intended to bridge each of these particular divides. Generally,
finance for innovation is said to be a problem in the UK by those close to
innovative business. This belies a view that good projects also find investors.
Signs of national and regional government co-funding as in Scotland's co-
investment fund and the new UK equivalent show that, in effect, public
goods subsidies are having to be deployed to incentivise risk investors – a
somewhat bizarre notion it may be thought. But to the extent it is successful, in future the hypothesized distinctiveness between strong and weak
high-technology cluster areas ought to become convergent. This is reinforced by evidence of so-called *decapitation* (R&D stays but the rest of
the business goes to the US or elsewhere) of Oxford, Cambridge and some
Scottish biotechnology firms owing to the poor risk investment climate in
the UK (Ward, 2005).

University–Business Relationships

In Oxford, Isis Innovation at Oxford University is the key knowledge transfer intermediary. What it can do is to raise substantially more money for
academic entrepreneurs than they can raise themselves. This has evolved
markedly since 1998 when a typical seed funding round raised some
£250 000 whereas by 2005 it had become normal for Isis to raise £1.6 million
for the first round. The Oxford Trust perception of university–business
links concurs with the survey finding reported earlier in this chapter,
namely that bioscience companies tend to have stronger links with universities, whether it is through formal science advisory boards, research collaborations or PhD studentships. Contrariwise, in general terms, ICT
companies are perceived not to have such strong links. In Cambridge, the
evolution of university–business networking has resulted in the establishment of I 10, a UK government-funded network forum of the Eastern
region's ten leading research institutes and universities. Such a consortium
need not be limited to thinking only of start-ups but may enter a large-scale
project with a multinational like Shell, possibly continuing with 'blue skies'
projects over the longer term. A mistake academe thinks that regional
development agencies (RDAs) have made is they may assume that, because
they have money and universities enter into collaborative projects, their
agendas are identical. This is a good example of the weakness inherent in
a generalized triple helix model that assumes consensus across distinctive
'epistemic communities' because basic interests are often not shared.
Academics may enter such projects to get more money then get other projects, whereas an RDA may be looking for successful outcomes from that
project but then to move on to other things.

In London, where complexity and scale have traditionally inclined university–business links towards the bilateral rather than multilateral, the London Technology Network represents financial support for, rather than establishment of, organizations. London is perceived to have high-grade resources and talent in its higher education base, with a capability to have a greater economic impact by stimulating the university community to work far more as a whole. So London's Technology Network seeks to encourage, through network funding, collaborative consortia leading to the launch of its 'Proof of Concept' fund, launched in March 2005, involving 24 London universities and four other consortia. In Wales the WDA Knowledge Exploitation Fund was grounded in the belief that moving people between industry and academia is the most important means of facilitating knowledge transfer. This somewhat underestimated the lock-in effects of contemporary evaluation and career-structure conventions in academe. There was admiration for the London Technology Network that buys out half a day a week of the academic time of an academic in any department to become a 'business fellow', with the role of marketing their department for business, helping seek out opportunities for collaboration with business, and attending organized monthly networking events for business where all business fellows meet people from business interested in such interaction. However, it is recognized, despite the aspiration in Wales to emulate London, that the likelihood that many leading research academics with 'star' status and knowledge would actually consider this is indeed remote. This is a yet further indication of the naivety of triple helix policy belief in the desirability of secondments and sideways mobility, something common in civil service and other administrative organizations, and their impracticability in the different epistemic community of academic practice also often involving sizeable research teams painfully put together in accomplished laboratory settings.

In Scotland, the Scottish Technology Forum is one of numerous initiatives that have sought to improve interactions across the university–business interface. The triple helix hype is perceived as forgetting a business community view that there is no incentive for academia to conduct commercialization work. It is thought by business in Scotland that even what are widely perceived to be successes emulated in London, such as Proof of Concept funding, do not incentivize the commercialization of the underlying concept because academics say, believably, they are incentivized internally by and to their institution. Thus, aspiring to move to a commercial model is somewhat fraught. While the Forum recognizes that in university commercial offices there is a mix of views, some perceive universities and research institutions erecting barriers around research such that start-ups and spin-offs may face greater difficulties to get through that barrier. Universities are viewed by the Forum often to have a problem working with large organizations in the co-development of technological innovation because universities are large organizations themselves. Equally, there is seen to be an opportunity cost in working with many small companies which makes it a comparatively more feasible option to work with large companies than to work with many small technology companies. Either way there are difficulties, but neither kind of partner is widely sought after. A perceived problem from the industry side is that although in Scotland it is believed, perhaps unreliably, that the majority of business owners are university graduates, they are nevertheless critical of universities for their distance from the commercial edge and their lack of commercial focus. So what you really have is a number of people who are all saying that 'those guys over there are impossible to work with'. In conclusion, the Forum expressed surprise at the animosity of the technology community towards interacting with universities in terms of access to research and support. The hypothesis of closer university–industry interaction is supported moderately but signs exist that it has begun dawning on both elements that established policy assumptions of easy cooperation have been naive, to say the least.

Geographical Proximity and Clustering

Cambridge Enterprise was of the view that critical mass is clearly important because a key feature of clustering is that when you have a large number of people who are extremely good at producing high-quality research then, inevitably, you are going to get a better chance of it being commercialized as well. However, this analysis is open to question because high-grade research institutions do not 'inevitably' develop innovative clusters, a fact which is perceived to be highly problematic in Europe, and also exists in the USA where many Ivy League universities and R1 (first-rank research) institutions like Johns Hopkins and the University of Chicago share this paradox. A different view is that the university has helped to encourage knowledge entrepreneurship such as the Cambridge Network and Library House, as active commercialization bodies that take research to market. Cambridge also has patent lawyers, management accountants and venture capital, essential ingredients for high-tech clusters. EEDA agreed that 8 per cent of EU venture capital money only is spent in Cambridge but that the fact that Cambridge is a city of only 120 000 people needed also to be taken into account in understanding its clustering capabilities.

A similar understanding of proximity and clustering is held by Oxford Innovation, the key factor being the research quality and volume in Oxford.

An estimated £150 million a year of top-quality research is conducted there. A high proportion is in life sciences and ICT, especially research which is a heavy user of ICT. Moreover some companies overlap, such as health care companies using large quantities of software and information technology. So the first ingredient is seen as the strong research base. The next most important ingredient is entrepreneurship with a good number of entrepreneurs present, combined with the amount of venture capital of various kinds, notably business angels. Related to that there are numerous lawyers and other professionals who are geared up to deal with start-up companies and spin-outs, and then there are science parks and industrial parks that have learnt rapidly over the past ten or fifteen years to accommodate new technology companies. Finally, there have been a few people with very great influence, such as Sir Martin Wood, the founder of Oxford Instruments and the Oxford Trust. He set up the Oxford Trust in 1985 to encourage the study and application of science and technology. The Trust has had a key role to play in cluster formation, in local economy thinking and in stimulating spin-outs. Oxford Innovation was itself a spin-out from the Oxford Trust and Oxford Innovation invested heavily in developing business angel networks and in establishing numerous business incubators. Oxford Innovation managed in 2005 12 business incubators with a total of 300 innovative start-up companies and of these, seven were in Oxfordshire with some 200 companies. There are many ICT companies in those centres and some biotechnology firms. But ICT, broadly defined, is the biggest activity, not least because it is favourable for innovative start-ups to become established therein.

The London Development Agency conducted a study inspired by Michael Porter's cluster work but, questionably, adopted a sector rather than cluster strategy. The sector-based approach identified seven key sectors that it was seen LDA could usefully be seen working with. Of these seven sectors, ICT is the only one where LDA by 2005 had not set up an explicit budget or programme of work. However, priority was beginning to be ascribed to it as one of the key sectors in London with the largest growth potential for the next ten years. The sector research, conducted in partial emulation of Porter's cluster methodology, identified key technology quasi-clusters and strengths in London, but the public sector in London as elsewhere tends to make decisions slowly because they are spending public money and there are various competing priorities. For that reason it is perceived to be a mistake for the LDA to proceed down purely technology lines because the LDA moves too slowly and the technology market moves too quickly. It is thought doubtful that the LDA would ever be able to keep up with technological change in the market given the way that technology markets evolve. A higher priority is to focus upon the demand base, and applications related to customer groups, such as ICT sector interventions for financial services, so the focus is on customers rather than technology because the customers are not likely to change as quickly as the technology. Hence, once more, financial intermediation looms large in London's expectations towards supporting new technology business – something dramatically at odds with the manner of proceeding in Cambridge and Oxford. But that is not to place a value judgement upon the London approach, which may be most appropriate for its type of internationally traded services economy.

In Scotland, as indicated, early adoption of a direct Porterian clustering model bought in 1992 from Monitor, Porter's consultancy, gave the Scots a lead in developing knowledge about the strengths and weaknesses of the approach. By 2003 the weaknesses were perceived to be outweighing the strengths and an ambitious ITI model of technology intermediation to speed up globally competitive innovation was initiated. This is meant to show the seriousness with which Scotland's 'developmental state' wishes to further innovation, not necessarily through cluster-building, a policy approach seen as being of questionable provenance given Scottish Enterprise's pioneering adoption of this economic development methodology. The post-cluster idea in Scotland is perceived to involve taking 'platform policy' actions that enable businesses to gain from complementarities, collaborations and knowledge spillovers, especially where related firms operate in geographical proximity. From some viewpoints this may smack of exclusivity or even construction of quasi-monopoly conditions where a group of firms enjoy 'club' benefits as with clusters. But from the ITI perspective, building the post-cluster in Life Sciences meant disrupting such monopoly conditions by introducing a competitor to an existing CRO. Unsurprisingly, this caused anxieties from incumbents, introducing as it did new knowledge of the problems of cluster-building but justified in terms of 'building the post-cluster' or *platform* rather than protecting established interests. In other respects, the cluster perspective, while being perceived as a worthwhile business development model, is also seen as a policy retro-model, meaning it is based upon 'a compartmentalized 1980s picture' rather than a future-shaping one. Thus from the ICT perspective software is seen as a pervasive platform technology. But TechMedia, the ICT-ITT, perceived software having no policy identity in Scotland's cluster strategy. Worse, software was subsumed under 'e-learning', considered a bizarre inversion of reality given 'e-learning' has, from a market perspective, zero profile. Platform technologies like software, displaying pervasive characteristics, do not fit the cluster idea as practised in regional economic development policy. It is more akin to a network metaphor. This lends greater credibility to the diverse interactions involved in complex, multi-use systems than more conventional sectoral-spatial notions such as clusters.

The Welsh Electronics Forum considered that at best there was only a small amount of clustering in Wales. Mapping sectoral concentrations had sought to draw together related companies to deduce in a given area of, for example, engineering and industrial software, existing groups of companies even though they may not be collaborating but possibly producing related products. These are good examples of agglomerations more than true clusters. Analysis suggests this is probably a result of a strong heavy engineering past in Wales, where adaptive survivor firms may have invested in use and production of new technologies. The WDA identified such traditional sectoral groupings and for innovative businesses implemented the incubator concept of Technium. This aggregates a number of start-up companies by sector in the hope that they have conversations and then engage in collaboration. In addition to that, the WDA provided business advice and support, helping companies in their formative years to set themselves up robustly. If this sounds like a conventional incubator model, it is because, more or less, that is what it is. The key to Techniums was seen to be that they are at least in some cases supported by universities – Swansea being the most loyal exponent. This is despite industry often perceiving university research or pedagogy as not quite at the cutting edge, nor working quickly enough or indeed close enough to the market, a similar perception to that held in Scotland. Nevertheless, it was envisaged that such communication would develop. As relationships developed it was presumed this would lead to all the technical and business support building blocks coexisting in an environment encouraging conversations and then collaborations. Perhaps this optimistic geographical determinism explains the demise of the WDA as clearly as anything could since most Techniums were, in 2006, five years after their introduction, either less than half-full or, indeed, empty. Compared with Scotland, Wales is more like London in its traditional adherence to sectors, the Electronics Forum stressing the undesirability of going against the principle of only bringing in related companies as a means to sustain collaboration. Otherwise there is a perceived danger that commercially driven Techniums would seek to fill rentable space with the wrong type of company. So finally, the proposed hypothesis, to the effect that greater proximity and clustering were gradually being mainstreamed was, on the contrary, perceived to be something of a source of scepticism. Of rising relevance to industry is the notion of platforms as a pervasive means by which innovative knowledge transfers into what we would term *related variety* industries. Only policy-makers and consultants remain committed to what industry perceives as a 'retro-model' – a condition analysed and critiqued at length in Asheim et al. (2006).

CONCLUDING REMARKS

This chapter has proposed empirical evidence that gives confidence for the following key observations. First, and in theoretical terms, it advances our understanding of the persistence and conceivable reinforcement of an asymmetric economic geography of prosperity and accomplishment. That is, partly through the quasi-monopolistic character of modern clustering, the scarce resource of knowledge becomes king in the economic development stakes. That clusters exist cannot be gainsaid but this only makes policy-makers everywhere envious and hopelessly emulative, especially when – as tellingly displayed in Table 7.9 – market and financial *conditions* have to be highly propitious to attract the necessary risk-capital to fuel them. Second, we saw that in an evolving and intensifying knowledge economy, science-driven and otherwise technologically sophisticated economic activity gives rise to demands upon industry organization that reinforce collaborative activity among smaller knowledge-intensive businesses. Then interaction between smaller, smart firms and university laboratories towards customer (and supplier) firms can ensue, even with transnational corporations.

Second, high-technology cluster regions perform relatively well, but more because of collaborative conventions than because of clusters. The latter may well have performance-inhibiting characteristics on financial outcome metrics. Nevertheless, on other metrics, concerning innovation inputs, clusters offer superior locations. This is important and original support for the thesis that *regional knowledge capabilities* increasingly determine the distribution of growth regions. Currently favoured are those that gain increasing returns from asymmetric knowledge distribution that assists in the construction of regional advantage in terms of talent recruitment and retention, spatial knowledge quasi-monopolies, and 'R&D outsourcing' or 'open innovation'. In UK ICT and biotechnology such features are pronounced with key bioregional capabilities attracting these advantages to clusters like Cambridge and Oxford, while for ICT, London and its satellites in the M25 and M4 corridors is the dominant market-led magnet.

Finally, this research has tended to find support for the superiority of collaboration in respect of a variety of performance indicators, and clustered cooperation for innovation being supported more by the collaborating part of the firm sample than the respondent group as a whole. This broadly applies in ICT and biotechnology, but as we have seen less regarding clustering for innovation activity by ICT than biotechnology firms, and much more for research interactions by biotechnology than ICT firms. Clearly, in the UK's biotechnology sector there is considerable asymmetry in the location of, for the most part, research-driven clustering. The fact that innovation attracts distant networks of cooperation for this sector is consistent with a characteristic feature of biotechnology, which is that R&D and innovation consistently involve externalized interfirm and firm–laboratory interactions. To that extent biotechnology conforms fairly well, judged from these UK data, to the standard narrative. Information and communication technology shares some of this character but in an inverse manner. Research is less of a cluster-driver than innovation activity, but the latter is not as pronounced as supply-chain innovation stretching globally, and intra-firm interactions. The one thing that appears to be almost transparent, especially in the ICT data, is the superiority for firm performance of collaborative knowledge exchange and innovation activity over stand-alone competition.

We conclude that clustering happens in ICT, mainly driven by supply-chain requirements; we could denominate it a 'Smithian Clustering' (Bottazzi et al., 2002) mainly based on division of labour. Where clustering occurs it is mainly software, R&D and services that cluster and are also higher collaborators. Conversely, the hardware industry tends not to collaborate even if they are located in clusters. So, as argued in Chapter 4, a clear contrast is revealed between ICT and biotechnology, confirming that crucial differences exist within science-based industry regarding clustering and that agglomeration economies exist and are relevant in science-based sectors (again, Bottazzi et al., 2002). The chapter also confronted different types of institutional settings that coexist in the liberal market economy of the UK. On the one hand, regions such as Eastern and South East England are characterized by entrepreneurial competitiveness little influenced in the past by regional public-led network-building. This is, to some extent, changing as regional agencies are now active in promoting regional advantage in the globalized, knowledge-based innovation economy. On the other hand, despite the liberal market nature of the UK, Wales and Scotland developed over a quarter of century strong regional agencies, aimed at developing their economies. Accordingly, they became the main vehicles for managing policies in support of innovation. This included coping with the demands of the knowledge economy yet with different outcomes. Scotland's agency has embraced knowledge economy principles closely, while Wales was so wedded to a 'public enterprise' mode of top-down stimulation that its development agency was closed. It was perceived to have failed to meet the need for new ways of supporting entrepreneurship and innovation. Contrariwise, London, despite its strong entrepreneurial tradition, is approaching its problems of scale and fragmentation in the knowledge value chain with rather traditional enterprise support arrangements that favour a sector-oriented approach. These have so far focused mainly on the London financial cluster.

NOTES

1. Estimating the level of employment in the ICT industry is not always straightforward since some definitions of ICT do not accord with the classification system used for employment statistics. For the purpose of the research conducted, ICT has been defined following the OECD ICT sector definition approved in 1998. This definition includes a combination of manufacturing and services industries that capture, transmit and display data and information electronically. A description of the sectors included in the OECD definition is reported in Chapter 6.
2. The ABI data set refers to data units. These do not readily correspond to the commonly used terms firms, companies or businesses by which employers are sometimes identified. They are roughly equivalent to workplaces but because of the way the data are collected two or more units can be present in the same workplace.
3. This geography lists the single-tier (unitary) authorities together with the upper-tier authorities in areas of two-tier local government. The constituent areas in the UK are: England: the City of London Corporation, London boroughs, metropolitan districts and unitary authorities (all providing single-tier local government) and counties (upper-tier in areas of two-tier local government); Wales: single-tier unitary authorities; Scotland: single-tier unitary authorities.
4. The research results on industry and cluster innovation interactions in ICT are based on a postal survey undertaken in spring 2004 and in-depth interviews conducted with firms and policy-actor informants starting in spring 2005.
5. It is worth noting that, within the ICT sectors, proximity factors are relatively of lower importance as the scale developed rated between 1 and 4, being 1 not important and 4 very important and for ICT all the different categories are below 3 – relatively important.
6. Insignificant non-collaborator numbers made this impossible to be performed for biotechnology.
7. The overarching aim of the research conducted was to identify the industrial location and organizational characteristics of knowledge-based clusters, differentiate these from other forms of industry concentration and ascertain the extent to which clustering and collaboration are beneficial to firm performance. Respondents to the survey were asked to provide their postcodes and these were matched with the LQs calculated on number of employees and data units. This exercise allowed us to group the respondents firms according to their location. One hundred and twenty-nine firms (52 per cent of respondents) were located in 'cluster'; eight firms (3 per cent of respondents) were located in district; 29 firms (12 per cent of firms surveyed) were located in areas identified as 'dominating firm' and 84 firms surveyed (34 per cent) in area characterized by no concentration. It is worth noting that the sample size may vary owing to the fact that location of some respondents is missing. Owing to the small number of respondents, analysis of performance was only possible for the sub-groups of clustered and non-clustered firms.
8. Despite its liberal market nature, the UK has had strong regional development agencies, especially for over a quarter of a century in Scotland and Wales where intervention was pronounced, evolving later in more associative directions. Since 1999, the nine English regions have also received regional economic development agencies. These agencies promote liberal market ideals but also tend to further, for example, entrepreneurship, innovation, clustering and talent formation through *associative* mechanisms. Hence, the UK government promotes networking and cooperation as an important driver in innovation. Successful firms also practise associative methods towards other firms and organizations with appropriate knowledge (see also, Cooke and Clifton, 2005).

REFERENCES

Amin, A. and Cohendet, P. (2004), *Architectures of Knowledge – Firms, Capabilities and Communities*, Oxford: Oxford University Press.

Asheim, B., Cooke, P. and Martin, R. (eds) (2006), *Clusters and Regional Development*, London: Routledge.

Bottazzi, G., Dosi, G. and Fagiolo, G. (2002), 'On the ubiquitous nature of agglomeration economies and their diverse determinants: some notes', in A. Curzio and M. Fortis (eds), *Complexity and Industrial Clusters*, Heidelberg: Physica-Verlag.

Cairncross, F. (1997), *The Death of Distance: How Communications Revolutions Will Change our Lives*, Boston, MA: Harvard Business School Press.

Caniëls, M. and Romijn, H. (2003), 'Localised knowledge spillovers: the key to innovativeness industrial clusters?', paper presented to Conference on Reinventing Regions in the Global Economy, Pisa, 14–16 April; published in P. Cooke and A. Piccaluga (eds) (2006), *Regional Development in the Knowledge Economy*, London: Routledge.

Centre for Advanced Studies (CASS) CASS ICT and Biotechnology Collective Learning Survey, Cardiff University ESRC (Economic and Social Research Council).

Cooke, P. and Clifton, N. (2005), 'Visionary, precautionary and constrained "varieties of devolution" in the economic governance of the devolved UK territories', *Regional Studies*, **39**, 437–51.

Department of Trade and Industry (DTI) (2001), *Business Clusters in the UK*, London: Department of Trade and Industry.

Department of Trade and Industry (DTI) (2004a), *Electronics 2015 Making a Visible Difference*, London: Department of Trade and Industry.

Department of Trade and Industry (DTI) (2004b), *Sector Competitiveness Analysis of the Software and Computer Services Industry*, London: Department of Trade and Industry.

Department of Trade and Industry (DTI) (2005), *DTI Sector Competitiveness Studies No.1, Competitiveness in the UK Electronics Sector*, London: Department of Trade and Industry.

Fontes, M. (2006), 'Knowledge access at distance: strategies and practices of new biotechnology firms in emerging locations', in P. Cooke and A. Piccaluga (eds), *Regional Development in the Knowledge Economy*, London: Routledge.

Nomis, http://www.nomisweb.co.uk/home/census 2001.asp.

Office of Science and Technology (OST) (2002), *Science Delivers*, London: Department of Trade and Industry.

Organisation for Economic Co-operation and Development (OECD) (1999), *S&T Indicators: Benchmarking the Knowledge-Based Economy*, Paris: Organisation for Economic Co-operation and Development.

Organisation for Economic Co-operation and Development (OECD) (2004), *STI Review*, Paris: Organisation for Economic Co-operation and Development.

Owen-Smith, J. and Powell, W. (2004), 'Knowledge networks as channels and conduits: the effects spillovers in the Boston biotechnology community', *Organization Science*, **15**, 5–21.

Roper, S., Love, J., Cooke, P. and Clifton, N. (2005), *The Scottish Innovation System: Actors, Roles and Policies*, Edinburgh: The Scottish Executive.

Shohet, S. (1998), 'Clustering and U.K. biotechnology', in P. Swan, M. Prevezer and
    D. Stout (eds), *The Dynamics of Industrial Clustering*, New York and Oxford:
    Oxford University Press.
Ward, M. (2005), 'Strategy: decapitation', *BioCentury*, **13**, A1–A4.

8. Austrian catch-up in knowledge-based sectors: research exploitation, spatial clustering and knowledge links

Austria has a highly competitive economy, but it is internationally not known as a location for high-technology activities. Its strength has traditionally been in medium-technology sectors such as speciality steel and materials, vehicles and engineering as well as in services, particularly in tourism. Under a coordinated market model of social partnership the country was able to achieve high productivity gains and growth rates. More recently, however, the Austrian economy has shifted more towards a knowledge economy, with increasing shares of sectors such as high-technology industries and knowledge-based services, including ICT and biotechnology. In the current chapter we analyse Austria's position as a latecomer in the knowledge economy and we ask to what extent we can observe a catching-up process in these industries (next section). In the second section and in greater detail we investigate the locational pattern of these knowledge-based sectors and any tendencies of spatial clustering. The third section then looks at innovation activities and analyses in particular the nature of knowledge interactions of firms in those sectors. We look at the most important innovation partners such as firms, universities and research organizations as well as at the types of relationship (market, network, spillover and milieu). The final section is devoted to the development, spatial clustering and knowledge links of the Austrian biotechnology industry. It is shown that there has been a dynamic development of this industry since the mid-1990s, partly related to the Austrian science base in relevant fields. The dominating Vienna biotechnology cluster as well as two more recent biotechnology locations (Styria and Tyrol) are investigated with respect to their structures and knowledge links.

AUSTRIA AS A LATECOMER IN KNOWLEDGE-BASED SECTORS AND ITS COOPERATIVE BUSINESS MODEL

The coordinated market economy of Austria is considered to be a latecomer regarding the development of knowledge-based sectors. Its economy is characterized by a dominance of traditional, medium-tech industries such as mechanical engineering or the fabrication of metal products. For many years, a strategy of technology adoption and modification instead of own R&D efforts and radical innovation was pursued, reflecting both the dominance of SMEs and policies supporting this development path. The cooperative model of social partnership (*Sozialpartnerschaft*) also seems to have contributed to a certain 'structure conservation', maintaining the status quo by favouring existing industries (van der Bellen, 1994; Nowotny, 2001) instead of promoting the emergence of new ones. Until 2000 relatively low levels of R&D and patent activities and a lack of venture capital were prevailing features, giving rise to the existence of a distinctive Austrian 'technology gap'. In spite of the specific specialization pattern in more traditional sectors, Austria displayed a good performance in terms of employment, level of income and growth over the last three decades. There are several factors explaining this, termed the 'Austrian paradox'. First, Austria benefited enormously from the fall of the Iron Curtain and the socio-economic transformation processes going on in its neighbouring countries in Eastern Europe. Second, a successful coordination of different policy fields (a coherent system of macroeconomic governance), harmonious relations between employers and employees (industrial relations), and continuous incremental innovations contributed to the international competitiveness of the country. In the mid-2000s, however, a dwindling of Austria's good growth performance became apparent. The low level of specialization in dynamic, high-technology sectors has been identified as a key structural deficit, threatening the country's long-term growth prospects (BMBWK et al., 2003). It reveals a limited capacity to break path dependence, brought about by specific configurations of the national and regional innovation systems that have not been ripe for promoting high-technology industries. In the emerging knowledge economy the strategy of technology adoption, adjustment and modification, which had for a long time been pursued successfully, had reached its limit.

In the recent past, however, a catching-up process could be observed, indicating a gradual move of the Austrian economy towards more knowledge-intensive activities. As shown in Chapter 2, the R&D intensity of the Austrian economy climbed above the EU-25 average and, even more importantly, grew quite strongly in the 2000s. The availability of highly qualified and technical workers and of venture capital, however, remain poor, signalling a persistent weakness of the Austrian innovation system. Accordingly, Austria had below average employment shares of high- and medium-high-technology manufacturing sectors and a comparatively weak performance in knowledge-intensive services internationally. With respect to productivity changes in the overall economy, as well as in ICT-related sectors, Austria has a good performance. Furthermore, there is strong productivity growth in ICT-using industries. It is clear that Austria's good overall productivity performance does not result from ICT *production* but is more the outcome of the *use* of ICT in a variety of sectors, as well as of other factors. Since 1997 Austria has considerably improved its position on R&D intensity, but it still suffers from a low share of highly qualified scientific and technical workers, a poorly developed venture capital market and a lagging knowledge-intensive sector, both in manufacturing and in services. In terms of scientific publications, patents and productivity growth Austria has shown a successful development in the recent past. The overall picture, thus, is that the Austrian economy has managed to benefit from knowledge application and innovation in many sectors, not just high technology, as well as from the use of ICT in services and other areas. Looking specifically at Austria's performance in ICT and biotechnology, which represent two key growth sectors of the emerging knowledge economy, enables us to get a better understanding of the development logic of high-tech industries in Austria and the dynamics of the catching-up process in these sectors.

According to a recent study the *Austrian ICT sector* is characterized by an average size and structure in international comparison (Schneider et al., 2004). It has technological strengths in communication technology, notably in 'television and radio transmitters and apparatus for line telephony and line telegraphy'. About 60 per cent of the R&D personnel in the ICT sector could be found in this field. The publication activities of Austrian research organizations in ICT show a good performance, also when compared internationally. It is in particular the fields of engineering mathematics, artificial intelligence, robotics/automation, and measurement and test engineering where Austria seems to have specific strengths. With respect to patents, Austria is clearly lagging behind the leading countries. Only about 20 per cent of the Austrian European Patent Office (EPO) patents have been in ICT, whereas countries such as Finland (57 per cent), the Netherlands (57 per cent), Sweden (38 per cent), Japan (44 per cent) or the USA (39 per cent) have shown larger shares by far. Between 1990 and 2000 new firm formation in the Austrian ICT sector was dynamic (up 25 per cent). It was particularly in the field of software development and consultancy that considerable growth rates occurred in this period (up 86 per cent). When compared with Bavaria or Germany, however, Austria has a poor performance regarding newly founded ICT firms. A more positive pattern can be found in R&D expenditures in the ICT sector as a percentage of GDP, and the employment growth of highly qualified ICT workers, where Austria is among the leading countries in Europe (Schneider et al., 2004). Looking specifically at telecommunications, Unger (2003a) identified the following strengths, niches and weaknesses of Austria. Voice processing and smart antennae constitute successful Austrian niches. Voice processing is one of the key activities of Siemens research in Austria. The geographic proximity to the East, know-how about trade with Eastern Europe, a high educational level, and cheaper labour costs than Germany are among Austria's main assets. The country, however, has weaknesses such as its small size, late restructuring, and a lack of venture capital. Furthermore, newcomers seem to use Austria as a sales department and not for research.

In *biotechnology* Austria has the status of a latecomer. Unger (2003a) compares Austria's performance in biotechnology with that of Finland, Germany and the Netherlands, and shows for the 1990s that Austria was clearly lagging behind these countries. This held true for public funding, patent activities and deliberate releases of genetically modified organisms (GMOs) (1992–2000). The small Austrian biotechnology scene is strongly specialized in 'red' or medical biotechnology, notably in cancer research, vaccines and blood products. The 'green' (agro-food) and 'grey' (environmental) variants of biotechology are almost negligible. According to Unger (2003b) this specialization is the result of political rather than economic reasons. In 1997 there was a *Volksbegehren* (petition) against gene manipulated food, signed by more than 1.3 million people, and as a consequence, applications in 'green' and 'grey biotechnology' plant releases have been forbidden. Other institutional factors, such as a lack of finance for start-ups, a non-adjusted technology policy, and low public funding have been main hindrances for the development of the sector (Unger, 2003b; for similar findings see Baier et al., 2000). Austria's long-delayed start in the development and commercialization of biotechnology is, thus, inextricably linked to various deficits in the Austrian innovation system. There are, however, also strengths, such as a history and reputation in bio-medicine, highly qualified scientists and skilled personnel (Unger, 2003b). Moreover, an international study has shown that Austria has a good performance in science. It was also noted, however, that it suffers from poor commercialization capabilities (Reiss et al., 2003). As shown below, in recent years a catching-up process in biotechnology has set in. This manifests itself in a dynamic process of new firm formation, largely driven by the rise of academic spin-offs. Such processes are observable in the Vienna biotechnology cluster in particular. The rebuilding of the national and regional innovation systems is a crucial aspect when it comes to explaining the emergence of biotechnology in Austria. Key factors in this context comprise the creation of new research and educational institutions, the establishment of supporting organizations specialized in fostering high-technology industries, as well as new policy routines such as new modes of governance and network facilitating instruments.

The Cooperative Business Model in Transformation?

As we have argued in Chapter 1, Austria represents an ideal type of an institutional regime that could be referred to as 'coordinated market economies' (Hall and Soskice, 2001). Austrian social partnership has constituted a key factor for the nation's excellent economic performance since the Second World War (Mesch, 1995; Nowotny, 2001). Being a comprehensive system of institutionalized cooperation and coordination between labour, business and government, it formed an essential 'institutional advantage', guaranteeing high growth rates in the post-war period. Since the late 1970s, however, the economic, political, and social conditions of social partnership action changed considerably (Unger, 1999). Globalization tendencies, the slowdown of economic growth rates, increasing unemployment, the erosion of homogenous interests, and so on are forming major challenges, threatening the once successful neo-corporatist arrangement. Austria's EU membership since 1995 limited its room for manoeuvre and the coalition of two conservative parties that came into power in 2000 challenged the dominance of the social partners, trying to reduce their strong influence in policy-making (Unger and Heitzman, 2003; Müller, 2006). Recently, Müller (2006) put forward the thesis that Austria is moving in the direction of a liberal market economy. In the past decade, he argues, Austrian capital has become less patient. One reason for this is the dramatic decline of the public sector due to the privatization of many state-owned enterprises in the 1990s. Furthermore, the capitalization of the Vienna stock exchange[1] has increased dramatically and Austria succeeded in becoming a major investor in East-Central Europe. Additionally, managers seem to have become more unconstrained and, finally, changes in collective bargaining and intra-firm power of labour can be observed, reflecting a considerable decentralization of industrial relations. Exemplifying these changes, in the following section we will deal with the Austrian coordinated market model from an innovation perspective.

SPATIAL PATTERN AND CHANGE OF KNOWLEDGE-BASED SECTORS IN AUSTRIA

Methodological Aspects and Database

The empirical findings presented below were collected in the context of the research project KNOWING[2] which focused on the innovation process in knowledge-intensive industries and was particularly interested in the prevailing mechanisms of knowledge exchange and their geography. It was based on the analysis of secondary data, a postal survey and in-depth interviews in selected industries. The analysis of secondary data focused on the size and growth of knowledge-based sectors (see definitions in Table 8.1) in Austria as well as on the spatial pattern and change of those activities (presented in the following section). The postal survey of Austrian knowledge based firms (in the third section) yielded basic insights into the process of knowledge generation and of the character of knowledge interactions. The personal interviews with representatives of firms, knowledge institutions as well as policy and support organizations were undertaken in the Austrian biotechnology sector in order to gain a deeper understanding of the knowledge processes and respective networks in this high-technology sector (final section).

The following analysis of the locational pattern and dynamics of the knowledge-based sectors in Austria builds on the data from the firm census (*Arbeitsstättenzählung*) for the years 1991 and 2001. Knowledge-based sectors were defined applying the OECD classification (OECD, 2001) and are investigated in the following both in a narrow and wider definition. For reasons of comparison we will refer in the following analysis also to:

- *medium-low-technology industries (MLT)*, comprising the NACE sectors Coke, Refined Petroleum Products and Nuclear Fuel (23), Rubber and Plastics Products (25), Other Non-Metallic Mineral Products (26), Basic Metals and Fabricated Metal Products (27–28) and Building and Repairing of Ships and Boats (351)
- *low-technology industries (LT)*, comprising the NACE sectors Food Products, Beverages and Tobacco (15–16), Textiles, Textile Products, Leather and Footwear (17–19), Wood, Pulp, Paper, Paper Products, Printing and Publishing (20–22) and Manufacturing n.e.c. (36–37).

Size and Growth of Knowledge-Based Sectors in Austria

Although the Austrian economy cannot be considered as technology intensive in international comparison, knowledge-based sectors have some

Table 8.1 Knowledge-based sector definitions

Narrow definition (NACE)	Wide definition (NACE)
High-technology industries (HT) Pharmaceuticals (244) Office, Accounting and Computing Machinery (30) Radio, TV and Communication Equipment (32) Medical, Precision and Optical Instruments (33) Aircraft and Spacecraft (353)	*High-technology industries (HT)* Pharmaceuticals (244) Office, Accounting and Computing Machinery (30) Radio, TV and Communication Equipment (32) Medical, Precision and Optical Instruments (33) Aircraft and Spacecraft (353)
	Medium-high-technology industries (MHT) Chemicals excluding Pharmaceuticals (24 except 244) Machinery and equipment n.e.c. (29) Electrical Machinery and Apparatus n.e.c. (31) Motor Vehicles, Trailers and Semi-Trailers (34) Railroad Equipment and Transport Equipment n.e.c. (352, 354, 355)
Technology-related knowledge-intensive business services (TKIBS) Computer and Related Activities (72) Research and Development (73) Architectural and Engineering Activities and related Consultancy (742) Technical Testing and Analysis (743)	*Knowledge intensive business services (KIBS)* Telecommunications (642) Finance and Insurance (65–7) All business consulting excluding real estate (71–4)

relevance. According to our data, in 2001 about 20 per cent of overall employment could be classified in a broad sense as 'knowledge based', applying the OECD definition (OECD, 2001). This includes the relatively large group of knowledge-intensive business services (KIBS) (13 per cent), as well as high-technology (HT: 1,7 per cent) and medium-high-technology manufacturing (MHT: 4.9 per cent; Table 8.2). In large parts of the following section we will look at a smaller subset of knowledge-based activities, however. These are the high-tech sectors in manufacturing (HT) as well as the more technology-based TKIBS, which are a subset of the more broadly defined KIBS (see above). In this narrow definition of knowledge-based sectors, which excludes also MHT sector, the knowledge-based

Table 8.2   *Knowledge-based sectors in Austria: employment and
             employment growth, 1991–2001*

	Employment abs. 1991	Employment shares in % 1991	Employment abs. 2001	Employment shares in % 2001	Employment growth 1991–2001
Total	2 933 438	100.0	3 420 788	100.0	0.17
Mf industry	735 862	25.09	663 339	19.39	−0.10
High-tech	53 407	1.82	59 295	1.73	0.11
Medium-high-tech	174 711	5.96	166 524	4.87	−0.05
Medium-low-tech	188 564	6.43	179 885	5.26	−0.05
Low-tech	319 180	10.88	257 635	7.53	−0.19
KIBS	288 812	9.85	450 171	13.16	0.56
TKIBS	53 637	1.83	103 981	3.04	0.94

sectors in Austria account for less than 5 per cent of employment (1.7 per cent in HT and another 3 per cent in TKIBS).

The knowledge-based service sectors were clearly the most dynamic sectors as the employment changes between 1991 and 2001 reveal. Employment in KIBS increased from 289 000 to 450 000 (55 per cent), whereas in the smaller subset of TKIBS employment almost doubled from 54 000 to 104 000 (+94 per cent). Employment in high-tech manufacturing (HT) expanded much more modestly from 53 000 to 59 000 (+11 per cent). This is low in comparison to the dynamic service sectors but a relatively good performance compared to the overall shrinking of the manufacturing sector (−10 per cent). Employment losses were severe in particular in the low-tech industries (−19 per cent) but they could also be observed in the medium-tech sectors (−5 per cent).

In international comparison, Austria's HT sector has been about 20 per cent lower than the respective EU average in 2001, but about 40 per cent below the levels of the leading countries, Sweden and Germany. Regarding knowledge-intensive business services (KIBS), the respective share in Austria is about 30 per cent below the EU average. In international comparison, thus, Austria has to be regarded as a latecomer to the knowledge economy. However, Austria caught up, as in the five years from 1996 to 2001 both knowledge-intensive services and the HT/MHT sectors displayed above average employment growth in a broader EU comparison. Knowledge-intensive business services employment grew by +6.3 per cent annually in Austria as against +3.2 per cent in the EU15; and HT/MHT employment increased by +1.4 per cent in Austria as against +0.5 per cent annual average growth rate for the EU15 (EC, 2003). In fact, the catching-up process continued after 2001 as more recent data revealed (see Chapter 3).

Spatial Pattern and Concentration of Knowledge-Based Sectors

Similar to the spatial pattern in other countries, knowledge-based sectors in Austria are strongly concentrated in geographical space. To a large extent they are an urban phenomenon: in 2001 43 per cent of Austrian HT employment was located in the Vienna region and 29 per cent in the other urban regions. An even stronger spatial concentration can be observed for the TKIBS: 47 per cent of their employment was to be found in Vienna and 33 per cent in other urban regions. This is much higher than the respective shares of overall employment, as the location quotients (LQs) reveal. In contrast, only 28 per cent of HT employment and 20 per cent of TKIBS employment were located in rural areas. This is much less than their respective employment share as the respective LQs reveal. The analysis of the sector-specific location quotients by type of region and by political districts reveals a more detailed picture.

High-tech manufacturing (HT)
Table 8.3 shows for Vienna an HT-LQ of 1.42 in 2001, indicating that its HT employment share is 42 per cent higher than its overall employment share in Austria and more than double its manufacturing employment share. Thus Vienna can be regarded as the most important HT location in Austria both in absolute and relative terms. Besides Vienna, HT employment is concentrated in a number of other urban centres such as Graz, Salzburg, Villach, Klagenfurt and Innsbruck and their surroundings. Looking at the spatial pattern of location quotients by political districts (Figure 8.1), we observe that in comparison to overall employment the cores of the cities have rather low shares of HT employment. This result is also supported by the fact, that the 'other urban regions' as a whole had an HT-LQ of 0.91 in 2001 (Table 8.3) indicating a below average HT employment share. This finding reflects a low share of manufacturing in general and a higher importance of the service sector in urban areas. High-technology employment, thus, is concentrated in urban regions with a certain manufacturing base such as Vienna, Graz, Leoben and Villach, but not in the other more service-oriented cities.

Not surprisingly, the rural regions in general have the lowest share of HT employment (LQ of 0.74 in 2001, Table 8.3). But Figure 8.1 also demonstrates that there are a number of rural, and partly even peripheral districts with high shares of HT employment as, for example districts in Southern Styria, Lower Austria or Tyrol. In many cases this pattern can be attributed

Table 8.3   Knowledge-based sectors: employment shares and location
quotients (LQ), 1991 and 2001 by type of region

Employment shares	Year	Vienna	Urban regions	Rural regions	Austria
Total employment	1991	31.6%	31.3%	37.1%	100.00
Total employment	2001	30.5%	31.9%	37.6%	100.00
High-tech	1991	52.6%	22.4%	25.1%	100.00
High-tech	2001	43.2%	28.9%	27.9%	100.00
TKIBS	1991	45.5%	35.1%	19.4%	100.00
TKIBS	2001	47.3%	32.8%	19.9%	100.00
Medium-high-tech	1991	25.9%	30.9%	43.2%	100.00
Medium-high-tech	2001	18.4%	33.9%	47.7%	100.00
LQ	Year	Vienna	Urban regions	Rural regions	Austria
High-tech	1991	1.66	0.71	0.68	1.00
High-tech	2001	1.42	0.91	0.74	1.00
TKIBS	1991	1.44	1.12	0.52	1.00
TKIBS	2001	1.55	1.03	0.53	1.00
Medium-high-tech	1991	0.82	0.99	1.17	1.00
Medium-high-tech	2001	0.60	1.06	1.27	1.00

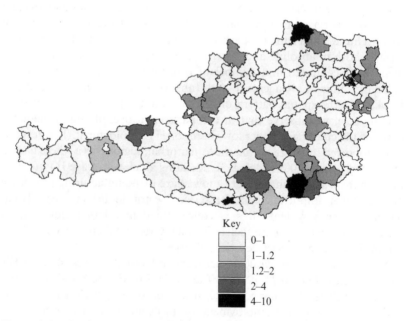

Key

0–1
1–1.2
1.2–2
2–4
4–10

Figure 8.1   High-tech location quotients, 2001

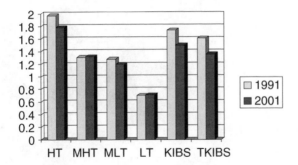

Note: Cofficients of variation based on sectoral employment by districts.

Figure 8.2 *Spatial concentration of knowledge-based sectors in Austria, 1999–2001*

to subsidiaries of a few well-known firms such as Siemens (in Deutsch-landsberg/Styria), Infineon (in Villach/Carinthia), Biochemie Kundl (in Kufstein/Tyrol) and others. We find here a situation of 'dominant firms' rather than 'clusters' (see below).

Medium-high-tech, MLT and in particular LT manufacturing sectors are clearly more dispersed than HT. In fact, we find a correlation between the technology intensity and the spatial concentration of the sub-sectors. The coefficients of variation (measuring the 'unevenness' or the degree of concentration of sectoral employment among political districts) decline from 1.78 for HT to 1.32 for MHT, to 1.20 for MLT and to 0.71 for LT (Figure 8.2). Thus HT is the most concentrated, and LT the most dispersed of the investigated sectors in Austria. The spatial pattern of employment measured by location quotients (2001) reveals that HT industries are con-centrated in suburban and intermediate locations (Figure 8.1), whereas MHT is located in and around the secondary cities of Graz, Linz and Salzburg (Figure 8.3). The lower technology sectors have a different pattern in comparison. The MLT sector (mainly basic industries and mechanical engineering) is often located in the old industrial areas of Styria, and of Upper and Lower Austria. Employment of the more labour-intensive LT sector is, in contrast, clearly more dispersed and spread across rural areas, often also in peripheral locations. In particular the LT sector has been subject to relocation towards low-cost countries and was declining in most regions of Austria in the past two or three decades.

Thus, overall we find a distinct spatial structure of manufacturing indus-tries according to their technology and knowledge intensity. For HT a strong concentration in a few urban regions such as Vienna, Graz and

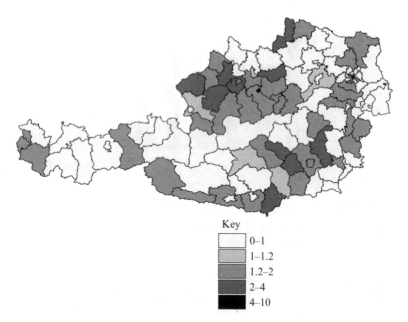

Key

	0–1
	1–1.2
	1.2–2
	2–4
	4–10

Figure 8.3   Medium-high-tech location quotients, 2001

Villach and their surroundings can be identified. But high-technology sectors are not an exclusively urban phenomenon, as there are also concentrations in intermediate and even peripheral districts. The latter are often due to subsidiaries or branch plants of larger firms rather than to cluster situations (see below).

Knowledge-intensive business services (KIBS and TKIBS)

The more broadly defined KIBS sector is strongly concentrated in space (coefficient of variation of 1.50). Different from HT, however, this sector is located exclusively in the cores of the major cities as the spatial pattern of employment (LQs) reveals. This central city location might be due to financial services such as banking and insurance, which often hold prestigious locations in the central business districts (CBDs). As has been stated already, the narrower defined TKIBS are even more concentrated in Vienna and other urban regions: Vienna's share of Austrian employment in this sector is 47 per cent and the respective LQ 1.55 in 2001. Also the other urban regions have an above average employment share of this sector (33 per cent of sectoral employment in 2001). But their LQ of 1.03 reveals that TKIBS are only slightly above the overall employment share of urban regions in Austria and, thus, not particularly strong.

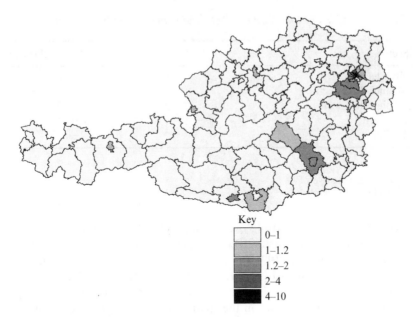

Key

	0–1
	1–1.2
	1.2–2
	2–4
	4–10

Figure 8.4   TKIBS location quotients, 2001

Disregarding the dominant position of Vienna, the overall spatial pattern of the TKIBS sector, however, is less concentrated (coefficient of variation of 1.36) than the KIBS (1.50) and the HT sector (1.78). This comparatively more dispersed pattern of TKIBS might partly be due to the high use of ICT in this sector which frees companies to some extent from central city locations. But their locational requirements (labour market, access to knowledge organizations, knowledge spillovers) still keep them within reach of urban areas. This is confirmed by the more detailed spatial pattern (Figure 8.4). In particular, Vienna and its surroundings to the south and the urban regions of Graz and other provincial centres are important locations for TKIBS.

Knowledge-intensive business services and TKIBS, thus, are more concentrated in urban regions than HT. Knowledge-intensive business services strongly favour the cores of those cities, whereas the TKIBS also spread to suburban locations. Good access to customers as well as knowledge spillovers and networks play an important role for these sectors, as investigated further below.

Spatial Dynamics of Knowledge-Based Sectors

So far we have observed a marked spatial concentration of knowledge-based sectors in the Vienna region and in other urban areas. In dynamic

Table 8.4   Correlation of LQ91 and employment growth, 1991–2001

Sector	Correlation coefficients
High-tech	−0.21[2]
Medium-high-tech	−0.25[1]
Medium-low-tech	−0.28[1]
Low-tech	−0.18[2]
TKIBS	−0.35[1]

Notes:
1 Significant 1% level.
2 Significant 5% level.

terms we have seen a strong overall employment growth in particular of KIBS (+55 per cent) and TKIBS (+94 per cent) and somewhat less in HT (+11 per cent) between 1991 and 2001. This contrasts with sectors having a lower technology intensity such as MHT and MLT for which a loss of employment has been registered (−5per cent each) and for LT (−19 per cent). Hence, what changes in the spatial pattern can be observed for the investigated sectors in the period 1991–2001? In general, and contrary to many cluster statements, there has been a tendency towards de-concentration for the knowledge-based sectors in the investigated period. Looking at the correlation coefficients (Table 8.3) we can observe a marked decline for all knowledge-based sectors, indicating a more even spatial distribution of employment in 2001 compared to 1991: for the HT sector the correlation coefficient has been declining from 1.97 to 1.78, for the KIBS from 1.74 to 1.5 and for the TKIBS from 1.61 to 1.36. The de-concentration trend is also reflected in a significant negative correlation between the LQs for the year 1991 and the employment growth of districts between 1991 and 2001 (Table 8.4). It implies that districts with low sectoral shares of knowledge-based sectors in 1991 have expanded their employment in knowledge-based sectors more strongly than those with high sectoral shares. This negative correlation is strongest for the TKIBS and for both medium-tech manufacturing sectors (MHT and MLT).

Regarding the spatial pattern of change for the *HT sector* we observe employment loss (−9 per cent) in the Vienna region (Table 8.3). This is a substantial reduction, but it is still small compared to Vienna's loss (−27 per cent) of total manufacturing employment in the same period. High tech, thus, has been part of the general de-industrialization process occurring in the Vienna region during the past three or four decades (Mayerhofer, 2006). The biggest winners of HT employment, however, are not the rural areas (+24 per cent) but the other urban areas (+44 per cent).

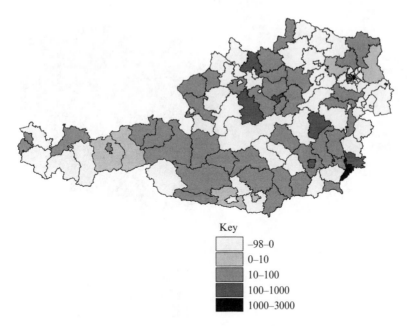

Key

	−98–0
	0–10
	10–100
	100–1000
	1000–3000

Figure 8.5   Medium-high-tech employment growth, 1991–2001

The location quotients for HT reflect those changes (Table 8.3): for Vienna it is reduced severely from 1.66 to 1.41, but it is still high. For the other urban regions it has moved up from a low 0.71 to a moderate 0.91, and for the rural areas it increased from 0.68 to 0.74. These results indicate that, despite some de-concentration from Vienna to other locations, spatial clustering in urban regions is still a relevant pattern for HT industries. The more detailed spatial changes for the HT sector (growth rates of employment) can be seen from Figure 8.5: strong employment growth can be identified in secondary cities such as Graz, Linz and Salzburg, but it can also be identified in a number of rural districts of Austria – some of them even in peripheral locations.

In the *TKIBS sector* we find a different picture. Overall, it is the sector with the highest employment growth. By type of region, the strongest growth is in Vienna; it more than doubled its employment in this sector (+102 per cent). But rural areas (+99 per cent) and the other urban areas (+81 per cent) also experienced high growth in this sector. The relative spatial changes are expressed by the location quotients for TKIBS: these increased for the region of Vienna from 1.43 to 1.55, remaining low and stable for rural areas at around 0.53 and declining for the other urban areas from 1.12 to 1.03. In the TKIBS sector we find, despite strong

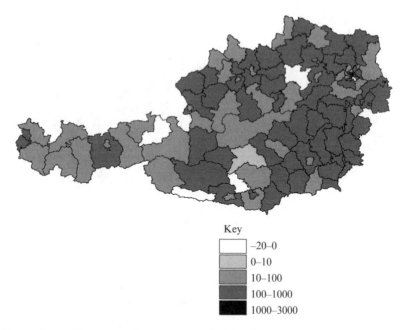

Key

☐	−20–0
☐	0–10
☐	10–100
☐	100–1000
■	1000–3000

Figure 8.6   TKIBS employment growth, 1991–2001

employment growth in most districts of Austria, an even stronger position of Vienna in 2001 compared with 1991. The more detailed spatial
changes for TKIBS employment can be seen in Figure 8.6. Strong growth
rates can be observed for Vienna and suburban areas of secondary cities
such as Graz, Linz and Salzburg, indicating a process of urban sprawl of
this sector. However, starting from a very low base, TKIBS have also been
growing in rural and more peripheral districts of Austria. Presumably, the
expansion of TKIBS is linked to the overall growth of jobs and population in those areas. It is probably also facilitated and supported by
the spread of modern ICT infrastructure and related technologies
(Kaufmann et al., 2003).

Regarding spatial changes in knowledge-based sectors we find partly
similar and partly different trends. The common pattern is one of overall
spatial de-concentration, as is indicated by the coefficient of variation. This
implies that districts with low employment shares of knowledge-based
sectors in 1991 have had a higher employment growth than districts with
high shares. However, there are also marked differences in the spatial development between HT and TKIBS.

Whereas in HT we observe a de-concentration from Vienna towards
the other urban areas, in the TKIBS sector we see the reverse pattern,

namely, that the position of Vienna has been further strengthened to the disadvantage of the other urban areas. Obviously, there is a process of specialization going on among the city regions of Austria with respect to knowledge-based sectors. Vienna becomes more specialized on TKIBS, whereas the other urban regions as well as some rural districts have been catching up, in particular in HT manufacturing.

Spatial Clustering of Information and Communication Technology Firms (ICT)

Two types of knowledge-based industries were researched in more detail, namely, information and communication technology and biotechnology, for this book. ICT was analysed through the firm census and the survey, and has thus been integrated into this and the following section. Biotechnology was investigated mainly through qualitative interviews, and the respective findings are presented in the penultimate section. The ICT bloc comprises both manufacturing and service industries, thus it combines specific subsets of HT and TKIBS (see Chapter 6). Regarding the SIC codes, we applied the same cluster definition as in the UK study (Chapter 7). Basically this includes the production of ICT-related goods such as office machinery and computers, telecom equipment, radio, television and electronic instruments as well as related services such as telecommunication services, software supply and consultancy, data processing and database activities, as well as R&D activities (see Chapter 6).

Overall in 2001 there were 13 841 ICT establishments in Austria employing 131 656 people. From 1991 to 2001 the number of establishments had increased more than threefold (up from 4256, or +325 per cent), whereas ICT employment had grown more modestly by +43 per cent (up from 92 341) in this ten-year period (Table 8.5). These numbers indicate both a strong growth of the ICT sector (driven mainly by the respective services) as well as a trend towards smaller establishments caused partly by the growth of small service firms and partly by the externalization of respective activities (vertical disintegration). In fact the average size of establishments has gone down from 22 to 10 employees. To what extent do ICT industries and related activities form spatial clusters in Austria? ICT activities are strongly concentrated in the Vienna region: in 2001 44 per cent of establishments and up to 51 per cent of employment were located there. Thus Vienna is clearly the dominant ICT location in Austria. The changes between 1991 and 2001 indicate a differentiated development. Rural areas are losing further ground in this sector: they display a below average growth of both establishments and

Table 8.5   ICT sector 1991–2001 by type of region

Regions type	1991 absolut	%	2001 absolut	%	% change
Establishments					
Rural	1123	26.4	3545	25.6	216
Urban	1372	32.2	4245	30.7	209
Vienna	1761	41.4	6051	43.7	244
Total	4256	100.0	13841	100.0	225
Coeff. Var.	1.02		0.99		
Employment					
Rural	18753	20.3	26214	19.9	39.8
Urban	25109	27.2	38414	29.2	53.0
Vienna	48479	52.5	67028	50.9	38.3
Total	92341	100.0	131656	100.0	42.6
Coeff. Var.	1.68		1.53		
Average size of establishments					
Rural	16.7		7.4		
Urban	18.3		9.0		
Vienna	27.5		11.1		
Total	21.7		9.5		

employment. Vienna has an above average growth of establishments (+244 per cent) but only a below average growth of employment. This is due to employment loss in production-related activities, particularly in large plants, and dynamic growth of smaller establishments particularly in the ICT services. The other urban areas show strong ICT employment growth but below average growth of establishments. This indicates that, similar to the development in HT, they are gaining larger, often production-related, activities. Overall these data indicate a trend towards clustering of ICT in urban regions of Austria, whereby Vienna has strengthened its position specifically in the ICT services and the other urban regions also in ICT production activities. Basically this is in line with the more general development analysed above.

Types of ICT concentrations

As already indicated in Chapter 6, we have distinguished between different types of spatial concentrations, using both the index of employment and the index of establishments (see Table 8.6).[3] Situations where the employment index is above 150, but the establishment index below 150 indicate the existence of few but large plants. We refer to such a situation as 'dominating firms'. A high index of establishments accompanied by a low

Table 8.6 Types of regional concentration

Types of regional concentration	Establishment index < 150	Establishment index > 150
Employment Index > 150	Dominating Firm	Cluster
Employment Index < 150	No Concentration	District

Notes:
Employment index, 100 = Austria (mean of the 121 political districts).
Establishment index, 100 = Austria (mean of the 121 political districts).

Source: Arbeitsstättenzählung, 2001.

employment index indicates the concentration of mainly small firms. Inspired by the literature on industrial districts we call this constellation 'district'. 'Clusters' combine both a relatively high number of establishments as well as of employment, and thus a combination of both small and large firms. We have to consider that these 'clusters' only constitute potential clusters, since interrelationships have not been taken into account in this typology. However, this aspect is included in the following section by analysing knowledge linkages.

From Table 8.7 we can see that, from about 120 administrative units in Austria, 18 are classified as 'clusters' according to the above criteria. Eleven of these locations are part of the Vienna region, the rest are mainly in secondary cities such as Graz, Linz, Salzburg, Klagenfurt and Innsbruck. This finding underlines the strong role of Vienna and some other cities as locations for ICT clusters. 'District' situations (that is, the concentration of mainly small ICT firms) we find in a few other locations within and around Vienna as well as in suburban areas of Linz and Innsbruck. The concentrations of ICT firms (clusters and districts) mainly in urban regions indicates that firms might benefit considerably from local knowledge spillovers and other externalities. The extent to which this is the case will be analysed in the following section.

In five administrative units we find a constellation of 'dominating firms', such as in Villach (Infineon), Deutschlandsberg (Siemens), Leoben (AT&S). There are good reasons to assume that in these smaller cities a strong role of 'hierarchy', that is, transactions and knowledge flows within those large companies and a considerable dependency of the regions on the respective firms in terms of employment and local development. Such a hierarchical situation might be an obstacle for knowledge exchange and collective learning of firms as well as for long-run development of the respective region as Storper and Harrison (1991) pointed out.

Table 8.7   *Types of regional concentration*

	Political district	Code
Dominating firm	Villach (Stadt)	202
	Deutschlandsberg	603
	Leoben	611
	Feldkirch	804
	Wien 20., Brigittenau	920
Cluster	Klagenfur (Stadt)	201
	Mödling	317
	Linz (Stadt)	401
	Salzburg (Stadt)	501
	Salzburg (Umgebung)	503
	Graz (Stadt)	601
	Graz (Umgebung)	606
	Innsbruck (Stadt)	701
	Wien 1., City	901
	Wien 2., Leopoldstadt	902
	Wien 3., Landstraße	903
	Wien 9., Alsergrund	909
	Wien 10., Favoriten	910
	Wien 12., Meidling	912
	Wien 19., Döbling	919
	Wien 21., Floridsdorf	921
	Wien 22., Donaustadt	922
	Wien 23., Liesing	923
District	Wien (Umgebung)	324
	Linz (Land)	410
	Innsbruck (Land)	703
	Bregenz	802
	Wien 4., Wieden	904
	Wien 6., Gumpendorf	906

INNOVATION AND KNOWLEDGE INTERACTIONS IN KNOWLEDGE-BASED SECTORS

What kind of innovative activities do knowledge-based sectors in Austria undertake and which knowledge sources and partners are they relying on in their innovation process? To what extent do various knowledge-based sectors differ in this respect? We will compare the different subsectors including the ICT sector on these aspects. The results which will be

Table 8.8   Sample for postal survey

Sectors (NACE 1995 Rev.1)	Sample	Return	Rate of return (%)
Medium-high-technology (24, 29, 31)	816	62	8
High-tech-technology (244, 30, 32, 33, 353)	593	56	10
TKIBS (72, 742, 743)	675	51	8
Research (73)	144	16	11
ICT (30, 31, 32, 33, 72)*	878*	79*	9
Not classified		4	
Total	2228	189	9

Notes:
Firm size: employees > 10; 244, 353 and 73 without size restrictions.
* The ICT group includes selected MHT, HT and KIBS industries.

presented in the following section rely on a postal survey undertaken in 2003.

Two thousand two hundred and twenty-eight Austrian firms were sampled from the Herold Professional Database and invited to fill in a questionnaire.[4] Approximately two-thirds of the firms belonged to the manufacturing sector and one-third to the knowledge-intensive service sector (data processing and engineering). One hundred and eighty-nine firms responded, yielding a rate of return around 9 per cent (Table 8.8). The sectors have been classified according the OECD classification (OECD, 2001) and grouped in four sub-sectors as described above:

1. High-technology industry (HT), comprising the NACE sectors Pharmaceuticals (244), Office Machinery and Computer (30), Radio, TV and Communication Equipment (32), Medical, Precision and Optical Instruments (33), and Aircraft and Spacecraft (353).
2. Medium-high-technology manufacturing (MHT) comprising the sectors Chemicals without Pharmaceuticals (24 except 244), Machinery (29), and Electrical Machinery and Apparatus (31).
3. Technology-related knowledge-based services (TKIBS) with a focus on Computer and Related Activities (72), Architectural and Engineering Activities and Technical Testing and Analysis (742, 743).
4. Research firms (R) with R&D (73).

In addition were included:

5. The ICT sector, consisting of selected HT and TKIBS activities (see above).

Nature of Innovation Activities

As to be expected from the literature on knowledge bases (see Chapter 3) there are clear differences with respect to innovation activities between the investigated sectors. In line with the high importance of an analytical knowledge base, high-technology (HT) firms and research (R), firms have higher numbers and shares of researchers in total employment and more patenting activities (Table 8.9). This is due to more continuous research activities both of a basic and applied nature (Table 8.10).

Given the knowledge and innovation orientation of TKIBS, it is interesting to observe that they have low research activities in comparison with the other sectors. In line with their reliance on a synthetic knowledge base, medium-high-technology (MHT) firms have less basic and applied R&D, but relatively more activities of design and market introduction instead. Information and communication technology firms are between the HT and MHT categories on most of these indicators, suggesting that ICT includes a variety of related activities from both analytic and synthetic knowledge.

The structure of innovation activities is reflected also in different types of innovation output (Table 8.11): HT and R firms have more innovations which are 'new for the market', whereas MHT and TKIBS rely more on modifications and technology adoption (innovations 'new to the firms') in order to maintain their competitiveness. Innovation strategies, that is, the attainment of competitive advantages via the introduction of substantial product innovations, thus, have clearly more prominence in the case of HT and R firms, compared with the rest. It is somewhat surprising to find that the TKIBS are only slightly more innovation oriented than the MHT firms.

Table 8.9   Firm economic and innovation data

	MHT	HT	TKIBS	Research	ICT	Total
Employment 2002	147	93	54	74	233	98
Turnover 2002 (million Euros)	25.07	16.84	9.01	4.84	57.87	16.59
Export ratio 2002 (%)	46.34	44.04	16.95	34.15	37.46	37.26
R&D ratio 2002 (%)	4.18	31.27	7.36	62.93	12.65	19.76
R&D department (%)	40	66	31	67	50	48
Researchers (%)	21	50	9	93	23	33
Number of researchers	4	14	39	32	274	19
Technicians (%)	97	91	100	86	97	95
Number of technicians	30	38	35	10	100	32
Patents (%)	39	57	8	47	34	36
Number of patents	5	3	10	4	24	4

Table 8.10   Innovation activity (percentage of responding companies)

	MHT	HT	TKIBS	Research	ICT
Basic research					
Yes, regularly	14.5	20.0	17.8	66.0	20.0
Yes, sometimes	12.7	26.0	20.0	26.7	21.4
No	72.7	54.0	62.2	6.7	58.6
Applied Research					
Yes, regularly	16.1	48.1	13.6	100.0	30.0
Yes, sometimes	26.8	19.2	40.9	0.0	31.4
No	57.0	32.7	45.0	0.0	38.6
Development					
Yes, regularly	60.7	72.7	55.3	93.3	72.4
Yes, sometimes	21.3	10.9	29.8	6.7	15.8
No	18.0	16.4	14.9	0.0	11.8
Design					
Yes, regularly	66.7	58.5	37.8	7.1	67.6
Yes, sometimes	18.3	7.5	31.1	57.1	12.2
No	15.0	34.0	31.1	35.7	20.3
Market implementation					
Yes, regularly	47.5	67.3	44.0	20.0	65.4
Yes, sometimes	34.4	20.0	40.0	73.3	25.6
No	18.0	12.7	16.0	6.7	9.0

Table 8.11   Type of innovation (percentage)

	MHT	HT	TKIBS	Research	ICT
Improvement of existing product	87.7	88.4	93.0	93.3	91.5
Innovation, new to the firm	70.2	65.1	69.8	66.7	66.7
Innovation, new to the market	60.0	77.6	66.7	92.3	74.6

This may partly have to do with the fact that the innovation process of service firms is not very well covered by existing innovation concepts. Partly it may reflect the fact that the Austrian TKIBS sector is for various reasons less advanced and sophisticated compared with leading European and US economies (Tödtling and Traxler, 1995; European Commission, 2003). The ICT range as a whole has high shares of all of these categories, indicating the importance of innovation in this sector in general. This indicates the rapid pace of technological change in ICT, covering product and process innovations as well as organizational changes. Thus innovation processes and activities clearly differ between the investigated sectors, a fact which has to be taken into account when investigating external knowledge sources and innovation partners.

Knowledge Sources

Also with respect to the *dominating sources of knowledge* there are clear sectoral differences (Table 8.12). While the most important knowledge sources for the MHT firms are other firms along the value chain (customers, suppliers) including competitors, for HT and R firms universities are a clearly more important knowledge source (for 58 per cent and 67 per cent of firms respectively). This finding underlines the importance not only of practical but also scientific knowledge for HT and R firms and demonstrates that these companies rely on a larger variety of knowledge types and respective sources than do MHT firms. Basically this is in line with the higher relevance of an analytical knowledge base and it is confirmed by other research on knowledge-based industries (Keeble and Wilkinson, 2000; Asheim and Gertler, 2005). For TKIBS and R firms, in addition, other service firms and commercial R&D are important knowledge sources. Thus we find, in accordance with Daniels (1995) and Moulaert and Tödtling (1995), frequent knowledge links within the service sector and fewer relations to other types of knowledge providers. Information and communication technology firms, in comparison, rely less on the group, commercial or non-profit R&D, but more on customers, competitors and service firms instead. Hence, their pattern of knowledge sources is more along and within the value chain and, accordingly, more similar to TKIBS and MHT than to HT and R firms.

Table 8.12   Importance* of knowledge sources (percentage)

	MHT	HT	TKIBS	Research	ICT
Own firm	96.7	100.0	95.8	100.0	98.7
Group	56.1	54.0	37.2	50.0	46.6
Customer	83.1	80.8	85.1	73.3	88.2
Supplier	79.7	51.9	58.3	46.7	63.6
Competitor	56.7	50.0	53.2	66.6	55.3
Service firm	28.3	25.0	43.2	40.0	28.9
Commercial R&D	20.7	33.4	37.8	40.0	19.7
University	29.3	57.7	32.6	66.7	32.9
Non-profit R&D	13.8	11.8	10.9	26.7	10.7
Technology transfer centre	13.8	17.6	20.0	40.0	14.9

Note: * Percentage of firms, rating knowledge source as important or very important.

Regarding the *role of the region* we can observe that knowledge sources from the region are clearly more important for all three kinds of knowledge-based sectors (HT, TKIBS and R firms) than for MHT firms (Table 8.13). This holds true in particular for universities and service firms, but to some extent also for customers and technology centres. As reasons for this, firms indicated that the contacts with knowledge sources from the region are more informal, faster and more appropriate for the respective purpose. Also the ease of contacts and the trustworthiness were mentioned by more than half the respondents. A less important reason, according to firms, is a higher security of regional information flows. In general, these results are in line with findings of Kaufmann et al. (2003) on similar questions. For the ICT sector the region has a similar high importance as for HT and TKIBS, whereby customers and to a smaller extent universities are, in particular, relevant.

Looking at the *spatial levels of knowledge sources* in more detail, we find that HT firms utilize regional, national and international knowledge sources (Table 8.14). Knowledge flows from clients, suppliers and competitors, in addition to intra-group knowledge flows, are highly internationalized (EU, USA). Relevant knowledge sources from the region are universities, technology centres and suppliers, but it is clear that knowledge sources from the region are in general less important than those from the rest of Austria and from Europe. Thus HT firms combine knowledge sources from the region with those of national and international origin in their innovation process. This result is in accordance with findings of Cooke et al. (2000), Sternberg (2000) and Bathelt et al. (2004). Basically this pattern also holds true for the other investigated sectors, with the

Table 8.13   Regional importance* of knowledge sources (percentage)

	MHT	HT	TKIBS	Research	ICT
Group	22.2	20.5	18.0	9.1	19.7
Customer	42.3	45.1	60.9	57.2	51.4
Supplier	42.4	26.5	35.5	35.7	32.9
Competitor	15.8	22.0	21.8	7.1	19.7
Service firm	29.6	43.8	39.6	35.7	35.8
Commercial R&D	20.0	32.6	24.4	35.7	27.3
University	38.1	65.4	56.8	78.5	50.7
Non-profit R&D	25.9	17.7	16.7	71.5	17.2
Technology transfer centre	27.8	23.4	37.2	57.1	24.2

Note: * Percentage of firms, rating the regional existence of the knowledge source as important or very important.

Table 8.14 Geographical location* of knowledge sources of high-tech firms (percentage)

High-tech	Region	Austria	EU	USA, Canada	Asia	Other
Group	8.9	14.3	41.1	25.0	1.8	3.6
Customer	16.1	44.6	51.8	26.8	19.6	8.9
Supplier	17.9	44.6	57.1	23.2	5.4	3.6
Competitor	8.9	21.4	48.2	39.3	10.7	3.6
Service firm	14.3	41.1	26.8	7.1	0.0	0.0
Commercial R&D	10.7	37.5	25.0	10.7	1.8	0.0
University	21.4	53.6	39.3	14.3	1.8	0.0
Non-profit R&D	14.3	21.4	19.6	7.1	1.8	0.0
Technology transfer centre	17.9	35.7	12.5	3.6	0.0	0.0

Note: * Percentage of firms, using a knowledge source at the relevant geographical level.

qualification that for the TKIBS and the R firms the region is a compara-tively more important knowledge space. A stronger role for tacit knowledge and a higher need for personal contacts with various knowledge sources and innovation partners are largely responsible for this pattern.

The knowledge sources correlate significantly with the types of innova-tion activities (see Table 8.15 and Tödtling et al., 2006): The use of com-mercial R&D, universities, non-profit R&D and technology centres as knowledge sources correlates positively with the performance of basic and applied research and development. This can be interpreted both from a demand and capability perspective. Regarding the first, we can argue that firms performing R&D have a higher need of various kinds of knowledge inputs from these different kinds of organizations. From a capability per-spective, we find that the performance of these functions enables firms better to interact with and exploit these various knowledge sources. Significant correlations can also be found between the types of innovation output and knowledge sources. Only the introduction of innovations 'new to the market' correlated strongly with the science and research related knowledge sources (universities, commercial R&D, non-profit R&D, tech-nology centres). Obviously, the more innovative products not only imply a more complex knowledge process, but they also require inputs from various kinds of knowledge organizations (Kaufmann and Tödtling, 2001). Products which are new to the firms only (adoptions) are significantly related to competitors and service companies as knowledge sources. The relation to competitors could be interpreted as a process of monitoring and imitation of rival companies with respect to new products (Malmberg and Maskell, 2002), whereas service companies contribute complementary

Table 8.15 Importance* of knowledge sources and innovation characteristics

	Basic research	Applied research	Development	Design	Implementation	Improvement of product	Product new for firm	Product new for market
Own firm	+++	++	+++					++
Group								++
Customer	– – –	++	++	+	++	+++		
Supplier								
Competitor							++	
Service firm	++		+				++	+
Commercial R&D	+++	+++	+++		++			+++
University	+++	+++	+++	+	+++	++		+++
Non-profit R&D	+++	++	++		++			+++
Technology transfer centre	+++	++	++	+++	++			++

Notes:
* Mann-Whitney U-Test.
Knowledge source more (less) important, significant 1% level +++ (– – –).
Knowledge source more (less) important, significant 5% level ++ (– –).
Knowledge source more (less) important, significant 10% level + (–).

knowledge relevant for the introduction and marketing of the new products. Product modifications, in turn, are significantly stimulated by knowledge inputs from clients. This supports the findings of Dosi (1988), von Hippel (1988) and Kaufmann and Tödtling (2001), that smaller changes and improvements of products take place continuously and are strongly stimulated and supported by relations to clients.

Knowledge Channels and Types of Interactions

To what extent do knowledge-intensive sectors differ not just in the knowledge sources but also in the types of interactions? We have tried to operationalize the investigated types of knowledge interactions (market, spillover, network, milieu: see Chapter 3) through the various channels firms use in getting their knowledge for innovation. Although there may be overlaps, we have related the channels to the four categories of interaction as in Table 8.16.

We find remarkable differences in the mechanisms and channels of knowledge transfer between the investigated sectors (Table 8.17): for MHT firms the most important channels of knowledge exchange are the buying of machinery and software (that is, market links). Also the places and institutions where trading partners and other people from the industry meet, that is, fairs and conferences, as well as informal contacts (milieu) are important channels. The buying of machinery and software represent a process of embodied technology transfer cited often in the economic literature on innovation (Coombs et al., 1987; Fagerberg, 2005). The participation in fairs and conferences, on the other hand, can be interpreted in the context of the monitoring of technologies, markets and other firms, and of informal knowledge exchange (Maskell et al., 2004). As in some other findings shown above, TKIBS are quite similar to MHT firms regarding their knowledge channels: fairs and conferences, informal contacts and the hiring of specialists are the dominant mechanisms. In the last case, knowledge is 'embodied' in and transferred through mobile qualified labour, an important mechanism of knowledge spillover as pointed out by Saxenian (1994) and Feldman (2000). Both MHT firms and TKIBS, thus, get their knowledge through a mix of market links, spillovers and milieu effects in the innovation process.

High-tech firms rely more on consulting, contract research, R&D cooperation and the joint use of R&D facilities in addition to intermediate goods and informal contacts. Since the former channels are usually based on more durable and reciprocal relations, we find a stronger overall importance of formalized networks for HT firms. This is in line with other studies on HT industries (Saxenian, 1994; Sternberg, 1995; Camagni and Capello, 2000;

Table 8.16   Types of knowledge interactions and channels

Types of knowledge interactions	Channels
Market links	Intermediate goods Licenses Consulting Contract research
Spillovers	Employment Reading of literature and patents
Formal networks/cooperations	Research cooperation Shared use of R&D facilities
Informal networks – milieu	Conferences, fairs Informal contacts

Table 8.17   Knowledge transfer channels (percentage)

	MHT	HT	TKIBS	Research	ICT
Employment	58.1	60.7	68.6	68.8	58.8
Intermediate goods	79.0	66.1	66.7	68.8	66.3
Literature, patents	66.1	67.9	60.8	75.0	62.5
Conferences, fairs	74.2	66.1	72.5	62.5	73.8
Informal contacts	79.0	71.4	74.5	81.3	72.5
Licenses	27.4	39.3	23.5	56.3	32.5
Consulting	59.7	62.5	52.9	75.0	58.8
Contract research	37.1	51.8	33.3	75.0	40.0
Research cooperation	40.3	60.7	35.3	75.0	42.5
Shared use of R&D facilities	33.9	50.0	19.6	75.0	30.0
Firm takeover	29.0	19.6	15.7	12.5	20.0

Bathelt, 2001; Powell and Grodal, 2005). Research firms, finally, get their knowledge through a variety of channels including reading of scientific literature and patents, contract research and research cooperations as well as informal contacts. Hence, these firms have the most distributed knowledge base (Smith, 2002). They draw on a large variety of knowledge sources and use various knowledge channels and interactions. Thus they combine all types of relations (market, network, spillover and milieu) in their innovation process. Information and communication technology firms as a whole use most of these channels less frequently than HT and R firms, with the exception of conferences and fairs. This pattern may be due to the large share of service firms with less R&D activities in the ICT sector.

Cooperations in the Innovation Process and their Spatial Levels

One key mechanism of knowledge exchange is cooperations, which belongs
to the network category. They constitute intentional and selective relations
to particular partners in the innovation process and they are more interactive and durable than market links. It is argued in the literature that
cooperations are of special importance for technology-intensive and
knowledge-based sectors because they can reduce uncertainties, provide
access to complementary resources and technologies, and speed up the
innovation process (Camagni, 1991; De Bresson and Amesse, 1991;
Hagedoorn, 2002; Fritsch 2003; Fritsch and Franke, 2004). In accordance
with this literature we find that firms in knowledge-based sectors (HT,
TKIBS, R firms) clearly cooperate more frequently in the innovation
process (49 per cent to 80 per cent) than MHT firms (34 per cent). In line
with the findings on the use of external knowledge sources we identify the
highest shares of cooperating firms among the research firms (80 per cent).
But also from the HT firms and TKIBS about half of the investigated companies cooperate (Table 8.18). Owing to their mix of activities, ICT firms
occupy an intermediate position: they cooperate more frequently than
MHT but less so than HT firms in comparison.

Regarding the objectives of cooperation, we find that for R, HT and ICT
firms the most frequent goals are product innovations 'new to the market'
and the opening up of new technical fields, that is, the entering of new technology paths (Table 8.19). Cooperations, thus can be regarded as an
important tool for accessing complementary knowledge, distribution channels or other innovation resources. Interestingly, for ICT firms, cooperations are less often undertaken with the intention to develop a product or
to publish jointly. Obviously these goals are not so important for them.

With respect to the areas of cooperation we can observe that R firms
most frequently cooperate in their own core activity, namely basic and
applied research and development (Table 8.20). High-tech firms cooperate
most frequently in the medium phases of the knowledge and innovation
process: applied research, development and testing are frequent areas of
cooperation. Technology-related knowledge-intensive business services
and MHT firms cooperate rather in later phases in comparison: development, testing and commercialization are their most frequent areas of cooperation. It becomes obvious that cooperations follow more or less the
dominant pattern of innovation activities identified above.

The spatial pattern of cooperations (location of partners) resembles
the geography of knowledge sources analysed above: Austria and the
European Union are also the most important cooperation spaces, the
region also has relevance, but seems to be less important in comparison

Table 8.18   R&D cooperation

Sector	%
Medium-high-tech	33.9
High-tech	49.0
TKIBS	48.9
Research	80.0
ICT	40.0
Total	46.5

Table 8.19   Objective of R&D cooperation (percentage)

	MHT	HT	TKIBS	Research	ICT
Improvement of existing product	77.8	81.8	84.2	90.0	75.0
Innovation, new to the firm	66.7	68.2	70.6	30.0	71.4
Innovation, new to the market	78.9	95.8	84.2	100.0	92.9
Patent development	55.6	50.0	13.3	55.6	16.0
Entering new technical fields	68.4	87.5	76.5	100.0	82.1
Joint publication	47.1	40.9	43.8	80.0	30.8

Table 8.20   Area of R&D cooperation (percentage)

	MHT	HT	TKIBS	Research	ICT
Basic research	25.0	47.6	29.4	100.0	32.0
Applied research	52.9	83.3	33.3	100.0	46.2
Development	95.0	83.3	95.2	90.0	93.3
Prototyping, testing	88.9	82.6	78.9	70.0	96.4
Commercialization	43.8	31.8	58.8	77.8	44.0

(Tödtling et al., 2006). Compared with MHT firms, HT firms cooperate more, both at the level of the region and internationally. In particular, universities (28 per cent) and commercial R&D (25 per cent) were frequent cooperation partners within the region. Among the R firms we found the highest share of cooperating firms. For them, the region is a more important cooperation space than for HT firms, in particular for partners from universities (50 per cent), technology centres (50 per cent) and non-profit R&D (42 per cent). But the R firms are even more cooperative internationally, with the EU and North America as the most relevant areas. For

HT firms, and for R firms in particular, we find a pattern of both regional and international cooperation links in the innovation process. Compared with HT and R firms the TKIBS and MHT firms cooperate much less frequently, and their spatial pattern of cooperation is more confined to the region and the rest of Austria.

BIOTECHNOLOGY IN AUSTRIA

Austria has the status of a latecomer in biotechnology.[5] In spite of the country's good tradition in medicine and biomedicine, the commercial exploitation of science and research in the key future field of biotechnology remained weak so far (Reiss et al., 2003). Austria's poor performance in the commercialization of biotechnology is due to a variety of hampering factors, including a missing critical mass of high-level biotechnological research, the lack of academic entrepreneurship, a culture which does not encourage risks and accept failure, a lack of venture capital, a weakly developed public support infrastructure, missing managerial competencies, and regional and national economic policies stuck in old routines. However, recently, a catching-up process in the development of biotechnology can be observed. In the following we will investigate this process for the Austrian regions of Vienna, Tyrol and Styria. These three regions differ enormously with respect to their preconditions for promoting biotechnology activities. Vienna is Austria's leading economic and scientific centre. Tyrol is mainly known for tourism, and the province of Styria has to be characterized as an old industrial area, which suffered from the downturn of its traditional coal and metal industries in the past. In the 1990s, a recovery of the region could be observed, brought about by a renewal of the steel sector and the growth of a new automotive cluster (Tödtling and Trippl, 2004).

Size and Structure of Biotechnology in the Investigated Austrian Regions

A key factor that distinguishes the three Austrian regions examined here is related to the presence of universities and other research organizations. In this context Vienna clearly has the lead. The region has an excellent science system which consists of five universities, several hospitals and other public and private research organisations. There are the Institute of Molecular Pathology (IMP) which is Boehringer Ingelheim's cancer research centre, the Novartis Research Institute (NRI), and the Antibiotic Research Institute Vienna (ABRI) which is owned by Biochemie Kundl (part of Sandoz [Novartis] R&D). A further strengthening of the local research base can be observed as the Austrian Academy of Sciences recently established two new institutes, namely, the Institute of Molecular Biotechnology (IMBA) and the Research Centre for Molecular Medicine (CeMM). Moreover, five cooperative research centres between university institutes and firms have been set up in the Vienna region. Finally, a technical college for biotechnology has also been created in order to improve the supply of specialized and highly skilled labour. The regions of Styria and Tyrol, in comparison, are only poorly endowed with knowledge-generating institutions. The scientific base in Tyrol comprises three universities, the Tyrolean Cancer Research Institute, and the Institute for Biomedical Ageing Research of the Austrian Academy of Sciences. There are, however, no cooperative research centres located in the region. The knowledge infrastructure in Styria is made up of three universities and two recently established cooperative research organizations carrying out bio-scientific research.

Until recently, the capability to reap commercial benefits from the presence of scientific expertise was rather weak. However, more recently evidence suggests that there has been a gradual catching-up by Austria in the biotechnology sector. An analysis of Austria's emerging business sector in biotechnology shows a strong specialization in medical ('red') biotechnology (Baier et al., 2000) and a clear tendency towards geographical specialization. Almost 70 per cent of all biotechnology-related firms are located in the Vienna region. Smaller agglomerations can be found in the provinces of Styria, Lower Austria and Tyrol (see Table 8.21).

The Vienna region hosts the largest Austrian biotechnology agglomeration. About 80 biotechnology-related firms are located in this cluster. Key activities comprise therapeutics and specialized supply (see Table 8.22).

Table 8.21 Proportion of biotechnology-related companies in different Austrian provinces

Region	Number of firms	Proportion of firms (%)
Vienna	77	67.0
Styria	10	8.7
Lower Austria	10	8.7
Tyrol	9	7.8
Upper Austria	4	3.5
Salzburg	4	3.5
Vorarlberg	1	0.8
Total	115	100

Source: BIT and LISA (2004), complemented by our own inquiry.

Table 8.22 *Main fields of activity of biotechnology-related companies in three Austrian clusters**

	Vienna		Styria		Tyrol	
	Number of firms	%	Number of firms	%	Number of firms	%
Therapeutics	22	28.6	2	20.0	3	33.3
Diagnostics	6	7.8	1	10.0	4	44.5
Specialized suppliers	19	24.6	4	40.0	1	11.1
Other	30	39.0	3	30.0	1	11.1
Total	77	100.0	10	100.0	9	100.0

Note: * We understand clusters in Porter's (1998) sense as agglomerations of related companies and supporting institutions specialized in a particular industry. It is obvious that in the case of Tyrol and Styria the biotech agglomerations are quite small. Nevertheless, we refer to them as clusters, but we are aware that they are rather 'potential clusters', far from being 'working clusters' in the sense of Enright (2003).

The industry structure is characterized by the existence of three subsidiaries of big pharma companies, which were attracted after the Second World War. These are Boehringer Ingelheim Austria (cancer research, biopharmaceutical production), its basic research subsidiary, the Institute of Molecular Pathology (IMP), Novartis (employing more than 3000 workers) and Baxter Austria. Moreover, there are about 25 dedicated biotechnology firms (DBFs) located in the cluster, such as Intercell (vaccines against oncological and infectious diseases), Igeneon (oncology), Austrianova (oncology, gene therapy) or Green Hills Biotechnology (oncology). About 40 per cent of these firms were founded after 2000 and many of them employ fewer than ten workers. The Vienna cluster also hosts about 20 specialized suppliers, such as producers of research agents (Bender Med Systems, Nano-S), bioinformatics providers (Emergentec, Insilico) and firms performing clinical trials services.

The Tyrolean biotechnology cluster is far smaller and still in its infancy. There are only three well-established companies, including the Novartis subsidiary Biochemie Kundl (production of generic medication), Gebro and Montavit (not included in Tables 8.21–8.23), which could be classified as traditional pharmaceutical firms. Tyrol only hosts seven DBFs. There are three therapeutic firms (Alcasynn, medication against pain; Ugichem, antisense, gene therapy against HIV and cancer; and Sentimmun, vaccines against kidney tumours) and four diagnostic firms (Amynon, tumour diagnostics; Vitateq, diagnostics and therapeutic products against infertility and neurodegenerative diseases; Biocrates, diagnostic instrumentation and

Table 8.23   Classification of biotechnology-related firms in three Austrian clusters

	Vienna		Styria		Tyrol	
	Number of firms	Proportion of firms (%)	Number of firms	Proportion of firms (%)	Number of firms	Proportion of firms (%)
Multinational companies	6	8.0	1	10.0	1	11.1
Dedicated biotech firms	25	32.4	2	20.0	7	77.8
Specialised suppliers	19	24.6	4	40.0	1	11.1
Other suppliers	10	13.0	3	30.0	0	0.0
Other firms	2	2.6	0	0.0	0	0.0
Sales and distribution firms	15	19.4	0	0.0	0	0.0
Total	77	100.0	10	100.0	9	100.0

analytical services; and Immumetrics, diagnostic immunological tests) present in the region. Almost all of these DBFs were established since 2000 and the large majority of them employs fewer than ten workers. Finally, there is the company Innovacell, a specialized supplier active in the fields of tissue engineering and stem cell therapy within urology.

The Styrian cluster hosts ten biotechnology-related companies. This agglomeration differs from the two other clusters as suppliers clearly dominate (see Table 8.22 and Table 8.23). This segment includes both larger incumbent firms such as Anton Paar (Laboratory Equipment and Instruments) and Roche Diagnostics (diagnostic instruments) as well as small and young suppliers of research agents (Pichem, Aurora Feinchemie) and firms specialized in clinical trials and clinical services (JSW Research and Molekulare Biotechnologie). There are only two dedicated biotech firms operating in the region including Eccocell (stem cells and genes) and Oridis Biomed (high-grade validated targets and optimised lead substances for chronic diseases of the liver and liver cancer). Styria's specialization in supplies and equipment is related to its associated variety in engineering and instruments.

Venture capital firms and business angels are a missing ingredient in all three clusters studied here. The main reason for this is the coordinated market, bank-dominated landscape, which coincides with a cultural preference for traditional credit instruments and a widespread adversity to risk-taking. Consequently, successful companies like Intercell or Igeneon had to attract external financing from international venture capitalists and funds. The gradual emergence of commercial biotech activities in Austria has been supported by conscious policy actions. It was, however, only at the end of the 1990s that biotechnology attracted the interest of policy-makers in Austria. Today, there exist several initiatives, both at the local and national level. The national policy level is of key importance, even if there are only two major policy initiatives which specifically focus on the promotion of biotechnology. These include the 'LISA-Life Science Austria' initiative (started in 1999) that mainly aims at promoting the foundation of new companies and the Austrian Genome Research Programme (launched in 2001), which has the goal of fostering the development of networks between universities and industry. Horizontal programmes, that is, policy measures which are not specifically designed to promote biotech exclusively but nevertheless have an impact on its development, also play an important role. Some of them have the goal of encouraging university–industry partnerships (such as the programmes K-plus and K-ind, Christian Doppler Laboratories), others support the establishment of centres that focus on the stimulation of new firm formation (AplusB programme). At the regional policy level new efforts were recently made to create more favourable conditions for the development of biotechnology clusters. In all three regions economic policy has undergone considerable changes in the recent past. A stronger focus on innovation and technology and the adoption of new policy approaches are clearly observable. All three regions have strategic policy priorities on life sciences and cluster initiatives in this field. This reorientation manifests itself in an intense process of institution building.

In Vienna new funding agencies have been established, which have special programmes for biotechnology organized as contests of proposals. The Centre for Innovation and Technology (set up in 2000) offers funding of R&D activities of high-technology companies and the Vienna Science and Technology Fund (founded in 2001) provides financial support to research organizations. Another new institution is INiTS (established in 2003), which aims at promoting academic spin-offs by providing counselling and assistance to scientists in the process of new firm formation. There are strong links between regional and national policy-makers and these are expressed in the recently started joint initiative 'Life Science Austria Vienna Region' which provides cluster management services to the local industry.

In Tyrol institutional learning is observable. A key player in the regional policy setting is the Tyrolean Future Fund, which has been set up in 1998 in order to promote the development of clusters, among others in the field of the health sector. In the year 2003 the policy initiative Life Science Cluster Tyrol was launched to foster commercial activities in the fields of biotechnology, pharmaceutics and medical device technology by stimulating local knowledge exchange, initiating collaborations and reporting on international trends and trade shows. The Kompetenzzentrum Medizin Tyrol (set up in 2001) acts as a management unit for this cluster and coordinates a research programme that is geared to encouraging university–industry partnerships. In 2002 the Centre for Academic Spin-Offs Tyrol was established in which start-ups are qualified, counselled and coached.

In the region of Styria a stronger focus on stimulating high-technology activities could be found. In 2001 the Styrian government created the Future Fund in order to support innovative projects, new high-technology activities and clusters. Even more interesting is the launch of the cluster initiative 'human.technologie.styria' in 2004. The Styrian Business Promotion Agency, which is a key actor in the regional institutional set-up, has accumulated several competencies in developing clusters since about 1995, particularly in the automotive industry. This expertise in implementing cluster policies is now being 'transferred' to the field of biomedical technologies. The cluster initiative 'human.technologie.styria' reflects a high level of coordination between key regional stakeholders, as it is organized as a formal cooperation between supporting organizations, firms and knowledge providers.

This is not the case for the two other regions investigated here. In Vienna we only found informal connections between the local supporting organisations. The Tyrolean cluster is characterized by a higher level of coordination than the one in Vienna, as the various supporting institutions have set up a formal network called Bio.Com.Net (Biotech Competence Network Tyrol) in order to coordinate their activities. There are, however, no firms or researchers involved as far as could be observed in Styria. Finally, it should be noted that the linkages between the three clusters are rather poor. There is almost no interregional exchange of experiences and know-how between the policy and supporting institutions in Vienna, Tyrol and Styria. Tyrol, however, has managed to build up such relations with the neighbouring German region of Bavaria, centred on Munich's biotechnology capability.

Spatial Dimension of Knowledge Flows

This section examines the spatiality of knowledge interactions in Austria's three biotechnology clusters. Building on the typology developed earlier in this book we distinguish between spin-offs (Audretsch, 1995; Fuchs and Krauss, 2003; Feldman et al., 2005), labour market recruitment and labour mobility (Keeble, 2000) and other knowledge flows, including spillovers (Feldman, 2001; Prevezer, 2001), market links (Zucker et al., 2002), informal networking (Keeble, 2000) and formal collaborations (Garmbardella, 1995; Powell et al., 2002; McKelvey et al., 2003; Owen-Smith and Powell, 2004).

Spin-offs
Spin-offs have become a key mechanism of knowledge transfer in the Austrian biotechnology agglomerations examined here, particularly in the Vienna cluster. An analysis of this process reveals three prevailing features (see Table 8.24). First, the establishment of spin-out companies is a recent phenomenon. The majority of these firms is rarely older than five years and employs fewer than ten workers. Second, new firm formation is highly localized in nature. A considerable share of all spin-off companies originated from parent organizations located in the region. Third, a clear dominance of universities as 'incubators' of these firms was found. Given the youth of the clusters, only a few second generation spin-offs have emerged so far.

Labour market recruitment and labour mobility
In all three Austrian biotechnology clusters, the local labour market seems to be crucial for most firms in order to get access to expertise and skills. A more detailed analysis reveals that it is to a great extent the local universities that are a key source of well-educated personnel. The movement of highly skilled workers between local firms is rather weak. There is a lack of locally available managerial competencies in all three Austrian biotechnology clusters. Academic entrepreneurs lack managerial expertise, and top managers with experience in the field of biotechnology or pharmaceutics are in short supply within the regions investigated here. Not surprisingly, a rather intensive inflow of international managerial know-how could be observed. Furthermore, the international recruitment of highly skilled research staff by firms also seems to be of relevance for all three clusters.

Consequently, the Austrian biotechnology clusters are highly dependent on successfully attracting international scientists and experienced managers. International recruiting is an explicit strategy as there is no talent available locally at excellent levels. This holds particularly true for the smaller biotechnology centres in Styria and Tyrol. Thus labour market recruitment and labour market mobility have both a local and an international dimension in the Austrian biotechnology industry. Owing to the youth of all three clusters, managerial competencies have to be 'imported'

Table 8.24 Characterization of spin-offs in the sample

		Vienna		Styria		Tyrol	
		Number of companies	%	Number of companies	%	Number of companies	%
Age of firm	Not older than 5 years	9	60	3	75	7	88
	Not older than 10 years	4	27	1	25	1	12
	Older than 10 years	2	13	0	0	0	0
	Total	15	100	4	100	8	100
Location of parent organization	Local	14	93	4	100	7	88
	International	1	7	0	0	1	12
	Total	15	100	4	100	8	100
Type of parent organization	Academic institution	11	73	2	50	8	100
	Firm	4	27	2	50	0	0
	Total	15	100	4	100	8	100
Firm size (number of employees)	1–10	8	53	1	25	7	88
	11–50	5	33	3	75	1	12
	More than 50	2	13	0	0	0	0
	Total	15	100	4	4	8	100

257

from abroad. Such a 'brain gain' strategy can be observed in the regions of Tyrol and Styria in particular, where the industry is still very young and small.

Spillovers, market links, informal networks and formal collaborations

External knowledge sources and innovation linkages with distant partners are crucial for the development of the Austrian biotechnology sector (see Figure 8.7 and Table 8.25). In particular Vienna and Tyrol exhibit a strong dependence on the inflow of international knowledge and expertise, as for these regions a considerable share of innovation links could be found at the international scale. But also the local level has to be regarded as an important interaction space, whereas national knowledge sources are almost negligible. It is only the Styrian cluster that seems to be embedded in the national innovation system. Local knowledge links play an important role for innovation in the Austrian biotechnology industry. In Vienna more than 40 per cent of all innovation interactions could be found within the cluster. Also, in Styria the local level clearly matters, as 37 per cent of all linkages are with regional partners. The picture is different for the Tyrolean biotechnology cluster. Owing to the small cluster size, only 31 per cent of all connections in this region could be found at the local scale.

Local interactions between firms and research organizations tend to be far more significant than local interfirm relations (see Table 8.25). The lack of a critical mass of firms in the clusters, particularly in Tyrol and Styria,

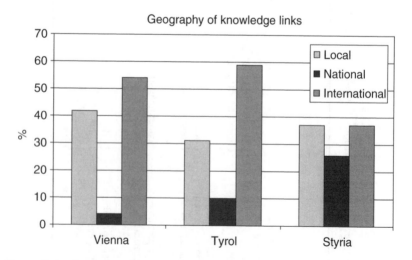

Figure 8.7   Geography of knowledge interactions in the Austrian biotechnology sector

Table 8.25   The geography of knowledge links in three Austrian biotechnology clusters

	Total number of links	With firms	Local			With firms	National			With firms	International		
			With RO*	Total	%		With RO	Total	%		With RO	Total	%
Vienna	172	23	49	72	42	2	5	7	4	52	41	93	54
Tyrol	81	9	16	25	31	3	5	8	10	23	25	48	59
Styria	51	6	13	19	37	6	7	13	26	11	8	19	37

Note: * RO = research organization (universities, clinics).

and the firms' highly specialized activities limit the potential for more intense knowledge exchange between local firm partners. Consequently, only a few examples of formal innovation networks and informal exchange of information about organizational, legal and funding issues have been found.

Local knowledge flows between research organisations and companies, in contrast, play a powerful role in all three Austrian biotechnology clusters. There is evidence of formal university–industry cooperations such as the publicly supported competence centre BioMolecular Therapeutics (BMT), the Austrian Centre for Biopharmaceutical Technology (ACBT) and several Christian Doppler Laboratories (CD Labs). Furthermore, our research results show for all three clusters a high significance of informal knowledge flows from universities and clinics to the local firms, signalling the relevance of local spillovers and milieu effects. Finally, there is clear evidence that most spin-off companies retain close links with their academic incubators, ranging from R&D collaborative partnerships, to the joint use of infrastructure, to the exchange of key personnel or to the buying of patents. National knowledge sources do not seem to play an outstanding role. Only the Styrian firms have built up some formal networks with national research institutions.

Distant knowledge links are a crucial mechanism for innovation in the Austrian biotechnology scene. Our research indicates that in the Vienna cluster more than 50 per cent of all knowledge linkages are with international actors. In the region of Tyrol, the inflow of international competencies and expertise is even more important, as almost 60 per cent of all knowledge interactions could be found at the international scale. This finding has to be interpreted against the background of the small size of the cluster and its geographical proximity to Germany, Italy and Switzerland. The biotechnology-related firms in Styria sustain fewer contacts with international knowledge sources than their counterparts in Vienna and Tyrol. For the Styrian companies international interactions are as important as local ones, amounting to about 35 per cent in both cases.

A closer look at the nature of these knowledge links shows the coexistence of market links, networks and spillovers. Interestingly, and contrary to the conventional wisdom in the literature (Jaffe, 1989; Bottazi and Peri, 2002), spillovers and informal networking are far from being overwhelmingly localized. International spillovers and informal networking at the global level undoubtedly matter. This can be observed for the Vienna biotechnology cluster and it seems to be even more relevant for the small agglomerations in Styria and Tyrol. A high importance of the reading of international scientific literature and patent specifications as well as of congresses and fairs for gaining new technical and managerial knowledge was found. Consequently, to access new competences and knowledge about technology, markets and rivals by reading international scientific literature and patent specifications and by attending international conferences and fairs seem to be of utmost importance for the Austrian biotechnology firms.

The analysis of knowledge interactions has also revealed a high significance of formal collaborations (such as R&D alliances and joint ventures) with global knowledge sources. This holds particularly true for the Vienna cluster, which seems to have strong connections with distant companies and research institutes. For Tyrol, in contrast, formal networks with international actors are rather unimportant. This finding has to be interpreted against the background of the youth of the cluster. The biotechnology-related firms in the Styrian cluster are more intensely inserted into cooperative networks with global partners, but they rarely sustain international market links. For the Vienna cluster, market links – mainly with international companies – seem to play a stronger role than for the Styrian firms. Finally, our research indicates that for the Tyrolean cluster market relations, principally with international firms which carry out more standardized activities such as the testing of arrays, are crucial.

SUMMARY AND CONCLUSIONS

Austria has to be regarded as a latecomer regarding the development of knowledge-based sectors. The country has specific strengths in medium-technology sectors and was able to achieve high productivity gains and growth rates under the cooperative model of social partnership (*Sozialpartnerschaft*) in the past. In the emerging knowledge economy, however, the prevailing strategy of technology adoption and modification has reached its limits. In the recent past, the Austrian economy seems to have shifted towards more knowledge-intensive activities, indicating a gradual catching up of the country in key future fields of high-tech industries and knowledge-based services including ICT and biotechnology. This seems to be accompanied by a move of Austria in the direction of a liberal market economy.

Looking at the *size of knowledge-based sectors in Austria* we found that in 2001 about 20 per cent of the overall employment could be classified in a broad sense as 'knowledge based' applying the OECD (2001) definition. This includes the KIBS sector (13 per cent), as well as high-tech (HT: 1.7 per cent) and medium-high-tech manufacturing (MHT: 4.9 per cent). Adopting a narrower definition that considers high-tech sectors in manufacturing (HT) as well as the more technology-based TKIBS, the knowledge-based sectors in Austria hold less than 5 per cent of employment (1.7 per cent in HT and another 3 per cent in TKIBS). The knowledge-based service sectors are the most dynamic ones. Between 1991 and 2001 employment in KIBS increased by 55 per cent, whereas in the smaller subset of TKIBS employment almost doubled (+94 per cent). Employment in high-tech manufacturing (HT) expanded more modestly (+11 per cent). In international comparison Austria's HT sector has been about 20 per cent lower than the respective EU average in 2001. With respect to KIBS Austria's share is about 30 per cent below the EU average. Austria seems to catch up regarding knowledge-based sectors, however. Since the mid-1990s both the knowledge intensive services and the HT/MHT sectors have shown an above average employment growth in a broader EU comparison.

Austria's knowledge-based sectors are *strongly concentrated in geographical space*. They are to a great extent an urban phenomenon. In 2001 not less than 43 per cent of HT employment in Austria was located in the Vienna region. Other important centres of HT could be found in urban regions such as Graz and Villach and their surroundings. Interestingly, a number of rural and even peripheral districts in Styria, Carinthia and Tyrol also exhibit high shares of HT employment. This pattern, however, is mainly due to the presence of subsidiaries or branch plants of large firms rather than to cluster settings. Knowledge-intensive business services and TKIBS are even more concentrated in urban regions than HT. Our analysis has revealed that KIBS can be mainly found in the cores of major cities. The geographical concentration of TKIBS is even stronger. Forty-seven per cent of the employment in this sector could be found in Vienna. Technology-related knowledge-intensive business services, however, have also spread to suburban locations.

Taking a more dynamic view by analysing the pattern of spatial changes of knowledge-based sectors between 1991 and 2001 we found indications for an ongoing specialization among Austria's city regions. For HT we could observe processes of de-concentration from Vienna to the other urban areas. Also, some rural areas and peripheral locations succeeded in catching up in HT manufacturing. For the TKIBS sector the picture is completely different, as Vienna has further strengthened its position to the disadvantage of the other urban regions.

Austria's ICT sector comprises almost 14 000 establishments employing more than 130 000 workers. Our analysis of the development pattern of the ICT industries between 1991 and 2001 has shown that the number of establishments had increased more than threefold (+325 per cent), whereas employment had grown more modestly (+43 per cent). The strong growth of the ICT sector seems to be driven largely by the dynamics of the services sector. Another important finding is the tendency towards smaller establishments which is caused by the growth of small service firms and by processes of vertical disintegration. Information and communication technology activities in Austria are characterized by a strong spatial concentration. In 2001 44 per cent of all establishments and 51 per cent of employment were located in Vienna. Looking at the changes between 1991 and 2001 we found that Vienna had strengthened its position in the ICT services, and the other urban regions in ICT production activities.

Our analysis of innovation activities has demonstrated that *knowledge-based sectors* such as HT and R *innovate in a different way* than the medium-technology sectors. Owing to a stronger reliance on an analytical knowledge base, they undertake more often activities of basic and applied research and development and their innovation output is to a higher extent oriented to more substantial and radical product innovations, as well as patenting. This differs from MHT firms and TKIBS, which follow the pattern of industries with a synthetic knowledge base. These sectors rely relatively more on activities such as development, design and market introduction, focusing in their output more on modifications and technology adoption. Information and communication technology firms in many respect are between these two groups. Owing to their heterogeneity of sub-sectors and activities, they obviously rely on both analytical and synthetic knowledge.

This pattern of innovation activity is also reflected in the predominant *knowledge sources* of those sectors. Whereas for MHT firms, TKIBS and ICT firms the most important knowledge sources are other firms along the value chain (customers, suppliers) including competitors, for HT and R firms universities and R&D organizations are clearly more relevant sources. High-tech and R firms thus rely on a larger variety of knowledge inputs including scientific and analytical knowledge than do MHT firms and TKIBS. Knowledge sources from the region, in particular universities and service firms, are clearly more important for all three kinds of knowledge-based sectors in comparison with MHT firms. However, it is obvious that HT firms rely even more on international knowledge sources than on regional ones. In particular, knowledge flows from clients, suppliers and competitors, in addition to intra-group knowledge flows, are highly internationalized. High-tech firms thus combine knowledge sources from the region with those of national and international origin in their innovation process.

There are significant relations between specific external knowledge sources and the type of innovation output. Innovations new for the market correlate positively with the use of specific knowledge sources such as universities, R&D organizations and technology centres. More fundamental innovations thus rely on a larger variety of knowledge inputs both from inside and outside the firms. On the other hand, the adoption of new products was more related to the monitoring of competitors and knowledge inputs from service firms, whereas the modification of products was significantly related to knowledge inputs from customers.

Not only the knowledge sources, but also the *transfer mechanisms and channels* differ: for MHT firms and TKIBS the most important channels of knowledge exchange are the buying of equipment and software, fairs, informal contacts and the hiring of specialists. 'Embodied' knowledge flows (both in people and equipment), the monitoring of markets and other firms (fairs) and informal knowledge exchange are dominant. Both for MHT firms and TKIBS we find, thus, a mix of market, spillover and milieu type of knowledge interactions in the innovation process. High-tech firms, in comparison, rely more on consulting, contract research, R&D cooperations and the joint use of R&D facilities. More durable and interactive relations, and thus networks, have more importance for these types of firms. Research firms use the full spectrum of channels to access external knowledge sources, including the reading of scientific literature and patents, contract research and research cooperations as well as informal contacts. They combine, thus, market, spillovers, network and milieu types of relations. Furthermore, the *goals and areas of cooperation* differ among the investigated sectors. For HT firms and R firms more fundamental product innovations and the opening up of new technical fields, and therefore strategic goals, are of particular importance. Research firms cooperate most frequently in their own core activity, namely, basic and applied research and development, whereas HT firms cooperate more in the medium phases such as applied research, development and testing. Technology-related knowledge-intensive business services, MHT and IKT firms focus in their cooperation relatively more on later phases such as development, testing and commercialization in comparison.

These empirical results indicate that innovation in knowledge-based industries is a complex phenomenon. Understanding these processes requires a detailed analysis of knowledge interactions since the firms rely, besides on internal R&D, to a large extent on external knowledge sources. They use various types of relationships (market, spillovers, networks and milieu) to acquire both codified and tacit knowledge from specific sources. Furthermore, it is obvious that innovation interactions between knowledge generation and exploitation actors are a multi-level phenomenon, taking place at regional, national and international levels. The sources, the kind of knowledge exchanged and the relationships differ between these levels, however: at the regional and national levels we often find universities, research organizations and service firms as typical knowledge sources, and tacit knowledge seems to be more often exchanged through informal relationships and spillovers. At the international level there is more exchange of codified knowledge, partly along the value chain, but also with competitors and research organizations. Formalized relations (market links, networks) are more important there. Regional, national and international innovation systems, thus, are strongly interconnected and play a complementary role for the development of knowledge-based sectors in Austria. This multi-level character is of particular importance for a small country and a latecomer in knowledge-based sectors such as Austria (see Trippl and Tödtling, 2007). In the following section we will analyse in more detail the catching-up process as well as the role of local and international knowledge interactions for the Austrian biotechnology sector.

Austria's catching-up process in biotechnology is a highly uneven phenomenon, revealing strong differences between the investigated regions. Vienna clearly has the lead regarding industry size and the existence of universities and other research organizations. Compared with Vienna, the regions of Styria and Tyrol are lagging behind. They are poorly endowed with knowledge-generating institutions and they still lack a critical mass of firms. These two embryonic biotechnology agglomerations differ markedly with respect to their pattern of specialization. In Tyrol the majority of companies is active in the fields of therapeutics and diagnostics. In the province of Styria, in contrast, suppliers dominate, indicating a stronger link of the biotechnology cluster to the region's economic and technological heritage. In all three areas considerable changes of economic policy have taken place in the past few years. A stronger focus on innovation and technology and the adoption of new policy approaches to transform the weak regional learning environments for biotechnology have been found. These policy learning efforts are inextricably linked to an intensive process of institution-building that becomes manifest in the creation of new funding and support organizations. Moreover, all three Austrian regions have launched cluster initiatives in order to promote the commercialization of biotechnology. The institutional solutions, however, betray notable differences. Vienna's cluster policy seems to be characterized by a top-down approach, a well-functioning interplay between national and regional policy-makers and an informal coordination between local supporting organizations. In Tyrol and Styria different paths have been followed. The cooperation with the national policy level is lower for both provinces. At the regional level, however, both in Tyrol and Styria more interactive modes of governance could be observed. In the region of Tyrol a formal cooperation between the key actors from the policy and supporting scenes has been formed and in the province of Styria a policy network has been created, comprising the main stakeholders from the policy system and the industrial and academic worlds.

Looking at the pattern of knowledge flows, we found that both *local and non-local factors and knowledge sources* are highly relevant. In all three regions, and particularly in Vienna, the development of the biotechnology industry is strongly localized in nature. This is true for the formation of new firms which are mostly local academic spin-offs, the mobility of highly qualified labour from universities to firms, as well as for other types of knowledge interactions. Such local factors and knowledge links are more significant in Vienna owing to the fact that the cluster here is relatively large and represents the most important biotechnology centre in Austria. The development of the biotechnology sector, however, was by far not just a local phenomenon. As we have shown, national policy initiatives played a key role for the sector's growth. Furthermore, knowledge links with international partners and knowledge sources turned out to be crucial. In fact, the most frequent knowledge links were with international partners. This was particularly the case for the region of Tyrol, the most embryonic cluster. International knowledge interrelations comprised a broad spectrum of links such as market links, formal networks, knowledge spillovers and informal relations. Through such links scientific knowledge, managerial know-how as well as venture capital and qualified labour are acquired. Our findings, thus, show that the cluster relations go beyond the misleading dichotomy of 'local buzz and global pipelines' (Bathelt et al., 2004), since we also found evidence for 'global buzz' (knowledge flows through international conferences and fairs) as well as 'local pipelines' (for example formalized R&D cooperations) for the regions investigated.

Finally, there remains the question of the implications of *Austria's cooperative business model* on the development of biotechnology. It has been argued in the literature that biotechnology as a risky high-technology activity flourishes better in an entrepreneurial or liberal business model (see Cooke, 2002; and Chapter 4). Relevant factors in this context might be less bureaucratic rules for the setting up of new firms, more venture capital, higher labour mobility, more university–industry links and spin-offs. The question, thus, arises as to what extent Austria's cooperative business model has hindered or supported the development of the biotechnology industry. From our analysis it appears that some of the country's weaknesses such as the lack of venture capital, the generally low levels of firm formation and the poor mobility of labour are indeed related to its cooperative business model. However, we also find supporting elements in the cooperative model for the development of biotechnology. These are, for example the cooperative R&D programmes at the federal level (initiatives K-plus, K-ind) which also support biotechnology; the networking between key actors both at federal (LISA) and regional ('human.technlogie.styria', Biotech Competence Network Tyrol) scales. Furthermore, we have to mention the relatively good coordination between federal and regional levels (Life Science Austria Vienna Region) as well as the integration of research and educational policies (setting up of new public research centres, Fachhochschulen) in this context. Then, it is important to state that Austria's cooperative model is not static but seems to be moving towards the entrepreneurial or liberal model to some extent. Indications for such a shift can be found in an increase of policy initiatives addressed to firm formation, spin-offs and seed finance, and an increasing role of competitive policy programmes as we have observed also for the biotechnology industry.

NOTES

1. 'In 2005 it was the best-performing stock exchange of the world. Shares in investment funds specializing in joint stock companies have replaced the savings account as the most popular form of investment of the little (and not so little) man' (Müller, 2006: 5).
2. The research project 'KNOWING – Collective Learning in Knowledge Economies: Milieu or Market?' has been financially supported from 2002 to 2004 by the Austrian Science Fund (FWF).
3. We have used a higher threshold index (150) than the UK (125: see Chapter 7) for identifying spatial concentrations in order to get a more adequate distribution of districts in the resulting matrix for Austria.
4. The questionnaire focused on the nature of firms' innovation activities within the past three years, the use of external knowledge sources and their location, the types and mechanisms of knowledge exchange as well as their cooperative behaviour and relationships.
5. The following findings are based upon results collected in the context of two projects: Collective Learning in Knowledge Economies: Milieu or Market? (2002–04) funded by the Austrian Science Fund; and Cluster development and policy in the Vienna biotechnology sector (2005–06) funded by the Jubilee Fund of the City of Vienna for the Vienna University of Economics and Business Administration. In sum, 72 interviews with representatives from firms, knowledge providers and the regional policy and supporting systems have been carried out, using semi-standardized questionnaires. Forty-three interviews have been with actors from the Vienna region, 13 with key players in the province of Tyrol and nine with actors from the region of Styria. Finally, we have interviewed seven representatives from the national policy and supporting system.

REFERENCES

Asheim, B. and Gertler, M. (2005), 'Regional innovation systems and the geographical foundations of innovation', in J. Fagerberg, D. Mowery and R. Nelson (eds), *The Oxford Handbook of Innovation*, Oxford: Oxford University Press, pp. 291–317.

Audretsch, D. (1995), *Innovation and Industry Evolution*, Cambridge, MA: MIT Press.

Baier, B., Griessler, E. and Martinsen, R. (2000), *National Case Study of Austria. European Biotechnology Innovation System*, Vienna: Institute for Advanced Studies.

Bathelt, H. (2001), 'Regional competence and economic recovery: divergent growth paths in Boston's high technology economy', *Entrepreneurship & Regional Development*, **13**, 287–314.

Bathelt, H., Malmberg, A. and Maskell, P. (2004), 'Clusters and knowledge: local buzz, global pipelines and the process of knowledge creation', *Progress in Human Geography*, **28**, 31–56.

BIT (Bureau for International Research and Technology Cooperation) and LISA (Life Science Austria) (2004), *Bio-Tech in Austria*, Vienna: BIT and LISA.

BMBWK (Bundesministerium für Bildung, Wissenschaft und Kunst), BMVIT (Bundesministerium für Verkehr, Innovation und Technologie) and BMWA (Bundesministerium für Wirtschaft und Arbeit) (2003), Österreichischer Forschungs- und Technologiebericht, Vienna.

Bottazzi, L. and Peri, G. (2003), 'Innovation and spillovers in regions: evidence from European patent data', *European Economic Review*, **47**, 687–710.

Camagni, R. (1991), 'Local "milieu", uncertainty and innovation networks: towards a new dynamic theory of economic space', in R. Camagni (ed.), *Innovation Networks*, London: Belhaven Press, pp. 121–44.

Camagni, R. and Capello, R. (2000), 'The role of inter-SME networking and links in innovative high-technology milieux', in D. Keeble and F. Wilkinson (eds), *High-Technology Clusters, Networking and Collective Learning in Europe*, Aldershot: Ashgate, pp. 118–55.

Cooke, P. (2002), *Knowledge Economies*, London: Routledge, p. 27.

Cooke, P., Boekholt, P. and Tödtling, F. (2000), *The Governance of Innovation in Europe*, London: Pinter.

Coombs, R., Saviotti, P. and Walsh, V. (1987), *Economics and Technological Change*, London: Macmillan.

Daniels, P. (1995), 'The locational geography of advanced producer services firms in the United Kingdom', *Progress in Planning*, **43**, 123–38.

De Bresson, C. and Amesse, F. (1991), 'Networks of innovators: a review and introduction to the issue', *Research Policy*, **20**, 363–79.

Dosi, G. (1988), 'The nature of the innovative process', in G. Dosi, C. Freeman, R. Nelson, G. Silverberg and L. Soete (eds), *Technical Change and Economic Theory*, London: Pinter, pp. 221–38.

Enright, M. (2003), 'Regional clusters: what we know and what we should know', in J. Bröcker, D. Dohse and R. Soltwedel (eds), *Innovation Clusters and Interregional Competition*, Berlin: Springer, pp. 99–129.

European Commission (2003), *Towards a European Research Area, Key Figures 2003, Science, Technology and Innovation*, Brussels: DG Research. Luxemburg: Office for Official Publications of the European Communities.

Fagerberg, J. (2005), 'Innovation: a guide to the literature', in J. Fagerberg, D. Mowery and R. Nelson (eds), *The Oxford Handbook of Innovation*, Oxford: Oxford University Press, pp. 291–317.

Feldman, M. (2000), 'Location and innovation: the new economic geography of innovation, spillovers, and agglomeration', in G. Clark, M. Feldman and M. Gertler (eds), *The Oxford Handbook of Economic Geography*, Oxford: Oxford University Press, pp. 373–94.

Feldman, M. (2001), 'Where science comes to life: university bioscience, commercial spin-offs, and regional economic development', *Journal of Comparative Policy Analysis: Research and Practice*, **2**, 345–61.

Feldman, M., Francis, J. and Bercovitz, J. (2005), 'Creating a cluster while building a firm: entrepreneurs and the formation of industrial clusters', *Regional Studies*, **39**, 129–41.

Fritsch, M. (2003), 'Does R&D-cooperation behaviour differ between regions?', *Industry and Innovation*, **10**, 25–39.

Fritsch, M. and Franke, G. (2004), 'Innovation, regional knowledge spillovers and R&D cooperation', *Research Policy*, **33**, 245–55.

Fuchs, G. and Krauss, G. (2003), 'Biotechnology in comparative perspective', in G. Fuchs (ed.), *Biotechnology in Comparative Perspective*, London: Routledge, pp. 1–13.

Garmbardella, A. (1995), *Science and Innovation*, Cambridge: Cambridge University Press.

Hagedoorn, J. (2002), 'Inter-firm R&D partnerships: an overview of major trends and patterns since 1960', *Research Policy*, **31**, 477–92.

Hall, P. and Soskice, D. (2001), *Varieties of Capitalism: the Historical Foundations of Comparative Advantage*, Oxford: Oxford University Press.

Jaffe, A. (1989), 'The real effects of academic research', *American Economic Review*, **79**, 957–70.

Kaufmann, A. and Tödtling, F. (2001), 'Science-industry interaction in the process of innovation: the importance of boundary-crossing between systems', *Research Policy*, **30**, 791–804.

Kaufmann, A., Lehner, P. and Tödtling, F. (2003), 'Effects of the Internet on the spatial structure of innovation networks', *Information Economics and Policy*, **15**, 402–24.

Keeble, D. (2000), 'Collective learning process in European high-technology milieux', in D. Keeble and F. Wilkinson (eds), *High-Technology Clusters, Networking and Collective Learning in Europe*, Aldershot: Ashgate, pp. 199–229.

Keeble, D. and Wilkinson, F. (eds) (2000), *High-Technology Clusters, Networking and Collective Learning in Europe*, Aldershot: Ashgate.

Malmberg, A. and Maskell, P. (2002), 'The elusive concept of localization economies: towards a knowledge-based theory of spatial clustering', *Environment and Planning A*, **34**, 429–49.

Maskell, P., Bathelt, H. and Malmberg, A. (2004), 'Temporary clusters and knowledge creation: the effects of international trade fairs, conventions and other professional gatherings', SPACES 2004-04, Philipps-University of Marburg, Germany.

Mayerhofer, P. (2006), *Wien in einer erweiterten Union – Ökonomische Effekte der Ostintegration*, Wien: LIT Verlag.

McKelvey, M., Alm, H. and Riccaboni, M. (2003), 'Does co-location matter for formal knowledge collaboration in the Swedish biotechnology-pharmaceutical sector?', *Research Policy*, **32**, 483–501.

Mesch, M. (ed.) (1995), *Sozialpartnerschaft und Arbeitsbeziehungen in Europa*, Wien: Manz.

Moulaert, F. and Tödtling, F. (eds) (1995), 'The geography of advanced producer services in Europe', *Progress in Planning*, **43**, special issue.

Müller, W. (2006), 'Towards a liberal market economy? Political economy and political forces of change in Austria', proto-paper prepared for the conference 'Austria as a Mirror for Small States in the European Union', Minda de Gunzburg Centre for European Studies, Harvard University, 7 April.

Nowotny, E. (2001), 'Das Wirtschaftssystem Österreichs', in E. Neck, E. Nowotny and G. Winckler (eds), *Grundzüge der Wirtschaftspolitik Österreichs*, 3rd edition, Wien: Manz, pp. 1–44.

Organisation for Economic Co-operation and Development (OECD) (2001), *OECD Science, Technology and Industry Scoreboard: Towards a Knowledge-Based Economy*, Paris: OECD, available at www1.oecd.org/publications/e-book/ 92-2001-04-1-2987/.

Owen-Smith, J. and Powell, W. (2004), 'Knowledge networks as channels and conduits: the effects of spillovers in the Boston biotechnology community', *Organization Science*, **15**, 5–21.

Porter, M. (1998), *On Competition*, Boston, MA: Harvard Business School Press.

Powell, W. and Grodal, S. (2005), 'Networks of Innovators', in J. Fagerberg, D. Mowery and R. Nelson (eds), *The Oxford Handbook of Innovation*, Oxford: Oxford University Press, pp. 56–85.

Powell, W., Koput, K., Bowie, J. and Smith-Doerr, L. (2002), 'The spatial clustering of science and capital: accounting for biotech firm-venture capital relationships', *Regional Studies*, **36**, 291–305.

Prevezer, M. (2001), 'Ingredients in the early development of the U.S. biotechnology industry', *Small Business Economics*, **17**(1/2), 17–29.

Reiss, T., Hinze, S., Dominguez Lacasa, I., Mangematin, V., Enzing, C., van der Giessen, A., Kern, S., Senker, J., Calvert, J., Nesta, L. and Patel, P. (2003), *Efficiency of Innovation Policies in High Technology Sectors in Europe (EPOHITE), Final Report*, Brussels: European Commission.

Saxenian, A. (1994), *Regional Advantage: Culture and Competition in Silicon Valley and Route 128*, Cambridge, MA: Harvard University Press.

Schneider, H., Mahlberg, B., Lueghammer, W., Erbschwendtner, J., Schmidl, B., Polt, W., Gassler, H. and Schindler, J. (2004) *IKT in Österreich-Grundlagen als Beitrag zur IKT-Strategiedebatte*, IWI, Joanneum Research, Vienna.

Smith, K. (2002), 'What is the "Knowledge Economy"? Knowledge intensity and distributed knowledge bases', United Nations University, Institute for New Technologies, Discussion Paper Series, Maastricht.

Sternberg, R. (1995), *Technologiepolitik und High-Tech Regionen – ein internationaler Vergleich*, Münster: LIT.

Sternberg, R. (2000), 'Innovation networks and regional development – evidence from the European Regional Innovation Survey (ERIS): theoretical concepts, methodological approach, empirical basis and introduction to the theme issue', *European Planning Studies*, **8**, 389–407.

Storper, M. and Harrison, B. (1991), 'Flexibility, hierarchy and regional development: the changing structure of industrial production systems and their forms of governance in the 1990s', *Research Policy*, **20**(5), 407–22.

Tödtling, F. and Traxler, J. (1995), 'The changing location of advanced producers services in Austria', *Progress in Planning*, **43**, 185–204.

Tödtling, F. and Trippl, M. (2004), 'Like phoenix from the ashes? The renewal of clusters in old industrial areas', *Urban Studies*, **41**(5/6), 1175–95.

Tödtling, F., Lehner, P. and Trippl, M. (2006), 'Innovation in knowledge intensive industries – the nature and geography of knowledge links', *European Planning Studies*, **14**(8), 1035–58.

Trippl, M. and Tödtling, F. (2007), 'Developing biotechnology clusters in non-high technology regions – the case of Austria', *Industry and Innovation*, **14**(1), 47–67.

Unger, B. (1999), 'Social partnership: anything left? The end of a dinosaur or just a midlife crisis?', in G. Bischof, A. Pelinka and F. Karlhofer (eds), *The Vranitzky Era in Austria*, New Brunswick, NJ: Transaction, pp. 106–35.

Unger, B. Giesecke, S., Rossak, S. and Oosterwijk, H. (2003a), 'Telecommunications and the Austrian Paradox', in F. van Waarden (ed.), *Bridging Ideas and Markets. National Systems of Innovation and the Organization of the Idea-Innovation Chain. Part II. Country-Sector Reports*, Final report of a project financed by the European Commission under the Fifth Framework Program (Targeted Socio-Economic Research), Utrecht, Utrecht University, pp. 24–73.

Unger, B., Oosterwijk, H. and Rossak, S. (2003b), 'Austrian biotechnology – where to find on the map?', in F. van Waarden (ed.), *Bridging Ideas and Markets. National Systems of Innovation and the Organization of the Idea: Innovation Chain. Part II. Country-Sector Reports*, Final report of a project financed by the European Commission under the Fifth Framework Program (Targeted Socio-Economic Research), Utrecht: Utrecht University, pp. 202–31.

Unger, B. and Heitzmann, K. (2003), 'The adjustment path of the Austrian welfare state: back to Bismarck?', *Journal of European Social Policy*, **13**(4), 371–87.

Van der Bellen, A. (1994), 'Sozialpartnerschaft – Ansätze einer Kritik', *Wirtschaftspolitische Blätter*, **41**(5–6), pp. 488–94.

Von Hippel, E. (1988), *The Sources of Innovation*, New York: Oxford University Press.

Zucker, L., Darby, M. and Armstrong, J. (2002), 'Commercializing knowledge: university science, knowledge capture, and firm performance in biotechnology', *Management Science*, **48**, 138–53.

9. Comparing the cases and lessons for knowledge-based sector policy

INTRODUCTION

Earlier chapters in the book have comprehensively highlighted how knowledge cannot be meaningfully analysed in isolation, hence a contextualized approach recognizing the characteristics of the regional knowledge ecosystem is necessary. A regional knowledge and innovation system has been defined as a dynamic and evolving constellation of actors shaped by the knowledge embedded in organizational systems and embodied in associated technological systems (Choo and Bontis, 2002). It has been argued that firms and research centres of expertise/excellence play a dual role within a region, both creating (or co-creating) knowledge and absorbing knowledge from outside the region. Optimizing the potential contribution to regional development of a region's knowledge stock, however, will require complementarity between the regional knowledge base and the requirements of regional firms (for example, Gunasekara, 2006). For instance, the evidence suggests that, in general terms, spillovers and productivity benefits are probably greatest from publicly funded basic research which contributes to the related public knowledge stock (for example, Guellec and van Pottelsberghe, 2004). Yet, research by Rodriguez-Pose (1999), Fernandez et al. (1996) and Jensen and Tragardh (2004) rather supports the idea that in an economy dominated by small and medium-sized firms with an intermediate technological and industrial base the returns are also greater from more applied research which is more easily absorbed by local firms (for example, Oughton et al., 2002).

To recall, we have also argued that knowledge-intensive industries make use of different types of knowledge links, distinguishable in market links, networks, spillovers and milieux; these knowledge interactions, introduced in Chapter 3, will have different spatial domains ranging from localities and regions to interregional and international domains, as the empirical chapters have shown. In Chapters 7 and 8 we also highlighted that what is typically missing from analysis of formalized R&D generated internally in the firm or university for transfer to market is the possibly greater impact upon innovative potential from accessing knowledge 'in the air' in clusters and even, in static externality terms, *agglomerations* where dynamic inter-activity is expected to be low.

The key point is that from the standpoint of its impact on regional development the nature of knowledge clearly cannot be considered in isolation. Instead, a more contextualized, systemic view is necessary, reflecting the supply of knowledge and its characteristics, the *absorptive* capabilities of knowledge users and the effectiveness of knowledge transfer/translation (for example, Cooke et al., 1997; Braczyk et al., 1998). The localized knowledge spillovers and proximity questions, which are by no means rooted in fantasy, also relate in interesting and complex ways to issues of specialization versus diversification as wellsprings of innovative capability. At issue is whether such localized knowledge spillovers as those discussed reside in the combination of effects that gives local innovation proactivity to the region or locale itself. Or does the locale derive its extra proximity capabilities from the combined effects of the firms that typify it? The latter is argued by Breschi and Lissoni (2001), who see no convincing evidence that non-pecuniary spillovers, the apotheosis of localized knowledge spillovers, have displaced Marshall's classic definition of 'external economies' in terms of pecuniary advantages.

For those seeking explanations for clustering and, at a higher scale, the formation of regional innovation systems, the existence of and contribution to an entrepreneurial social infrastructure of *localized knowledge spillovers* is a priority interest. To recall, as we said in Chapter 4, this is an area of hot debate, the outlines of which were explored by Caniëls and Romijn (2003) and the proponents of the view that proximity offers innovation advantages beginning in relatively recent times with Jaffe et al. (1993). Chapter 4 reviewed this debate, however here it is our aim to stress that R&D often constitutes a public good in locations where it concentrates. Furthermore, the 'market' dimension of such growth impulses may be overstated since, if free markets in such knowledge really operated, proximity would not be necessary. This also reinforces our hypotheses that clusters are more like 'clubs' than commodity exchanges. Thus the Caniëls/Romijn critique is that most 'knowledge spillover' perspectives attribute too much influence to regional *milieu* and too little to *knowledge entrepreneurship*. Clusters may be expected to have 'knowledge entrepreneurs', possibly in abundance, whereas 'institutional' regional innovation systems often try, through regional development agencies, to substitute for them, not necessarily with the appropriate expertise. In the course of the book it has also become clear that the idea of regional learning alone is an inadequate way of evoking regional economic development because of selection problems such as the composition fallacy, utilizing samples of one, learning legacies and delays and impossibility of repeat experience. So, necessarily, to offset regional imbalances responsible agencies have to explore solutions endogenously in greater measure. This means constructing regional advantage, not an easy thing to do, by integrating and exploiting a range of assets from economic strengths to knowledge assets, good governance and creativity. This means, of course, stepping further than firms alone in the analysis, to include epistemic communities that are *professionally* distinctive associations, gaining identity and economic status precisely from their distinctiveness. Good epistemic boundary crossing practices (for example, from science or culture to commercial output) are key to innovation and commercialization processes.

'Related variety' as discussed by Boschma (2005) assists this. He offers good evidence that key to innovation-related regional growth are economies possessing 'related variety' of the kind that on a micro-scale is narrowly expressed in clusters. As we see later in the chapter, building on that, we hypothesize regional economies that operate with related variety will take the form of 'economy platforms'. For example a regional economy with key expertise in sensors or software will, with related variety, be capable of diffusing such capabilities into numerous potential applications *inside the region* as well as beyond. Unlike the old days when regional economic 'diversification' was advocated, actually as a *defensive* means to insure against decline of old sectors, the 'related variety' thesis represents a modern, growth-oriented 'economy platform' that may forge ahead 'offensively' by rapid diffusion of innovation through related-variety channels. Of great importance in this is seeking to promote 'related variety' among economic activities. Single innovations diffuse swiftly across technology 'platforms' into related industries because absorptive capacity is high among them. The key methodology in constructing regional advantage is designing appropriate policy platforms that mix variable policy instruments in an integrated and judicious manner. This chapter maps out a theoretical approach as a means of understanding 'policy fitness' in the knowledge economy era enabling development to be accomplished. In the context of the book, our aim is to compare the two liberal and coordinated market economies of the UK and Austria, in respect of the high-technology industries studied, and to elicit broad policy implications and lessons learnt. Thereafter, Chapter 10 penetrates deeply into methodologies and exemplars, suggesting how this new 'post-sectoral' approach works in diverse contexts.

LIBERAL KBS MARKETS: SWIFT UPSIDE AND DOWNSIDE

The UK, as argued in Chapter 2 and Chapter 7, has become more and more of a knowledge economy, its strength being mainly the knowledge-intensive services, biopharmaceuticals and the ICT producing and using services. It also has a high share of highly qualified scientific and technical workers, and a comparatively good venture capital base. Two somewhat distinctive geographies appeared in Chapter 7 when analysing location quotients for the ICT sub-sectors: a concentration of manufacturing in peripheral regions, on the one hand, and a concentration of services and R&D in the South East, Eastern region and London, on the other. Information and communication technology concentration is found around London, university towns such as Oxford and Cambridge, and extends towards the M4 corridor, all locations that are within reach of urban areas and main transport facilities.

Biotechnology is the UK's fastest growing industry: in 2004 the UK also had some 200 'pipeline' products in various trialling stages. A high proportion of the UK's DBFs are located within a 50-mile radius of London but, unlike Vienna, most are concentrated in university towns like Cambridge and Oxford, and in the Guildford area. The university towns host what are officially denoted 'biotechnology clusters', whereas Guildford is less well known as a 'biotechnology agglomeration'. Whereas leading edge bioscientific research and talent support 'academic entrepreneurship' in the university towns, Guildford is a case of 'localization economies' spatial concentration due mainly to experienced technical labour and excellent transportation links to London and internationally.

The key characteristics of the UK biotechnology sector can be summarized as follow:

- Although geographical clustering is pronounced, distinctive kinds of interactivity by firms exist. This involves a combination of looser *collaboration* activities and tighter *cooperation* activities.
- This apparent paradox involves looser yet proximate collaboration for *knowledge* access and tighter but more distant cooperation interactivity for *innovation*. Thus biotechnology firms make clear distinctions between knowledge and innovation and the latter occurs most in partnership with global partners.
- The most important reason given by UK firms for locating in proximity is access to university research, followed by talent and other specific services expertise required in genomics. This is significantly more important than proximity to other kinds of R&D (for example, private or foundation) or other business transaction relations.
- *Regional* universities ranked high but not highest among the location of universities' collaboration; research links elsewhere in the UK came first, and in third position were R&D links in North America. Low R&D links are found from UK cluster firms to collaborators in the EU but relatively higher links measured in numbers of R&D

collaborators who are customers for the research, suppliers of inputs for the research, or competitors of the R&D practising firm.

● United Kingdom firms either have few direct competitors in their *region*, or they do not collaborate significantly with them if they are present. Contrariwise, they collaborate substantially with competitors elsewhere. The fact that numerous firms have collaborations with competitors in the UK may suggest the former interpretation is more likely. If so, it casts an interesting sidelight upon the 'spatial knowledge capabilities' thesis. That is, it indicates biotechnology firms have no desire to conduct R&D with local competitors because they already know its likely content due to 'open science' and due to localized knowledge spillovers among firms competing in highly specific local niches (Caniëls and Romijn, 2003; Owen-Smith and Powell, 2004).

The UK ICT research showed interesting similarities with the UK biotechnology sector. Collaboration in general is remarkably and similarly high for both UK ICT and biotechnology. Specifically cooperating on innovation is equally high while cooperating on innovation in the home region is comparably low, as it is in the EU and with global collaborations. However, cooperating nationally is much higher in UK ICT than biotechnology. Notably clustering, even with informal collaborators, is also much lower for ICT than biotechnology. Thus, on key *proximity* indicators for ICT we have a picture of a large majority of firms cooperating on *innovation* with other domestic but not particularly local firms and organizations. While for biotechnology, proximity to universities for accessing *research* was the first imperative, for ICT universities are ranked medium as 'proximity partners'. Geographic proximity to 'customers' ranks lowest in biotechnology but highest for ICT.

Thus a picture is relatively easily formed of ICT and biotechnology having rather opposite rationales for proximate interaction in research and innovation. Whereas biotechnology firms cluster around universities and other public laboratories for research knowledge and related interactions, meanwhile interacting distantly with customers many of which are pharmaceuticals transnationals, ICT firms prefer to cluster close to customer firms, keeping research at a distance. This is an original finding for both industries and tells us much about the nature of and differences between them relating to distinctive innovation and knowledge-transfer practices.

Finally, observing ICT *R&D* collaboration practices we can argue that, for ICT, the picture of R&D collaboration is significantly more national in orientation. Thus most UK ICT collaboration in R&D occurs nationally, with the host region some way behind, but relatively more engaged for universities and consultants. 'Suppliers' and 'competitors' are relatively important to ICT R&D collaboration in both the EU and North America, as indeed are customers. Europe is also the location where some collaboration with other public R&D laboratories takes place. Thus a picture forms of UK ICT firms engaged somewhat in transatlantic supply chains bolstered by UK and fewer regional R&D collaborations with a wide range of support actors.

Hence, it can be postulated that ICT and biotechnology clustering in the UK are driven by different imperatives – research for biotechnology, innovation for ICT – which also sees firms intimately involved in interacting collaboratively with customer firms with whom they engage for purposes of conducting 'open innovation' and/or 'R&D outsourcing' kinds of collaboration. Finally, a further elaboration is a greater valuation by UK biotechnology firms of *functional* or *relational* over *geographical* proximity for innovation through distant networks.

As more fully elaborated in Chapter 7, these results stress the distinctive features of clustering by UK ICT and healthcare biotechnology firms. The 'knowledge spillover' results, largely for collaborators, tend to support the 'club' rather than the pure market aspect of clustering, into which non-collaborators seek entry for reasons that were not obvious initially but which turn out to yield much of importance for theory and practice. They also point to important advantage being accrued by firms from related 'platform' variety. Performance indicator survey data on firms also show that firms that collaborate perform better on all performance indicators than firms that do not. Collaboration thus gives to firms in these industries an added competitive advantage. Second, geographical (cluster) proximity for non-collaborator UK ICT firms is important, as shown by the number of these that consciously co-locate in a cluster (56 per cent; this is also found in biotech where the proportion of non-collaborating firms located in clusters is 55 per cent). This appears to arise from ICT and biotechnology firms' conscious decision to access knowledge spillovers from the interaction effects and knowledge 'free-riding' opportunities available to firms within earshot of other incumbents with whom they have no intention of collaborating. However, as shown in Chapter 7, the 'cluster effects' may help innovation but may also have some performance-hindering characteristics.

The possibility that recruitment of talent is an element of knowledge spillover advantages being sought by such firms had also to be taken into account. However, illumination of the precise nature of knowledge spillovers and their variation in these two key innovation-inducing sectors proved to be relatively mundane. In interview, the attraction of clusters to such 'isolates' was 'in the air' knowledge about competitive tendering opportunities and, if needed, talent-spotting. These are relatively static 'localization' knowledge spillovers or externalities. The knowledge quasi-monopoly or 'club' hypothesis about the cause of clustering thus gains further credence, since not being in the cluster, even as a 'freeloader', equates to potentially lethally low absorptive capacity. One notable feature of free-riders is their significantly smaller than average employment size as non-collaborators in clusters, which may even suggest accessing talent is not their primary interest in comparison to picking up tendering gossip. The 'constructed advantage' of the knowledgeable cluster thus derives from its local linkages and conveys degrees of competitive advantage directly and indirectly to its collaborators and non-collaborators alike.

Bringing our attention to policy support, as highlighted in Chapter 7, we can argue that regions such as Eastern and South East England are mainly characterized by entrepreneurial competitiveness which had little influence in the past from regional public-led support. Nevertheless, the Eastern and South East regional development agencies are currently becoming more active players in promoting regional advantage. On the other hand, as also highlighted, strong regional agencies in Wales and Scotland became the main vehicles for managing policies in support of innovation, including coping with the demands of the knowledge economy, yet with different outcomes. Contrariwise, London showed that, despite its strong entrepreneurial tradition, problems of scale and fragmentation in the knowledge value-chain are approached in more traditional enterprise support arrangements that favour a sector-oriented approach, focusing mainly on the London financial cluster. It follows that these imperatives to further innovation and entrepreneurship can have profound influences on policy as we shall see later in the chapter.

COORDINATED MARKETS: SLOW UPSIDE BUT SOFT DOWNSIDE

The coordinated market economy of Austria has been regarded as a late comer vis-à-vis the knowledge economy. Nonetheless, Austria is showing signs of a catching-up process. On the one hand, despite the fact that Austria has a far smaller high-technology and knowledge-intensive services industry than the UK, it displays a good productivity performance resulting to some extent from ICT use in a variety of industries. On the other hand, in recent years, both knowledge-intensive services and high-technology manufacturing have displayed an above average employment growth in a broader EU comparison, as argued in Chapters 2 and 8.

To a large extent, ICT activities are considered an urban phenomenon, with stronger concentration in the Vienna region; ICT manufacturing[1] concentration is also found in other urban regions such as Graz, Leoben and Villach. Concentration of ICT manufacturing is also found in a number of rural and more peripheral districts such as Southern Styria, Lower Austria or Tyrol – mainly due to the presence of subsidiaries of larger firms such as Siemens, Infineon and Biochemie Kundull, to name but a few. Concentration of ICT services, on the contrary, is exclusively found in the cores of major cities, mainly due to the presence of financial services such as banking and insurance. Research and development firms concentrate in Vienna and its surroundings to the south, the urban regions of Graz and other provincial centres, thus locations that are within reach of urban areas, which favour access to customers as well as knowledge spillovers and networks. Whereas in ICT manufacturing we observe a de-concentration from Vienna towards the other urban areas, R&D is following a reverse pattern, in which Vienna has further strengthened its position, showing that a process of specialization is taking place in the Vienna region in favour of ICT services and R&D. The ICT firms surveyed are of a relatively larger size than those surveyed in the UK (mean employment and turnover being respectively 233 and $69 million in Austria and 89 and $26 million in the UK) but tend to spend less in R&D (13 per cent in Austria compared with 16 per cent in the UK). Austria, as highlighted, has a distinctive science and technology infrastructure and associated career path for scientists. The key role is played by the Austrian Research Centre network of national technological institutes, where research is conducted and knowledge transferred to firms. Other research is conducted in firms and universities. For industry-related research, Austria's technical universities are key. Frequently, these also have technology parks and business start-up - programmes.

Similar to the UK, ICT manufacturing and R&D firms tend to have more patenting activities and it is also these groups of firms that consider universities a more important knowledge source (as shown in Table 8.12 in Chapter 8). It can be concluded that the ICT sector in Austria includes a mix of various activities relying both on analytic and synthetic knowledge and shows a propensity to produce innovations that are both incremental and new for the company and the market. This indicates the importance of innovation in the ICT industries and its rapid pace of technological change. Information and communication technology firms rely less on commercial or non-profit R&D, but more on customers and competitors. Service firms are also important, although less important than they are for the UK ICT firms. This pattern of knowledge sources is more along and within the value chain as argued for ICT in the UK; however, Austrian universities are

Table 9.1   Importance of ICT knowledge sources (% of firm responses)

	Austria	UK
Own firm	98.7	96.9
Group	46.6	51.4
Customer	88.2	88.5
Supplier	63.6	67.5
Competitor	55.3	55.0
Service firm	28.9	34.0
Commercial R&D	19.7	15.9
University	32.9	25.5
Non-profit R&D	10.7	10.1
Technology transfer centre	14.9	7.4

Note: Percentage of firms rating knowledge source as important or very important.

considered more important knowledge sources than the UK universities (as shown in Table 9.1). The regional level plays an important role when the location of knowledge sources is considered. This is particularly true for knowledge sources such as universities, suppliers, public R&D and to some extent customers (see Tables 8.13 and 8.14, Chapter 8). Knowledge-flows from clients, suppliers and competitors are highly internationalized.

When comparative data are juxtaposed for Austria and the UK, they reveal that the regional and national levels play a relatively more important role in Austria. This holds true in particular at local level for knowledge sources coming from universities and other R&D; within the UK, relatively more important regional knowledge sources were local competitors and consultants as shown in Table 9.2. Presumably, for reasons of smaller economy scale, Austrian firms rely more on knowledge coming from other European firms and institutions. Reinforcing this, European value-chain relationships are relatively more important than national ones. This depicts Austrian firms being relatively more open than the UK to Europe and North America but variably so to Asia and the rest of the world.

Austrian ICT firms are more engaged in opening up external knowledge sources and in international networking, a result which may partly be due to its lower internal R&D and knowledge-generating capabilities (this is also shown in Chapter 2 for Austrian companies in general). As a consequence the Austrian economy developed a tradition of fast and intelligent search and selection followed by acquisition and adoption of external knowledge. In part this is probably also due to the coordinated market business model which, as we have seen, has been challenged and in retreat recently, but is far from being abolished.

Table 9.2   Location of ICT knowledge sources (% of firm responses)

	Local/ region	National UK (AU)	EU	North America	Asia	Rest of World
University	15.4	32.7	8.8	5.9	1.5	1.5
	(21.4)	(53.6)	(39.3)	(14.3)	(1.8)	(0.0)
Consultant	18.0	39.3	8.1	4.8	1.5	2.6
	(14.3)	(41.1)	(26.8)	(7.1)	(0.0)	(0.0)
Supplier	14.3	61.8	23.2	17.3	10.3	8.1
	(17.9)	(44.6)	(57.1)	(23.2)	(5.4)	(3.6)
Other R&D (public)	7.7	21.7	4.4	1.1	0.7	1.1
	(14.3)	(21.4)	(19.6)	(7.1)	(1.8)	(0.0)
Customer	17.3	62.1	36.0	26.8	17.6	18.4
	(16.1)	(44.6)	(51.8)	(26.8)	(19.6)	(8.9)
Competitor	11.8	50.4	29.4	26.1	9.9	12.5
	(8.9)	(21.4)	(48.2)	(39.3)	(10.7)	(3.6)

Note: Figures in brackets relate to the Austrian sample.

Industry coordination remains grounded in many institutions (collaborative approaches between state, firms and knowledge providers such as competence centres, CD laboratories, and so on) and emphasizes networking among relevant actors from business, state and research. Furthermore, the result that universities, research organizations and technology transfer are more important (while value-chain and service links are less important in comparison to the UK) may be due to the smaller and less complete clusters in Austria and its less sophisticated ICT industry. Public and non-profit organizations (universities, research organizations, technology transfer) take up some of the activities which are provided by firms in the UK system.

Austrian ICT firms get their knowledge sources through a variety of market links, spillovers and milieu effects in the innovation process. However, there are significant differences in the frequency and importance of channels used by sectors: research contracts and reading of scientific literature and patents are most commonly used in research firms, conferences and fairs mainly by services industries, while manufacturing ICTs tend to rely more on consulting. Most firms, however, consider informal contacts important for knowledge exchange (Table 9.3). A similar trend is found in UK firms. However, interestingly, the UK ICT firms predominately use informal contacts as preferable knowledge source channels and less so market links and formal networks. Also employment, although regarded as important, is used less frequently due to the small size of the respondent firms.

Table 9.3　Knowledge transfer channels in ICT (%)

	Austria	UK
Employment	58.8	29.3
Intermediate goods	66.3	29.3
Literature, patents	62.5	51.2
Conferences, fairs	73.8	51.2
Informal contacts	72.5	80.5
Licences	32.5	21.9
Consulting	58.8	46.3
Contract research	40.0	21.9
Research cooperation	42.5	24.4
Shared use of R&D facilities	30.0	9.8
Firm takeover	20.0	4.9

Austrian ICT firms tend to cooperate significantly in the innovation process, as distinct from knowledge transfer; however not as much as their UK counterparts. In the UK 73 per cent of firms conduct innovation activities in cooperation with other firms or institutes (in Austria the percentage of cooperating firms varies from 49 to 80, averaging 65 per cent).

Cooperation, as argued, is regarded as an important tool for accessing complementary knowledge, distribution channels or other innovation resources. Austrian ICT firm cooperation is less often undertaken to develop a product simply new to the firm or to develop a patent, but more to develop an innovative product both new to the company and the market (Table 9.4). A similar trend is also found in the UK, although the main reasons why firms cooperate on innovation activities is to access a complementary platform of technologies and knowledge to develop products that are new to the market and the firm.

Moving now to compare the biotechnology sector in the two countries, previous chapters have shown that Austria has to be characterized as a latecomer in the commercialization of biotechnology. There is a long-term presence of international pharmaceutical companies as well as early spin-offs but the Austrian dilemma has its roots in weak incentives and conditions for commercializing research, and a lack of tradition and culture for risk-taking. However, a catching-up process can be observed, mainly supported by policy efforts, which have intensified over the past decade. At the national level several initiatives like the 'Impulsprogramm Biotechnologie' or the 'Genome Research Programme' have been launched and the regions of Vienna, Styria and Tyrol have recently started to promote local biotechnology clusters. Hence, a change in attitude in public research organizations, partly endorsed by new policy programmes promoting the formation

Table 9.4   Objective of ICT R&D cooperation (% firm respondents)

	Austria	UK
Improvement of existing product	75.0	75.8
Innovation, new to the firm	71.4	64.0
Innovation, new to the market	92.9	81.1
Patent development	16.0	19.0
Entering new technical fields	82.1	85.9

of enterprises, has also brought a gradual rise of academic entrepreneurship within the sector.

Austria specializes in medical biotechnology; the sector is unevenly distributed in space featuring a strong spatial concentration in four centres that include Vienna and the regions of Lower Austria, Styria and Tyrol. Among these, Vienna clearly has the lead with more than 60 per cent of biotechnology firms being located in the area. Innovation in the Vienna biotechnology cluster is neither a result of informal and local milieu effects nor is it predominately shaped by formal global market links and networks. It is more the result of a complex interplay of those various types of relationships both at local and global scales. The regions of Styria and Tyrol, on the other hand, have younger and smaller biotechnology agglomerations than Vienna and, as a consequence, less local but more external knowledge links.

The national level plays only a minor role, owing to the strong concentration of Austria's biotechnology industry in Vienna and its relative youth, which reduces the likelihood of intense networking. Within even this cluster, there is a lack of managerial know-how. Academic entrepreneurs, who are gradually rising in importance, often have limited managerial competencies, and top managers with experience in the field of biopharmaceuticals are in short supply and mainly recruited in the international labour market. As there is low interfirm mobility within the cluster and within the Austrian economy in general, this essential type of knowledge-flow is not prominent. The significance of the region results to a considerable extent from a rather intense informal networking between some local companies and research organizations within the cluster. Personal relationships and trust among local actors have been supported by policy actions, such as the organization of 'Life Science Circles' and other meetings which have brought local companies together, stimulating an informal exchange of ideas and experiences. The Vienna cluster illustrates the importance of non-local connections as source of innovativeness in the industry: international universities, clinics and other research institutes are important partners. In order to access top scientific knowledge and expertise some local biotechnology firms have forged cooperative relations with internationally renowned knowledge centres like Harvard Medical School and the Scripps Institute in the USA or the Karolinska Institute in Sweden.

Furthermore, biotechnology firms in the Vienna region are also inserted into various innovation networks and R&D cooperations with foreign multinational pharmaceutical companies, such as Merck, Aventis or Johnson & Johnson, and young biotechnology firms. The case of the Vienna biotechnology cluster shows that knowledge spillovers are only partially geographically bounded. Although spillovers have some relevance, a considerable body of expertise used by the local companies in their innovation process is not untraded or 'in the air'. A lot of knowledge is generated and exchanged in an interactive way in a formal, more 'closed' network in the cluster. For biotechnology in Austria formal cooperations and spillovers (Bathelt et al., 2004) play a role at both the local and global level. Austrian and, in particular, Viennese biotechnology firms are not just benefiting from spillovers and milieu effects, but are also engaged in formal regional cooperations and networks. This seems to be partly due to the nature of the biotech industry and Zucker et al.'s (1998) finding that too much is at stake in this industry and for this reason knowledge is 'not freely in the air'. The stronger role of formal regional networks *in Vienna* is partly also due to some of the networking programmes such as the competence centres (K-plus, K-ind, K-net), the Christian Doppler laboratories and the Lisa and GENAU (Genome Research Austria: see Chapter 8) initiatives. Furthermore biotechnology firms in Austria maintain and benefit from international informal knowledge links in addition to formal R&D networks and other cooperations. This reflects the need of actors in young and small clusters with relatively small internal knowledge bases to access international knowledge hubs and meeting places of this industry (conferences, fairs, and so on). This need seems to be smaller for UK firms given its own strong knowledge base.

Earlier chapters have highlighted two major weaknesses affecting the Austrian knowledge and innovation system: on the one hand, Austria lacks sufficient higher qualified, scientific and technical workers; on the other hand, another important input missing for moving towards a knowledge economy is the weakness of the venture capital sector. Austria compares relatively poorly with other European countries and the USA in this respect and venture capital for the expansion phase is a rarity with companies having to deal with more traditional credit instruments, which are by no means risk adverse. Austria demonstrates a high network propensity among firms and knowledge institutions. However a problem that emerges is that a region such as Styria, which has excellent technical universities, producing numerous start-ups from its graduating PhD candidates, housing them in technology parks and linking them through collaborative activities connected to development agency cluster programmes, produces innovators that are often focused in medium-technology rather than high-technology sectors.

POLICY FOR EUROPEAN KBS: SWIFT UPSIDE AND SOFT DOWNSIDE

The research analysed within the book shows differences and similarities in the knowledge exploration to innovation exploitation process in the distinctive industry platforms of biotechnology and ICT. It also highlights the varying governance contexts that exist between the two countries and their knowledge and technology-intensive regions. We argued that the UK is a markedly 'European' variety of liberal market institutional capitalism with its weak commercialization performance, while Austria is moving away from its stylization as a coordinated market model. Both show solid regional institutional support for innovation, yet differ in policies and outcomes as we have seen.

Before comparing differences in policies, it is necessary to introduce the new concept we briefly hinted at earlier, namely, the idea of 'platform policies'. In the past, if policies were not regional, hence redistributive, to places of low accomplishment, they were sectoral and aiming to support lowly accomplished industries. Sometimes the two were combined in an attempt to diversify a failing regional economy by stimulating different industries to enter the regional space. But this was always a defensive measure, taken to respond to the decline of once dominant regional industry. Accordingly, it undermined regional 'related variety' among industries as well as regional 'absorptive capacity' because of cognitive dissonance between sectors and skills. Nowadays, however, it has become clear that the more accomplished regional economies tend to possess 'related variety' in their economic structures. We might think of the design-intensive clusters of north-central Italy, the related variety of semiconductors, microprocessors, computers, software, search engines, bioinformatics and biotechnology in northern California's Silicon Valley, or the varieties of printing machinery, machine tools, automotive and electronics engineering industries in southern Germany as exemplars of related variety. On a smaller scale, similar 'platform technology' industries are found in Cambridge, Massachusetts, and Cambridge, UK, or Leuven, Belgium, and Rehovot, Israel, each of which combines numerous strands of ICT and biotechnology with world-class research institutes or universities such as Harvard,

MIT, Cambridge University, KU Leuven or the Weizmann Institute at its cluster heart.

If this industry and technology 'platforms' analysis is accepted as a reasonable proposition about the conditions of contemporary growth, it has implications for two other dimensions of 'platform' thinking. First, it involves 'stakeholder platforms' and, second, it involves policy platforms. This arises from the key feature of industry platforms, which is that a single significant innovation may travel swiftly across the platform, earning surplus profits as it is applied to different but related technologies by 'communities of practice' sharing a high degree of 'absorptive capacity' across the platform strands. Of course, as in network theory more broadly, innovations tend to get 'repaid' by other platform actors and the whole becomes much more than the sum of the parts as innovations cascade through the platform over the years. Innovation productivity and returns remain higher than normal under such circumstances. At the micro-level this is one of the mechanisms sustaining growth by clusters: at the meso-level it is capable of being sustained across a broader range of related variety in regional innovation systems. Incorporating good governance and creativity into the brew as the following example illustrates, means that constructed regional advantage has been achieved:

> when the system into which a learning and change disposition is to be integrated is an externalized inter-organizational one rather than a corporation, the task is more difficult. Some attention is therefore devoted first to US practices in building 'economic communities'. Henton *et al.* (1997) argue for strong leadership to maximize, for profit, social capital effects and cite cases like the latest revival in Silicon Valley's fortunes through the association called Joint Venture Silicon Valley, the revival of Cleveland and the transformation of Austin, Texas. Austin's evolution from sleepy campus and government town to one of America's leading New Economy clusters is presented, as are the others, as cases of learning clusters based on cooperative advantage. The *New York Times* eulogized the place in January 1999 as:
>
> > '. . . at once the least Texan and the most Texan of cities, with a burgeoning hi-tech industry, a University population of over 50,000, the endless carnival of Texas statehouse politics, and a music and restaurant scene that would be envied by a city twice Austin's size. Austin is one of those cities like Seattle and Santa Fe that gets so much praise you wish you could hate it.'
>
> Economic growth was running at 9–10 per cent annually in the 1990s and 30,000 new jobs were being added each year, some 200 start-up technology companies were founded there each year and it had a relatively recently established specialist business support system, largely private in origin. This was propelled forward by the actions of community entrepreneurs, particularly George Kozmetsky, founder of the Austin economic model, known subsequently as the 'triple helix' of government, industry and university interaction (Etzkowitz and Leydesdorff, 1997). This led to a successful cluster policy in which all three

actors connected also to entrepreneurs, sources of technology, venture capital and an innovation infrastructure of lawyers, accountants and business incubators. The policy key is leadership on each part of a strategy to retain and attract existing business and to grow new ones, and each leader is drawn from the 'economic community'. Luck plays a part, but even bad luck can be parlayed into future benefits. Thus the University of Texas at Austin is the second richest after Harvard in the USA, hence the world. But it started with a handicap in that it could not be a federal land grant university as Texas contained no federal land in the 1870s when it was established. Instead it received 5,000 acres of cotton and forest property, plus building plots. So successful was the cotton crop in the first year that the South Texas legislature re-appropriated the land and gave the university in perpetuity 2.2 million acres of wasteland in exchange. Not a very auspicious start, except that beneath the wasteland were discovered enormous resources of oil. Now the University of Texas' endowment stands at $7 billion. Some of this windfall was used to attract 'anchor' projects like the semiconductor research consortia Sematech and MCC, which gave Austin a local high technology edge. This attracted 3M, Dell, IBM and Motorola to set up applied research facilities. Dell's just-in-time production system means that, as well as its 20,000 Austin employees, it sustains an equivalent number in its co-located supply chain. But New Economy clusters are not immunized against economic downturns as Dell's cut of 1,500 jobs in 2000, and expectations of lower rates of city job growth 2000–2002, down to 3.7 per cent from 5.2 per cent signify. (Cooke, 2002: 14–15)

We attempt to signify these relationships between stakeholder platforms, industry platforms and policy platforms in Figure. 9.1. In Chapter 10 the policy methodologies and configurations for constructing regional advantage are delineated. Thus far, policy experience has evolved first with 'stakeholder platforms'. That is, ideas of public-private partnership governance of innovation policy have become commonplace. Moreover, industry experience is even further evolved, notably through ideas such as 'technology platforms' as in biotechnology, sensor and software support initiatives; 'policy platforms' are almost unheard of. Nevertheless they exist, and in both the vertical and lateral dimensions. In the vertical, we can say most countries and some regions have a series of linked support policies for firms moving from the exploration through the examination to the exploitation phases of their evolution. They are exemplified in descriptive accounts of various pre-competitive supports as 'proof of concept' and 'seedcorn' funds through to 'research link' and 'innovative idea' grants to 'collaborative R&D' and 'knowledge transfer networks' subsidies. It is noticeable that these are often front-loaded at the pre-competitive end of the firm's evolutionary cycle – one reason why commercialization is often so difficult for firms facing the 'valley of death' afterwards. Regionally constructed advantage should assist in addressing this by the evolution of more lateral, interactive innovation support among 'related variety' investors, who thereby spread their risk by means of 'related portfolio' investments

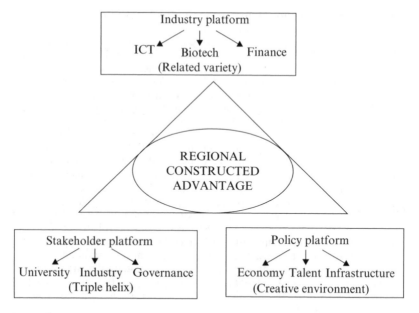

Figure 9.1   Regional constructed advantage: industry, stakeholder and
*   policy platforms*

boosted by the surpluses earned from single, swiftly diffusing innovations. This is good practice as conducted by leading venture capitalists
(Cooke, 2001).

What Figure 9.1 means is that as an aspiration a regional innovation
system integrates its industries, stakeholders and policies around ensuring
as far as possible regional advantage is created by judicious support where
required. A geographical information system (GIS)-based external knowledge management system such as those deployed in Scottish Enterprise,
Norway's SIFA knowledge support system and Sweden's VINNOVA
innovation systems agency would be inescapable adjuncts of this regional
policy approach. Following the model of 'precision farming' (Pedersen and
Pedersen, 2006) where Global Positioning System (GPS) satellite naviga-
tion interacts with an on-board GIS laptop to inform the tractor to broad-
cast exactly the required density of fertilizer and seed according to
micro-variations in natural soil humidity and fertility, policy resources
would be similarly disbursed.

Accordingly, regional advantage accrues from the precision application
of supportive inputs designed to optimize efficiency, while effective outcomes accrue from capabilities in swift germination and diffusion of innovations through related variety platforms with high absorptive capacity.

HOW TO ACHIEVE THAT HAPPY STATE?

In the introductory chapter, we argued that countries that resemble coordinated market economies operate a 'social partnership' model of economic organization in which interaction occurs by negotiation across boundaries between otherwise divergent interests associated with government, industry and labour. At a lower scale than the national economy, this resonates with the 'networking propensity' associated with regional milieux of innovative small firms. Contrariwise, a liberal market regime of economic governance might be hypothesized to have few of these milieu effects, being more associated with economic individualism and a competitive ethos than the 'networking propensity'. As highlighted for the Austrian and UK economies, both are engaging in more collaborative activities and firms' networking propensity is high in both countries, contributing to enhanced competitive advantage and firm performance.

We have also argued that the two models, coordinated and liberal, are to some extent converging. As the research has shown, where there is convergence this occurs in response to the necessities and constraints for firms based on the demand for continuous innovation due to a constant increase in global competition. This shows in the following three ways. First, innovation demands more use of a variety of knowledge types, ranging from research to commercialization. This forces firms to be more open in the search process, different from the more secretive corporate knowledge management of the past. This applies equally in both liberal and coordinated market. Second, firms have to be more willing to engage in selling their products to larger markets, involving more export-oriented activities. Finally, these two input and output imperatives, stimulating and stimulated by the innovation impulse, mean that there is a tendency towards convergence in industrial organization, with regional related variety of industry becoming more prominent again. Variety persists among countries in their growing capacity to engage in networking of many and diverse kinds. More to the point, we found variety in economic governance and entrepreneurship models between regions within countries, as with say, Scotland compared with Eastern England.

Thus against notions of global 'flattening' of innovation opportunity, there is substantial regional imbalance in the 'economy culture' and modes of economic governance of contemporary economic activity, in both countries studied in this book. The research undertaken in the UK shows a division between the 'entrepreneurial' regions of southern England, including London, and the rest, represented in the reported research by Scotland and Wales, but aspects of their 'institutional' dependence in seeking to construct regional innovation capabilities are shared elsewhere in 'outer UK'.

Similarly, Austria shows some regional imbalances with Vienna and the other urban regions, where related varieties is at its highest, performing relatively better than rural regions; regional policy is also responsible for providing the ground to facilitate informal networking and knowledge access, and that can provide the means to overcome the problems high-lighted in the Austrian knowledge and innovation systems. We explore a type of policy platform for modern rural development in Chapter 10.

The combination of factors that underpin innovation advantages for firms are now better understood. In both countries one feature was of outstanding importance, notably, collaboration. But the ways non-collaborators may benefit from access to valuable knowledge in cluster-settings were not clearly understood, indeed scarcely broached in the literature. This was because of a misplaced notion that clusters were simply spatially concentrated markets. Our findings suggest they are spatial know-ledge quasi-monopolies. Even 'isolates' are willing to pay a high price for entry – as eavesdroppers. To the extent they are not over-specialized and contain related variety, clusters may, for example, be superior locations to cities unless the latter also have related variety. Most cities have some ele-ments of this but policy-makers may often operate in the possibly mis-guided belief that economic diversification is generally superior to specialization. Regions in the UK with clusters, at least in ICT and biotech-nology, seem to perform markedly better than those without. So the region is by no means an inadequate category for understanding innovation. To the contrary, the region may be the essential entry level for policy, since cities seldom have this policy function while the national level is too remote to capture localized nuances of the kind discussed above. On this, finally, it is arguable that the related variety of growth industries – a notion under-pinning new thinking about economic growth, queries the narrow notion of 'sector' as an instrument for policy to engage with. What may be rising is a notion, by contrast, of 'platform technologies, products or services' capable of flexibly interacting and diffusing in innovative ways into neigh-bouring 'sectors' giving some UK and Austrian regions 'constructed advantage' over newly rising competitors. This may be partly due to depth of academic knowledge, including knowledge about how to rethink the targets of policy. Thus a regional policy analysis focused on innovation might begin by examining the deep structure of a region's knowledge capa-bilities to assess the potential for constructing related variety in the regional economy. A stylized question of relevance might be: what is the key regional capability with respect to 'exploration knowledge' (from within as well as distant from the region) in basic research – chemistry, medical biology, physics or some combination of these, and what are the present or future 'exploitation knowledge' interactions of these for commercial innovation? What are their talent, investment, and global market implications? This does not mean a nullifying redistribution of excellence is called for, rather it means evolving a better – post-sectoral – methodology for more fully understanding regional capabilities and challenges for all regions, then developing appropriate integrative policies connecting talent formation, policy governance and enterprise support where they are especially needed. The full 'constructed advantage' approach also involves building 'creative regions and cities' that are good-quality cultural environments in which innovation is increasingly fostered.

CONCLUDING REMARKS

This chapter, summarizing and comparing the findings of the research conducted in Austria and the UK, has aimed at highlighting differences and similarities among the knowledge-intensive sectors of biotechnology and ICT between the two countries. Earlier chapters postulated that coordinated economies were more likely to engage in networking activities due to their 'social partnership' model and less networking propensity was expected to be found in liberal market economies. However, the research has highlighted that convergence between the two regime categories is happening to a noticeable extent in high-technology industry. The demands of higher intensity and quality of global competition and the quest for innovation are stimulating convergence. Nevertheless, as highlighted in this chapter and in many other parts of the volume, convergence is happening but not as one trend but as many and diverse kinds. Similarly, interventionist economic governance actually to promote clusters in Scotland and Styria compared favourably, with some policy feebleness in London and Vienna as cases in point. Interventionist governance is embedded in policy efforts to overcome regional imbalances. The models emerging, which from an innovation viewpoint we term the entrepreneurial and the institutional regional innovation systems, suggest that the key in constructing regional advantage is designing appropriate policy platforms that mix variable policy instruments, key stakeholders and industries in an integrated manner.

NOTE

1. To allow comparison with the UK case study, high-tech manufacturing is described as *ICT manufacturing*, knowledge-intensive services are defined as *ICT services* and technology business services are defined as *R&D*.

Empirical findings

REFERENCES

Bathelt, H., Malmberg, A. and Maskell, P. (2004), 'Clusters and knowledge: local buzz, global pipelines and the process of knowledge creation', *Progress in Human Geography*, **28**, 31–56.

Boschma, R. (2005), 'Proximity and innovation: a critical assessment', *Regional Studies*, **39**, 61–74.

Braczyk, H.-J., Cooke, P. and Heidenreich, M. (eds) (1998), *Regional Innovation Systems*, London: UCL Press.

Breschi, S. and Lissoni, F. (2001), 'Localised knowledge spillovers versus innovative milieux: knowledge "tacitness" reconsidered', *Papers in Regional Science*, **80**, 255–73.

Caniëls, M. and Romijn, H. (2003), 'Localised knowledge spillovers: the key to innovativeness in industrial clusters?', paper to conference on 'Reinventing Regions in the Global Economy', Pisa, 14–16 April. Also published in P. Cooke and A. Piccaluga (eds) (2006), *Regional Development in the Knowledge Economy*, London: Routledge.

Choo, C. and Bontis, N. (2002), *The Strategic Management of Intellectual Capital and Organizational Knowledge*, Oxford: Oxford University Press.

Cooke, P. (2001), 'Regional innovation systems, clusters and the knowledge economy', *Industrial & Corporate Change*, **10**, 945–74.

Cooke, P. (2002), *Knowledge Economies*, London: Routledge.

Cooke, P., Gomez Uranga, M. and Etxebarria, G. (1997), 'Regional systems of innovation: institutional and organisational dimensions', *Research Policy*, **26**, 475–91.

Etzkowitz, H. and Leydesdorff, L. (eds) (1997), *Universities & the Global Knowledge Economy*, London: Pinter.

Fernandez, E., Junquera, B. and Vazquez, C. (1996), 'Government support for R&D: the Spanish case', *Technovation*, **16**, 59–66.

Guellec, D. and van Pottelsberghe, B. (2004), 'From R&D to productivity growth: do the institutional settings and the source of funds matter?', *Oxford Bulletin of Economics and Statistics*, **66**, 353–78.

Gunasekara, C. (2006), 'Reframing the role of universities in the development of regional innovation systems', *Journal of Technology Transfer*, **31**(1), 101–13.

Henton, D., Melville, J. and Walesh, K. (1997), *Grassroots Leaders for a New Economy*, San Francisco, CA: Jossey-Bass.

Jaffe, A., Trajtenberg, M. and Henderson, R. (1993), 'Geographic localisation of knowledge spillovers as evidenced by patent citations', *Quarterly Journal of Economics*, **108**, 577–90.

Jensen, C. and Tragardh, B. (2004), 'Narrating the triple helix concept in weak regions: lessons from Sweden', *International Journal of Technology Management*, **27**, 515–28.

Oughton, C., Landabaso, M. and Morgan, K. (2002), 'The regional innovation paradox: innovation policy and industrial policy', *Journal of Technology Transfer*, **27**, 97–110.

Owen-Smith, J. and Powell, W. (2004), 'Knowledge networks as channels and conduits: the effects spillovers in the Boston biotechnology community', *Organization Science*, **15**, 5–21.

Pedersen, S. and Pedersen, J. (2006), 'Innovation and diffusion of site-specific crop management', in J. Sundbo, A. Gallina, G. Serin and J. Davis (eds), *Contemporary*

Management of Innovation: Are We Asking the Right Questions? Basingstoke: Palgrave Macmillan.

Rodriguez-Pose, A. (1999), 'Convergence or divergence? Types of regional responses to socio-economic change in Western Europe', *Tijdschrift voor Economische en Sociale Geografie*, **90**, 363–78.

Zucker, L., Darby, M. and Armstrong, J. (1998), 'Geographically localised knowledge: spillovers or markets?', *Economic Inquiry*, **36**, 65–86.

10. Reflections on the research and conclusions for policy

INTRODUCTION

In the course of researching and writing this book, three key matters of considerable importance have happened for our understanding of the evolutionary economic geography of high-technology development. The first of these is that, while embarking on a methodology rooted in the notion of sectors, we found they exerted less and less leverage on the reality we were describing. As some of our respondents said, sectors are a fiction. They, by contrast, were interested in how *pervasive* innovations and any associated technologies could be. Some were not particularly complimentary about the policy of cluster-building operating in some locations, seeing it – correctly – as a 'retro-model' itself rooted in narrow sectoral concerns. Academics have been critical of ideas connected with clustering in the active policy sense, for some time now (Asheim et al., 2006) critical also of the fascination they have for economic development policy-makers worldwide.

But none until now have seen with the clarity of the experienced entrepreneur who temporarily held a key intermediary role at the knowledge transfer interface between the public and private spheres, that clusters are a nineteenth-century model of industrial organization. No one denies they exist, for instance, as echoes of a certain kind of agglomeration economy recognized by Alfred Marshall as *industrial districts* but with a peculiarly Italian turbo-driven and collaborative form of small-firm interaction often, for example, in mature, luxury sectors associated with the 'Made in Italy' label. Nor that knowledge-intensive variants could be found rooted in liberal market US and UK biotechnology and ICT clusters often attached to university research. Yet industry sees clusters not as revved-up industrial districts but as *platforms* of related variety industry and technology that are sometimes in geographical proximity, but often not. So our first reflection is that policy-making had better wake up to this reality and begin tackling the less rigid conditions associated with platforms, and by way of examples we offer below some real and some stylized methodologies for the achievement of platform policies. In doing this we were forced to discover actual platform policies in practice or envisioned. This quest took us beyond the confines of the Austria–UK data and even, to an extent, beyond the ICT or biotechnology fields. But so useful was that extra search that we considered it worthwhile to share its results with readers.

The second reflection concerns the findings from the Austrian and UK results that the differences in political and economic regulation of industrial organization that is captured in the further stylization of economies into liberal-market organization (USA, UK) and coordinated market organization (Austria, Germany, Nordic countries) breaks down quite severely, especially in respect of the manner in which high-technology industry of the kind researched in this book organizes itself. This rather goes against the view of Casper et al. (2001) who conclude that coordinated market economies such as Germany face too many institutional and regulatory barriers to be able to innovate in the ways the USA regularly does. We found coordinated market Austria introducing reforms to accommodate precisely the kind of innovative firm practices that high-technology industry habitually displays in terms of networking, flexibility in the workplace, interfirm relations and seeking risk capital. Perhaps large industry, even in these 'platforms', remains highly coordinated but it seems that certain practices like outsourcing, offshoring, and divesting non-core competences nevertheless occur with accelerating frequency by German and Nordic pharmaceuticals and ICT firms. Siemens and Bayer, for example, both active in Austria though German-owned, have done all three with gusto in recent years. So it seems these platforms themselves exert pervasive effects on firms large and small. But we cannot be definitive about the former since they were not directly researched in this study, nor were large or small firms in other industries so we remain silent on that.

The third reflection arising from the research takes us back to clusters and our core question as to whether firms perform better inside them than outside them. We now know they exert a 'discreet charm', as Steiner (1998) put it, upon certain kinds of firms that solves some puzzles in the literature on geographical proximity (Rallet and Torre, 1998; Boschma, 2005). For example, it was argued by Breschi and Lissoni (2001) that clusters offered no more than pecuniary externalities, hence advantages to their incumbents. But we found two separate features that undermine that viewpoint from widely divergent angles. First, we found that collaboration for many kinds of externality, including knowledge spillovers is pronounced in these industries but, crucially, collaboration in clusters was by no means the predominant form taken by such externalities. In many cases 'distant networking' (Fontes, 2006) was common, especially in biotechnology.

Moreover it was collaboration more than clustering that was associated most positively with superior firm performance on the appropriate indicators. Second, however, somewhat to our surprise, we found evidence, confined to ICT since there are few biotechnology firms not located in clusters of some kind, that non-collaborators could show superior turnover increases than collaborators in clusters. On further inspection we put this down to two things: costs of R&D, innovation and patenting all of which were high among collaborator firms inside and outside clusters; and diseconomies of agglomeration as measured in higher land rents and wage costs in clusters than in locations some 30 kilometres away with appropriate accommodation and labour. We are driven to conclude that profitability may be inhibited for firms in clusters and that is partly because of the high cost of doing business there. However, such is the discreet charm of the cluster location and postcode that, again to our surprise and again confined to ICT firms, clusters attract numerous consciously and actively non-collaborating firms. Why do these isolates locate in clusters? To access knowledge of possible supply chain contracts and availability of 'poaching' opportunities for suitable human capital or 'talent' (Florida, 2002). Florida, it will be recalled, noted that members of his so-called 'creative class', who are found inordinately located in cluster-type cities and university towns, were unlikely to move to a new job unless they could already see the opportunity for a further possible job-move in the locality. Non-collaborator cluster firms are free-riding on knowledge of these practices. So a possible negative as well as a probable positive pecuniary advantage from locating in clusters is our estimation of the empirical situation. Do they cancel each other out? Not perfectly since even our clustering non-collaborators do less well in pecuniary terms than do ICT non-collaborators in general.

In what follows we attempt to organize these and other findings into sets of analytical and policy relevant points. We concentrate, first, on some reflections and recommendations of what seem to us to be superior policy instruments to clusters, namely, regional innovation systems, and, second, on outlining – with illustrations – what we term Innovative Policy Platform (IPP) methodologies as our contribution to more modern approaches to 'constructing regional advantage' as it is termed in the European Commission (2006) report. Because of our finding that innovative, knowledge-based industry dominated by SMEs tends to impose upon its practitioners in liberal and coordinated markets alike a considerable degree of convergence in high-technology industry organization – rooted in networking, flexibilization, quest for venture capital and interactive innovation – we have relatively little further to say on this conclusion.

REGIONAL INNOVATION SYSTEMS AND CLUSTERS

In an era when theory and policy are agreed that productivity drives growth while innovation drives productivity, the insights of the leading scholarship on innovation, which are those of the evolutionary or neo-Schumpeterian school (Lundvall, 1992; Edquist, 1997) will remain influential. From a spatial perspective rooted in evolutionary economic geography, regional innovation systems (RISs) have played and will continue to play a strategic role in promoting the innovativeness and competitiveness of regions. More is said in the following section on the problem-solving innovation the RIS approach has brought to neo-Schumpeterian economic analysis and the way it has begun to have serious impacts in national and regional policy formulation. Essentially, the RIS approach has strengthened policy by the attention it directs towards the need – perceived by policy-makers at OECD, EU-member state and regional levels – for constructing regional advantage. The regional innovation system can be thought of as the knowledge infrastructure supporting innovation in interaction with the production structure of a region. It is necessary to think in post-sectoral or 'platform' terms to capture the full flavour of this contribution.

This, fundamentally, is how the RIS approach solved problems arising in the still mostly sectorally driven analyses of the national innovation system (NIS) school. Thus, when the following two subsystems of actors are systematically engaged in interactive learning, it can be argued that a regional innovation system is in place:

- the regional production structure or knowledge exploitation subsystem, which consists mainly of firms, often displaying clustering tendencies, and
- the regional innovation support infrastructure or knowledge-generation subsystem, which consists of public and private research laboratories, universities and colleges, technology transfer agencies, vocational training organizations, etc.

The interacting knowledge generation and exploitation subsystems of an RIS are, moreover, linked to global, national and other regional systems (Cooke, 2004). From this it follows that clusters and RISs can (and often do) coexist in the same territory. But whereas the regional innovation system by definition may host several clusters, a cluster is never isomorphic with an RIS. This is usually because of scale, but more analytically an RIS governance structure is normally a formal, meso-governmental body with policy responsibility and resources for animating and facilitating system-coherence. Clusters may have some kind of 'governance' but it is likely to be informal or a private 'cluster association' rather than more formally governmental in character. This is not to say there is no role for informal, associational involvement in RIS governance, for by definition that is precisely what the governance concept captures, including institutional context (that is, norms, trust and routines) in which interactive firm and organizational learning may take place. It is precisely such dynamic and complex interaction of a more or less formal kind that constitutes what is commonly labelled a system of innovation (Lundvall, 1992; Edquist, 1997), where systems are conceived as interaction networks (Tödtling and Sedlacek, 1997).

THE NATIONAL AND THE REGIONAL SYSTEMS OF INNOVATION GOVERNANCE

Moving on, and in common with the national innovation systems (NIS) perspective leading authors have espoused, it is entirely complementary to a *sectoral* analysis – indeed the concept of 'sectoral innovation systems' (SIS; Breschi and Malerba, 1997) continues as a minority interest in that conceptual domain. As with the sister concept of 'technological innovation systems' (TIS; Carlsson, 1995), it was a necessary de-construction of the hugely complex empirical reality confronting NIS analysts. The SIS and TIS approach strengthened NIS analysis by systematically linking *national* phenomena to new thinking about *global* value chains and globalization more generally.

However, by rooting NIS analysis in a non-spatial, sectoral perspective it weakened this neo-Schumpeterian school's understanding of the platform-like nature of emergent and soon to become dominant general purpose technologies like ICT, biotechnology and, most recently, nanotechnology (Helpman, 1998). As mentioned, sectors are constructs with little basis in contemporary business reality.

The perspective upon innovation systems that saved the neo-Schumpeterians from flirting with a dangerous irrelevancy was that which discovered and sought to analyse the great variety of regional innovation systems. Founded on a learning and innovation perspective that had regional science not industrial economics lineaments (Cooke, 1992; 2001) this field had, by 2006, become the leading bibliometric field in the innovation systems literature (Carlsson, 2007). Moreover, in policy terms, RIS building strategies have been adopted by countries as diverse as Sweden and South Korea and have become a mainstay of development policy advisory work by the likes of the United Nations Industrial Development Organization (UNIDO; Cooke and Memedovic, 2003; Cooke, 2006), the

OECD and the EU (European Commission, 2006). The latter institution had pioneered regional innovation policy, renaming its Regional Technology Plan instrument of 1994 Regional Innovation Strategy later on.[1] But these too suffered from a sectoral fibrosis, typified in the early, narrow technology emphasis, subsequently retained despite the softer innovation focus. DG Research (European Commission, 2006) committed to RIS in its desire for a new methodology for constructing regional advantage (CRA; Foray and Freeman, 1993). Regional innovation systems are the key to achieving CRA since they support the post-sectoral concept of platforms with all that implies for policy formation. Ironically, while the CRA concept was introduced by NIS adherents it was never investigated and it was only fully articulated eventually by RIS analysts (de la Mothe and Mallory, 2003; Cooke and Leydesdorff, 2006). The important points about the multi-level relationship between systems of innovation include the following:

- A national innovation system (NIS) guides and defines strategic policy especially at the input points in the 'linked chain' by which government assists firms and organizations to innovate.
- Thus science and technology (S&T) policy priorities and associated budgets are strategic for a country.
- Representation of these to international bodies and other national governments are NIS responsibilities.
- Research funding priorities are strategically set at NIS level, including NIS research institutes.
- University funding and assessment criteria are strategic NIS functions.
- Large-scale investments in S&T facilities and equipment (for example, synchrotrons) are NIS decisions.
- Finally, informing and engaging industry, research and talent stakeholders in support of and engagement with the above is a NIS function.

As may be seen in the following, RIS activities are complementary to these but are closer to the innovation outputs points of the innovation process.

- Innovation is commercialization of new knowledge. Hence an RIS arrangement is close to the market, assisting firms to translate knowledge into marketed products and services.
- Regional innovation system activities are often animated by a regional development agency (RDA) or if there are provincial or regional parliaments, the relevant Enterprise and Innovation ministry. Regional stakeholder advisory boards are involved.

- Of key importance is that the RDA or ministry designs, evaluates and reports on strategy supporting institutional elements that relate positively to each other for matters of consequence to innovation performance.
- Promoting an RIS it thus makes grants, loans or equity investments to assist firm innovation, especially where there may be some degree of market failure in such provision. Examples would be public venture capital or building of business incubators.
- Regional innovation system requirements may wish it to promote entrepreneurship to further innovation, subsidizing a spin-out programme or one that places a graduate in an SME.
- It may develop regional and interregional supply chain programmes to enhance innovation, possibly with partner regions in other countries.
- It may seek to attract innovative foreign firms to upgrade demand for innovation from indigenous firms.

The 'innovation system' concept can also be understood in both a narrow as well as a broad sense. A narrow definition of the innovation system primarily incorporates the R&D functions of universities, public and private research institutes and corporations, reflecting a top-down model of innovation. Such constellations traditionally resulted in regionalized national innovation systems, which constitute a top-down, supply (science push)-driven model. This type basically represents a large firm-driven model, where demand factors determine the rate and direction of innovation.

A broader conception of the innovation system includes 'all parts and aspects of the economic structure and the institutional set-up affecting learning as well as searching and exploring' (Lundvall, 1992: 12) and, thus, has a weaker system character. This broad definition incorporates some elements of a bottom-up, interactive innovation model but is more suited to a national system of innovation perspective, especially as found in, for example, Nordic countries where much 'knowledge scanning and borrowing' is conducted externally due to scale problems. It is not well suited to operating in European RIS arrangements which tend to be rather 'institutional' (that is, dependent on public intervention). But occasionally in Europe and possibly more frequently elsewhere, where entrepreneurship is strong, such 'entrepreneurial regional innovation systems' may be found. Even there huge public research budgets may have fuelled an ICT or biotechnology cluster, knowledge arising from which is nevertheless exploited by entrepreneurs, academic, serial or otherwise.

Yet a third 'hybrid' model can be found, which combines elements of both the foregoing in a network set-up that operates especially effectively at the meso-level of the region, state, *Land* or equivalent (Cooke et al., 2004). Certain southern German and Austrian regions such as Bavaria and Styria have this character and produce high innovation and other economic performance indicators. The networked system is commonly regarded as also the most efficient form of RIS: a related variety of regional clusters supported by a regional supporting institutional infrastructure. Emilia-Romagna has this character due to related variety of mechanical engineering clusters in luxury automotives, motorbikes, biomedical instrumentation, food packaging, machine tools and agro-food machinery, among others. These systems have a more 'coordinated' character, and a stronger, more developed role for regionally based R&D institutes, vocational training organizations and other local organizations involved in firms' innovation processes as in the case of Austria. In Styria a *Land* government policy to build a regional innovation system supporting clusters through networks among regional stakeholders has been successfully implemented during recent years. Upper Austria, centred on Linz, is evolving a comparable regional institutional structure. Vienna, as we have seen, benefits from the agglomeration economies, related variety and associated knowledge spillovers from proximity that large cities can offer.

There are some key differences and some key similarities between clusters and RIS set-ups. These are represented in Table 10.1. It immediately will be noticed that a cluster is normally geographically smaller than a region but an RIS is highly focused upon innovation, whereas a cluster need not be based on innovation – especially innovation of the formalized science and technology utilizing kind. Thus, many clusters function in localities without universities or formalized R&D of any kind, whereas it is unusual in countries that have evolved or consciously developed RIS arrangements for those regions to be lacking advanced R&D of some kind,

Table 10.1   Comparison of cluster and RIS concepts

Cluster	RIS
Firms	Knowledge
Markets	Markets and social use
Narrow variety, e.g. wine, biotech	'Platform' variety, e.g. engineering
Research/design	Exploration
Commercialization	Exploitation
Associated institutes	Intermediaries
Cooperation	Sub-system interaction
Competitiveness	Innovation

particularly university based. We may begin the elaboration of Table 10.1 by reference to given definitions of cluster and RIS after Porter (1998) and Cooke (2004). It may be seen that clusters highlight firms, an RIS highlights knowledge.

[clusters are] geographical concentrations of interconnected companies, specialised suppliers, service providers, firms in related industries, associated institutions (for example, universities, standards agencies, and trade associations) in particular fields that compete but also co-operate. (Porter, 1998: 197)

a regional innovation system consists of interacting knowledge generation and exploitation sub-systems linked to global, national and other regional systems for commercialising new knowledge. (Cooke, 2004: 3)

Thereafter, it is mainly the *focus* of a cluster that draws attention in comparing it with the standard RIS. In typical examples a cluster is not simply reducible to a localized sector. It is more than that, being comprised of firms, suppliers who may be from different sectors, 'associated institutions' including those that may supply research (exploration) knowledge or design (symbolic) knowledge of a kind that need not derive from research but possibly 'consultancy'. However, the scope of the typical cluster tends to be narrow and supportive of the lead industry (for example, wine, biotech, furniture, shoes, spectacle frames, ski boots, alloy-head golf clubs, and so on). In any case it is the lead firms in the lead industry that are primary and their various supports that are secondary. There may be competitive lead firms in clusters.

In the typical RIS, by contrast, there might be no clusters or there might be many. In the former case there might be oligopolies, duopolies and different kinds of agglomeration, supply chain or value network and industry mix that are, nevertheless, not necessarily wholly diverse and differentiated. However, it can be more diverse than that found in the cluster as stylized above – although its diversity may be underpinned by high absorptive capacity based on 'platform variety'. That is, as in the wealthy Italian region of Emilia-Romagna, there are diverse industries that are unified by an underlying professional 'community of practice' or even professional 'epistemic community' consciousness rooted in mechanical engineering. Regional economic growth status was earned from the region's engineering capabilities, exercised in advanced automotive sub-sectors such as sports cars (Ferrari, Maserati) and motorbikes (Ducati) but also biomedical devices, packaging machinery, machine tools, ceramics ovens and agro-food machinery. Thus appropriate innovations in one of these fields – each of which continues to operate in distinctive 'industrial districts' or 'clusters' – nevertheless diffuse rapidly because of the high lateral 'absorptive capacity' of the engineering 'culture' and associated learning by interaction from the evolutionary concept of 'related variety' (Cohen and Levinthal, 1989; Boschma, 2005). To an even stronger degree such advanced engineering capabilities typify Germany's Baden-Württemberg RIS, accounting for Mercedes, Porsche and Audi in automotives and excellence in printing machinery and machine tools as well as clusters of surgical instruments and, historically, clock manufacturing (Cooke and Morgan, 1994; 1998). The key unifying theme in RIS set-ups is the platform of networks and nodes by means of which firms access knowledge to enable them to innovate. Innovation is thus the defining feature in an RIS, whereas it need not be in a cluster. Vienna, Styria, Eastern England and the Thames Valley are like this, whereas Scotland and Wales are less so. Policy-makers perceive London's economy to be fragmented, but with the demise of its manufacturing base its related variety has risen around financial and creative platforms and it has prospered accordingly (DTI, 2006).

However, apart from scale and focus the two organizational models show many comparisons as well as contrasts in their remaining features. Thus a cluster – for example, focused in new media – might not access university research laboratories for key knowledge, nevertheless its members will seek symbolic knowledge in the form of new designs or other creative content. Moreover, varieties of research, mostly of a non-scientific kind, will have a role to play in assisting firms to compete. In an RIS, research of consequence to innovation is highly prized, whether coming from specialist suppliers in a regional or even global supply chain, from research laboratories inside or beyond the region, or from internal company R&D facilities. Again, we may compare and contrast Emilia-Romagna and Baden-Württemberg. In the knitwear cluster of Carpi, near Bologna, cluster firms lacked design capability to upgrade their output in order to escape imitation by cheaper competitors. The regional development agency, ERVET, was made aware of this and, utilizing European Social Fund resources from the EU, targeted workforce training, established CITERA, a design centre with computer-aided design-computer-aided manufacture (CAD-CAM) knitting equipment that entrepreneurs could make use of, learn about or purchase from the National Research Laboratory, which was also located in the region, where it was designed. Baden-Württemberg is well endowed with multi-level engineering-related technology centres and research laboratories as well as university laboratories specializing in exploration, or in the Max Planck Institutes. Examination and applied research occur in the Fraunhofer Institutes and exploitation in the Steinbeis Foundation whose technology centres assist SMEs to innovate. Thus it may be seen that knowledge for innovation is important in both clusters and RISs but that the distributed networks of externalized research capabilities offer greater related variety of knowledge choice than a cluster. As in the case of Emilia-Romagna, a cluster may lack necessary knowledge institutions, needing to call on the external RIS in order to overcome that deficit. Except for the limited case of science-based clusters that agglomerate in proximity to universities, clusters seldom sustain major research facilities.

Nevertheless, clusters do quite often have associated institutes, whereas RISs may not. Thus many biotechnology clusters have associations, councils or networks that work for, represent and provide common services for members. Hence the Massachusetts Biotechnology Council performs these functions from its base in Cambridge, Massachusetts, while the Eastern Region Biotechnology Initiative (ERBI) does the same from its base in Cambridge, UK. The UK-based motor sport cluster centred on Oxford has the Motor Sport Industry Association and in 'Plastics Valley' at Oyonnax in France Plasturgie Centre-Est performs the equivalent function. Incidentally, clusters are far more likely than RISs to have brand names. Many are 'Valleys', like – possibly the first – Silicon Valley south of San Francisco, now joined by the multinational Bio Valley linking France, Germany and Switzerland, the Netherlands' Food Valley at Wageningen, Wireless Valley at Stockholm and Language Valley in Flanders, Belgium. The last named is a cluster of firms specializing in speech recognition. This example has the added advantage of showing how a new industry was formed without the close presence of external research laboratories or key modern technologies but, rather, the common human skills of recognizing, reproducing and translating language. In some ways, for clusters 'The Valley is the Vision'.

Envisioning in local and regional economic development began with the early 'clustering' projects, some of the first European examples of which were conducted by Monitor in the early 1990s for the Basque Country and Scotland (Cooke and Morgan, 1998; Cooke, 1999). Described at the time as a 'vision, monitor and manage' approach which echoed the policy learning models pursued in management texts like that of Parsons (1995), it also chimed somewhat with the non-linear innovation systems approach, then in its infancy (Lundvall, 1992; Nelson, 1993). In common with the national innovation systems (NIS) perspective these and other authors then espoused, it is entirely complementary to a *sectoral* analysis – indeed, the concept of 'sectoral innovation systems' (SIS; Breschi and Malerba, 1997) continues as a minority interest in that conceptual domain. As with the sister concept of 'technological innovation systems' (TIS; Carlsson, 1995) it was a necessary deconstruction of the hugely complex empirical reality confronting NIS analysts. The SIS and TIS approach strengthened NIS analysis by systematically linking national phenomena to new thinking about global value chains and globalization more generally. But, as we have seen, it was not until the RIS perspective was introduced that fundamental problems associated with sectoral thinking could be fully addressed.

REFLECTIONS ON THE PROBLEMS OF CLUSTER POLICY

The *envisioning* perspective was a genuine improvement on the old 'wishlist' planning approach centred on 'goals achievement'. But it was deployed in cluster policy-envisioning on too narrow and too ambitious a front in which 'global competitive advantage' was a typical driver. Thus the best is the enemy of the good. This necessary diversion into the context in which early local and regional envisioning emerged brings us to discussion of that methodology. It is based on the Monitor response to the request by Scottish Enterprise for advice on its embryonic cluster programme. To explain, the first step is an *initiation* phase, conducted at the political-policy formulation interface. It involves public and private organizations and actors of consequence to the strategy in question. Following initiation, come two parallel steps, one of which is a *scoping* exercise, in which the range of serious candidate sectors for development of a regional clustering and global value-chain policy is determined. In the specific case in question, the development agency proposed a list of – ultimately – 16 sectors and the consultants advised on which had some element of global competitive advantage based on their global competitiveness methodologies. As the birth, in Scotland's Roslin Institute, of Dolly, the first transgenic animal, coincided with the launch of Scottish Enterprise's cluster programme in 1997, it is not surprising that biotechnology had been one of the first six 'cluster candidates' to be selected.[2] In parallel with the scoping exercise the other requirement was the assembly of necessary *resources* to achieve objectives. This latter activity involves pooling funds and personnel from a variety of public and non-public sources of consequence to the strategy in question. The initiating, scoping and resourcing activities lead to an envisioning process where the desired state is clearly envisioned in relation to existing assets, linkage possibilities and learning requirements. At this point, stakeholders from business, trade associations, knowledge centres, intermediaries such as 'agents', consultants, investors and lawyers, as well as specialist government agencies, are assembled, encouraged to engage with the process and collaborate with each other. From this process of networking, *leadership* must emerge and a monitoring process relating the vision to internal and external realities, supported by the governance and learning support system, is set in motion. So a *cluster leadership group* is formed from which individual leaders are drawn to implement specific agreed actions. At this point, not before, statistical exercises, benchmarking, and scenario-building in relation to global trends are conducted. Following this, actions are planned and implemented, drawing down appropriate and justified resources. Thereafter, implementation is *evaluated*, visions are updated, further resources are mobilized and new leaders identified to take the process forward. This, in short, is the envisioning and related cluster-building methodology in action, focused on a sectoral linkage target, mobilizing and leveraging indigenous assets, and engaging in a learning process to strengthen regional clusters to integrate with global value-chains.

Many recent policy assessments have called into question the way regionalization of innovation through cluster policy has been implemented (Roper et al., 2005; Asheim et al., 2006). Some of these are summarized in Table 10.2 and they point to disaffection with 'sectoral' thinking, 'picking winners' and even the utilization of 'retro-models' such as 'clusters' when, as noted, industry increasingly thinks in lateral 'platform' ways encompassing notions of 'knowledge spillovers' and 'related variety' of the kind discussed in this book so far. Other weaknesses of clustering and its associated narrow, sectoral focus, emphasizing only cluster-building in sub-sectors with global competitive advantage are described in what follows and broadly summarized with industry preferences in Table 10.2.

A frequent finding of research that focuses on innovation in firms and organizations is that the public development agency or administration is perceived as one of the least influential actors assisting firms in accessing the knowledge that helped bring forth the innovations. Yet this problem has to be overcome by designing an inclusive, consensus-based and envisioning strategy capable of successful implementation. This means it must avoid the negatives in Table 10.2 and 'make sense' to industry. This in turn means that pet political projects must not be allowed to determine action.

Table 10.2   *Problems with cluster policy for regional development: industry perspective*

Critiques from industry	Industry thinking
– Tick box mentality	– Platforms
– 'Picking winners'	– Pervasive applications
– 'One size fits all'	– Variety and flexibility
– 'Key sectors' policy rigidities	
– Policy 'retro-models' (e.g. clusters)	

Source: Roper et al. (2005).

Innovation policy prioritizing 'e-learning,' for which firms see a nearly non-existent market, over software, which firms see having pervasive, platform markets is a case in point (Roper et al., 2005). The most common thing firms typically state they value from public agencies is the disbursement of capital grants and tax reliefs that assist purchase of equipment and new premises or extension to existing buildings.

This is puzzling for three reasons: first, public agencies are major funders of basic research from which many innovations ultimately arise (Goozner, 2004). Second, as we have seen in Chapters 7 and 8, numerous government programmes exist to assist innovative firms to conduct early-stage technological development. Such pre-competitive technology schemes are often numerous, and total schemes supporting enterprises can usually be numbered in the hundreds for most advanced economies. Third, apart from schemes per se, most countries and even regions have a panoply of services and advice to offer firms through economic development agencies and the like. Yet firms seem to undervalue all this public largesse or worse, not even acknowledge it. Why is this? Is it a case of forgetfulness? Partly, but interviewers of such firms usually remind them and they still rank it low compared to the role of customers, suppliers, consultants and, for some industries more than others, university research laboratories (Cooke et al., 2000). Or is it that firms genuinely do not consider that public agencies of the third type generate policies that firms value? It seems that the latter explanation may be the stronger. Reconsider Table 10.2 and the explanations for low valuation of standard policies aimed at supporting enterprise and innovation.

THE FIRST REGIONAL DEVELOPMENT PLATFORM IMPLEMENTATION IN LAHTI, FINLAND

Thus we have identified the complexity problem at the heart of policy formulation. But of course the important question is how, short of a misguided neoconservative impulse to dismantle all instruments of enterprise and innovation support, to improve upon the inherited position? Rather than think this out *in vacuo*, it is easier to observe critically a live example that sought to overcome the kind of criticisms sketched in Table 10.2. We refer here to the single known case of the deliberate evolution of an innovative platform policy (IPP) – in a de-industrializing region in this case – as elaborated for Finland by Harmaakorpi (2006). Though this approach is rather general, it is consistent with that outlined above, it is theoretically and methodologically informed and it has the precious advantage of having been implemented in the Lahti region, though too recently to be *ex post* evaluated yet. There are thus three key segments expressing these phases – theory, methodology and implementation – outlined as follows.

Regional Development Platform Theory

Regional trajectories are 'path dependent' over time, with specific asset and capability configurations that relate to each other historically and, potentially, regional capabilities renew or atrophy, which are collective not individual processes and outcomes. Constructed advantage is seen to rest on valuable, scarce, non-imitable and non-substitutable resources – for example specific combinations of talent, environment, knowledge and so on. These may be referred to as bundles of knowledge capabilities. Regional dynamic capabilities are processes or routines that integrate and reconfigure assets to meet or create market opportunities, hence key regional process assets are knowledge, innovation, networking, leadership and envisioning capabilities.

The organizational basis for integrating these process or routine capabilities is a regional innovation system. As we have seen, this connects organizational and enterprise assets in a collective manner with a view to exploiting and renewing trajectories over time by reconfiguring dynamic capabilities. This process is subject to 'governance' of a facilitative rather than determinative kind, relying upon vision and leadership that express the dynamic capabilities of the region's actors, routines and processes.

Regional Development Platform Methodology

Within a framework of regional innovation systems (RISs) and evolutionary economics the key aim is to explore market potentials of regional IPP assets and capabilities.

- These explorations define regional development platforms in conceptual terms.
- Regional development platforms are asset configurations evolved from path dependencies aimed at examining and exploiting market potential.
- Regional platform actors include firms, intermediaries, knowledge centres, research, cultural and talent formation organizations.
- Regional platforms comprise industry, knowledge and contextuation in, for example, techno-economic paradigm change or 'global megatrends' of consequence to the specific region.

Among the guiding principles in the Lahti case were the following: understanding the 'drivers' of paradigmatic global change; understanding specific regional trajectories, especially regarding agglomeration; avoiding regional lock-ins; defining competitive asset configurations; forming multi-actor innovation networks to exploit such configurations; improving absorptive capacities among network members; utilizing and enhancing regional social capital; promoting leadership and foresight; and engaging multi-level governance in the exploitable platforms. These activities are performed by, *inter alios*, analysts, expert panels, scenario designers, system designers and knowledge intermediaries or entrepreneurs. All this activity is directed at identifying (with 'related variety' to the fore) the existing and future core competencies of the region (Harmaakorpi and Melkas, 2005).

The region of Lahti was explored and found to be weak in modern talent, knowledge and educational performance. Locked in to a declining mass-market furniture manufacturing profile, the region's leaders sought to envision a new future by analysis of RIS potential related to existing regional expertise. Expert panels graded regional entrepreneurship, growth potential, firm-size balance, global perspective, innovativeness, knowledge intensity, leadership capability, educational quality, research and technology-transfer expertise. The resulting industry grades were compared and leaders identified, then 'related variety' was stimulated by innovation system building around potential 'platforms'. Platforms integrated global megatrends to 'related variety' as opportunities for integrating materials, mechatronics and nanotechnology capabilities. As Harmaakorpi (2006) shows, the key actors diversified away from a declining furniture industry by identifying future advantage in the health-care industry. The new platform included firms in plastics, construction, furniture and metal, public health-care organizations, higher education and research organizations, the 'governance' leader – the regional development agency, incubators and talent development centres.

What were the effects and process evaluation results? The effects included changed assumptions by key actors and raised awareness of the key importance of RIS institutions and IPPs. Old industrial policy mentalities and measures were transcended. A previous perception, that absence of clusters and dynamic lead industries or firms (for example, Nokia) doomed the region, was transformed by application of policy platform methodology to exploit 'related variety'. Finally, the existential power of multi-actor networks was discovered and entrenched in the region. What problems had to be overcome? Foremost, 'knowledge' access, flow and integration asymmetries were overwhelming. Second, creative, imaginative thinking about the future lacked specificity to enable platform-envisioning to evolve. This was solved by formulation of a 'rye-bread'[3] model of knowledge creation

(Harmaakorpi and Melkas, 2005) which adapts Nonaka and Takeuch's (1995) SECI approach to knowledge creation in which tacit/codified internalization/externalization of knowledge processes stimulates creativity *inside* the firm. In the externalized world of the IPP far greater knowledge depth and complexity must be handled to enhance 'visualization' and 'potentialization' for the region. Finally, this new process proceeded raggedly and in time-consuming ways that subsequent experience may show can be foreshortened. The novelty of terms and concepts makes it initially difficult for outsiders to understand that clear communication is vital.

Constructing Regional Advantage

In Chapter 9 we called the broad RIS-IPP approach to be taken 'constructing regional advantage' (CRA) as evolved in the European Commission's (2006) deliberations about improving the elaboration and delivery of regional innovation policies. This embodies key concepts such as 'related variety' among sectors, innovating in the neighbourhood of current assets not 'leapfrogging' into generically fashionable new industries, regional policy platforms, and a vision-led implementation methodology – here the balanced score card (BSC) approach is preferred over, for example, strengths, weaknesses, opportunities and threats (SWOT). The reason is the former is vision led while the latter is not. Moreover SWOT can produce perverse results. An example is where a firm or region aspires to the opportunity to enter an attractive market – which has a high internal rate of return, possibly because of collusion which keeps prices artificially high and, thus, attractive. The entrant might expect to be collusive too. Then, a typically related aspiration might be to achieve competitive advantage in that market, which would destroy the collusion and make it a less attractive market. In other words, SWOT assumes variables are additive when they might be self-defeating.

Interestingly, the IPP approach has now also been commenced in neighbouring Estonia (Figure 10.1) with comparable results and a comparable set of envisioned directions to help diversify the economy away from dependence upon wood products (Tiits et al., 2006). In this they had 'conversations' around a number of scenarios:

- Scandinavian periphery
- vigorous modernization of traditional sectors
- venturing into new high-tech economic sectors.

Not favouring a 'Scandinavian periphery' scenario, hence illustrating the way the second scenario could be assisted towards linkage with the third,

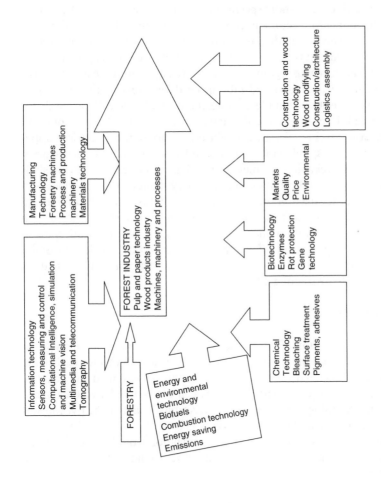

Source: After Watkins and Agapitova (2004).

Figure 10.1   Related variety platform: forestry diversification

advice from the World Bank on economic diversification in neighbouring Latvia was utilized. This shows the role of new technology evolving from a traditional sector towards new markets with new, stronger criteria of excellence, such as sustainability and high quality at affordable price.

A further evolution of this in the Estonian, as in the Finnish, context is towards health-care, or more specifically medicine. Seen as one of the most knowledge-intensive sectors it can boost economic diversification and provide cross-border health-care 'tourism' markets. Thus we clearly see an emergent candidate vision that may begin a CRA 'conversation' about an emergent platform with regional variety as well as sectoral 'related variety'. It is relevant to economies faced with the need for diversification away from a staple such as the example of forestry by building IPPs around, for example, cellulose, plastics and related variety industry in automotive industry components, electrical engineering, household electrical appliances, aeronautics, telephony, optics, toys, eyewear, hair accessories, fashion goods packaging, cosmetics and pharmaceuticals, involving skills such as design, mould production, machinery manufacture, polymer processing, finishing, decoration, recycling, and industrial and commercial logistics. Connecting to this type of vision will be those associated with improving and integrating talent, entrepreneurship and innovativeness in line with the expectations of a 'knowledge economy'. In less favoured regional economies where inward investment has dried up this is an alternative way of 'constructing advantage'.

Balanced Scorecard

An important question concerns the best way to ensure, in so far as possible, the implementation of the strategy-driving vision. As hinted earlier, the most closely associated with an envisioning approach to managing change is the 'balanced scorecard' (BSC), a management system based on a measurement methodology that enables organizations to clarify their vision and strategy and translate them into action.

The BSC suggests that we view the organization from four perspectives, and develop metrics, collect data and analyse it relative to each of these perspectives:

- Learning and growth perspective – 'To achieve the vision, how do we improve?'
- Governance process perspective – 'To satisfy stakeholders what processes must we excel at?'
- Stakeholder perspective – 'To achieve our vision what is our image?'

- Financial perspective – 'To succeed financially, what performance indicators?'

Each of the four perspectives is integrated to and driven by vision and strategy. Answers to each question might involve the following mission statements:

- Learning and growth – We improve by adopting an economic diversification vision and strategy that builds IPPs from a core industry towards 'related variety' industries.
- Governance – We must excel at identifying achievable actions that fit the vision strategy and identifying committed leaders to implement achievable action lines.
- Stakeholder – We must project the image of competent, innovative change managers who are designing an innovative economic future for the region.
- Financial – Key performance indicators will be growth rate improvements (to targets) in employment, talent, entrepreneurship, productivity, economic variety and competitiveness in the region.

Each perspective is measured in terms of:

- objectives
- measures
- targets
- initiatives

These are linked back to the governance process by 'learning' or feedback loops to see if:

- strategy needs adjusting to achieve vision
- vision needs adjusting to facilitate strategy.

This is because, as J. Edwards Deming (the well-known US expert whose business advice to the Ministry of International Trade and Industry [MITI] was substantially responsible for Japan's post-war economic success) saw, variety is created at every step in a governance process. Unanticipated consequences, excluded interests and 'wicked' problems are cases in point. So the causes of 'negative' variety need to be identified and fixed. These should be tackled as they arise and to establish a process to achieve that means governance has to be part of a system with feedback loops. The feedback data should be examined to determine causes of variety, which processes give rise to problems, and to focus attention on fixing that subset of processes.

Key 'associative' governance characteristics that lie between the liberal and the coordinated market policy models normally emphasized under this 'envisioning' regime are:

- stimulate civil society
- encourage associations
- governance to be inclusive
- governance to be accessible
- governance to be transparent
- create consensus on vision
- allow non-governmental leadership of specific actions within 'platforms'
- monitor and learn from implementation processes.

These are the characteristics that suit a BSC envisioning decision process for regional economic development most effectively.

A PLATFORM POLICY FOR REGIONAL HIGH-TECHNOLOGY INNOVATION

For illustrative purposes on what is, in effect, an IPP approach for high-technology development, we may turn to the region of Flanders in Belgium. Here, stakeholders reached an 'associative' consensus on a vision to generate innovative industry and employment from its scientific and technological base. Building on an earlier, local decision by its leading scientific and technological university (KU Leuven) to compete internationally by establishing a Microprocessor Engineering Centre of Excellence (IMEC), a special development association and agency charged with 'academic entrepreneurship' was established in 1998 and a management team appointed. This team works to defined measurable objectives with targets and initiatives to take in order to progress with achieving them. Among these are – following discussions with experts in the leading microprocessor technology areas – to establish, over a ten-year period six 'related variety' clusters each commercializing different varieties of the core technology in a set of laterally connected clusters, or a 'related variety IPP' as we would now term it.

Platform policy support instruments and investments in these vary according to the specific needs of actors and appropriate budgets. Thus 'knowledge centres' would be unlikely to attract economic development funding for scientific activities. But they would secure such funding from the NIS and supranational S&T bodies. Entrepreneurs would get start-up 'seed money' and later introductions to 'capital markets'. Infrastructure (incubator and science park) expansions would attract substantial public funding. Cluster policy would attract regional funding support from multi-level governance resources. Quality of life would be supported by appropriate public funding. International companies might be attracted with some public grants. Funds would be from various budgets at different governance levels as well as private–public mixes. Objectives, measures, targets and initiatives are embedded in the process. These are examples of accumulating and targeting distinctive platform-member needs and requirements. It is worth reflecting that the 'academic entrepreneurship' strategy here only began in 1998, although the 'mother centre', IMEC, was established in the 1980s. Hence, from a BSC perspective, in the six years to 2004 15 500 jobs were created, and the 20 000 jobs target for 2010 had already been achieved by 2006. Hence the constant need for monitoring and learning, and the present need for BSC-inspired vision steering and associated bringing forward of science park expansions and new developments (Hinoul, 2005).

A PLATFORM POLICY FOR RURAL DEVELOPMENT: ENVISIONING PRESELI, WALES

A different kind of health care than that supplied through pharmaceutical biotechnology or medical technology is that associated with preventative health care from relaxation, exercise, healthy eating and purchasing authentic products. Rural, agricultural or maritime touristic areas may not have the resources or desire to attract or develop high technology. Much of the available evidence points to that being a feature of certain 'talented' urban agglomerations. Nevertheless such regions may also benefit from seeking 'constructed advantage' from their existing and projected assets, for example, high oil prices mean the large oil producers are investing in production infrastructures for biofuels. Biofuels are produced by farmers as plant waste which can be transformed into cellulose ethanol; other sources include soya beans or corn (McNulty, 2006). In the stylized account below, which is drawn from an advisory mission to an economic development association (Cooke, 2007), biofuels are presented as part of a constructed advantage strategy linking related variety and policy platform thinking for agriculture in a rural region. In what follows from that account, other 'related variety' innovations – many of an organizational kind – are focused upon.

Preseli, Pembrokeshire, is on the Atlantic coast of south-west Wales (UK). This high-quality rural environment can be a 'lighthouse' for modern rural policy in support of economic development in many of the kind of regions perceiving themselves bypassed by the knowledge economy. Not many are fortunate enough to share Preseli's combination of attractions, such as the world's first Coastal National Park, neolithic monuments in abundance, hiking hills and beaches, but many have some of them. Preseli also has some disadvantages: it is perceived by residents and others to be 'remote'. In reality Preseli is not that remote nowadays, though it desperately needs a better road to the faster network that starts 50 kilometres away. Remoteness is relative, often in the mind, but sometimes it does involve movable barriers. Envisioning 'centrality' is an exercise in joined-up thinking required to make 'peripheral' areas like Preseli perceive themselves as not remote but at the heart of some things. This means engendering a conversation about how such places become those where residents have the choice of not necessarily leaving in order to get on, places which have their own intrinsic interest that forms a 'platform' on which they construct economic and other kinds of 'advantage' that is sustainable in all its many meanings. The vision of an evolved future where enterprise is possible and may thrive is required, with role-models assisting the process if possible. Thus, examples of farmers diversifying their activities assisted by EU rural diversification support (for example, agro-tourism) would be a case in point. Arguments about the diseconomies of urban agglomerations in relation to quality-of-life issues would conceivably be broached. Innovative policy thinking about 'poly-nucleated' settlement patterns that may be superior, as also advocated by the EU European Spatial Policy Directive (ESPD), might be explored.

A different kind of innovative reflection would take in a different kind of relevant role-model from a buoyant region. Clearly food production is an important element in the rural economy platform. One innovator who helped transform the conventional wisdom about food is Northern California restaurateur, Alice Waters. Her philosophy is as follows:

> If you choose to eat mass-produced fast food, you are supporting a network of supply and demand that is destroying local communities and traditional ways of life all over the world – a system that replaces self-sufficiency with dependence. And you are supporting a method of agriculture that is ecologically unsound – that depletes the soil and leaves harmful chemical residues in our food.
> But if you decide to eat fresh food in season – and only in season – that is locally grown by farmers who take care of the earth, then you are contributing to the health and stability of local agriculture and local communities.

Notice that today, farmers are having to turn away from ploughing and practise 'conservation tillage' because of the damage such mono-agricultural practice has wrought on the earth. Bounds (2006) reported that in Flanders – once again – what was previously necessary technology only in the Americas, where agro-food overexploitation of the soil has been endemic, innovation is necessitated due to climate change.

> As less rain falls, more of it is used for irrigation while what drops from the sky runs straight off the denuded land, taking chemicals and nutrients with it. Algae blooms in lakes, rivers silt up and chemicals enter the water supply. Europe's 25 member states lose 250 million tonnes of soil a year to erosion.
>   Conservation tillage is minimal ploughing, keeping the soil as intact as possible to prevent erosion and rotating crops to restore mineral balance. Instead of churning soil over with a plough to plant seeds deep . . . small drills . . . drop them into place without scarring the land. Supported by the National Trust and RSPB, savings are €40–60 per hectare annually. Ploughing costs up to €170 a hectare. In the UK, studies showed three times as many skylarks under conservation tillage as on ploughed land.

Although not such an arable agricultural economy as Flanders, Preseli nevertheless produces for high-grade and early vegetable and flower markets. In any case, it may be both losing money from its questionable practices and failing to reap the benefits from innovation under changed climatic conditions. Platform thinking allows the option not to cease full-scale farming and replace it with an agro-tourism monoculture, but to sensitively recapture an old heritage, spot and act on such trends towards organic and functional foods, by now spreading swiftly among consumers, as Alice Waters spotted in 1971 on the edge of America's San Joaquin Valley, home of the US agro-food monopolies. A further extension of Preseli's quality, authentic and sustainable food offered is the new kind of specialist organic food farm shop epitomized by the past Top Organic Food Shop in Britain winner, Daylesford in Gloucestershire, which attracts customers from far and wide, many of them easily two hours' drive away –'footprint alert' activated, but how else to bring the market to a deepest Cotswold farm – and overcoming any sense of 'remoteness' with its 'must have' produce?

'Related variety' linking to tourism is key, involving good quality holiday accommodation, visual arts and authentic, local fresh and seasonal Preseli cuisine. Local farmers re-discovering sustainable methods of producing such quality food, if they ever lost them, and restaurants and sustainable organic farm and delicatessen shops networked around the county and district, are at the heart of this constructed advantage 'platform' of excellent, modern consumer offerings that will bring the market for quality of life and sustainable tourism to Preseli. But it must be carefully distributed and appropriate light-touch support from policy given where required to maintain the delicate fabric of the world-class setting and cultural distinctiveness

Table 10.3   Regional and local constructed advantage: Preseli

What are our knowledge advantages?	What to innovate?
High quality agriculture and landscape	Branding organic and other food in demand
Diverse tourism offer	Growing and creating biofuels
Maritime resources	Gastronomy, art/cultural economy
	Fisheries, port logistics, maritime bioresearch
	Construction talent
	Connectivity in Wales and to global knowledge centres

of the stretches from its mountains, ancient geology, neolithic archaeology, Welsh language and landscape beauties south-westwards to the Atlantic shore.

The foregoing gives a flavour of how related variety platforms can be got under way and supported, linking by 'conversations' stakeholders from tourism, farming and food retailing in a 'joined-up' way that would even envisage close interactions with infrastructure and built environment planners for policy evolution (Table 10.3). Related variety is a broader but not totally diversified set of economic advantages. Under these circumstances, inter-linkages between key segments of the economy can be fostered, innovative ways of doing business, creating and selling new products or improving processes, like using cooperation for common purchasing where appropriate, can be effected. This is because the 'absorptive capacity' of entrepreneurs about the nature of the economic activities in neighbouring 'platform bundles' is high. Whereas the absorptive capacity between a silicon-chip maker and a pig farmer is likely to be low, that between a pig farmer and a supplier of organic pork to the hospitality industry is likely to be high.

So far we have not explored another contributor to the quality agro-food elements of the 'related variety' bundle, one which connects in important ways with it. This is the 'creative industries' grouping of activities. Here, Preseli is already in one or two instances succeeding in quality local and non-local markets. Quality-intensive and other woollen blanket and related products are produced on a small scale but sell in top international stores. This is not a common feature, yet Preseli has an advantage that has already been lost from well-known centres where the last of the larger-scale UK blanket manufacturers closed some years ago. Preseli's offer is authentic not generic, design intensive not standardized, and modern not traditional. Accordingly, though small, it has survivability through quality from fitting into niche, quality markets. But, importantly, the talent to work on design-intensive applications and be entrepreneurial in business terms must also be part of the platform linked to specialized training offered in colleges and, perhaps, schools. Marketing and other organizational skills and talents must also be available vocationally in relation to the 'creative industries' part of the Preseli platform. This is what 'constructing advantage' is fundamentally about – knowledge creation and knowledge exploitation in the market. Notice the manner in which modest support is given for talent formation in the creative industries in Northern Ireland. The fields covered are extensive, as shown below:

- advertising
- built environment (architecture, interior, landscape)
- designer-maker (ceramics, glass, furniture, textiles, metalwork and jewellery)
- design (product, graphics and communication)
- digital media (leisure software, games, interactive content)
- fashion
- moving image (film, television and animation)
- fine art (painting, sculpture, printmaking, mixed media)
- music (classical, popular)
- performing arts
- photography
- publishing
- radio.

Rural arts and crafts are also creating new markets elsewhwere. Being 'ahead of the curve' on quality, authentic and sustainable trends in consumer markets for services and products requires talented, foresighted people – usually entrepreneurs – but also support from special knowledge services, perhaps a 'competence centre' supported by the creative industries, business associations and public investment who would secure the necessary timeliness of creative innovation being offered to online as well as normal consumer and tourist markets.

BROAD IPP RECOMMENDATIONS

This section is intended to summarize the main recommendations arising from this analysis of evolutionary approaches to achieving local and regional economic development and growth across related variety industry platforms and the regional IPPs in their support.

Talent

There is a high priority with respect to human capital or talent and one way to do this is to raise the 'education platform' (Eliasson, 2000) through stimulating diversity in the uptake of university and other tertiary level education by diverting talent to 'tougher' courses in mathematics, science and engineering, design and languages. This showed through particularly strongly in the worked cases of rural economic development and high-technology urban, university-focused economic development. Talent has clearly been successfully mobilized in the accomplished regions that have constructed advantage through ICT and biotechnology platforms in Austria and the UK. In all cases the key elements enabling envisioning to produce realistic chances of success involved nurturing talented people and the institutions that train them.

Creativity

As countries move more and more into a knowledge economy, talent is increasingly attracted to creative and culturally rich locations in university towns and attractively designed cities. Mainly urban parts of the OECD occupy high positions on various indices of knowledge economy and GDP per capita associations. These countries have cultural variety that makes them attractive destinations for business and tourism, with related variety in entertainment, cuisine and the performing arts that such assets often imply. Hence quality-of-life criteria must play an important part in designing platforms to support regional futures as knowledge economies – bolstered by improvements to the scientific and engineering workforce. Modern evolutionary economic geography explains how investments in culture and creative industries is now a key contributor to local and regional economic growth (Florida, 2002).

Entrepreneurship

A further feature of successful knowledge economies is that they have high rates of entrepreneurship. As well as the services supplied direct to consumers and indirectly through material value chains, the research showed how a new category of 'knowledge entrepreneurs', distinct from traditional 'technology entrepreneurs' (Shane, 2004), some in developing countries, has come into existence. Knowledge entrepreneurs do not require many natural or material resources but, rather, knowledge and methodologies with which to find solutions, conduct research or create intellectual property. Enterprises are created that commercialize aspects of language and technology as well as creative and cultural specialisms enhanced by technology. Such entrepreneurs may or may not be academic entrepreneurs, they may be pure entrepreneurs who know how to create a successful business and some may be serial entrepreneurs who do this with successive enterprises. They may have to be recruited either from a returning migrant pool or simply from the global entrepreneurial talent pool. But all regions must become less unfavourable institutional climates in which to be entrepreneurial. Entrepreneurs of this kind may well thrive in clusters but, as the research showed, being entrepreneurial means exercising 'collective entrepreneurship' in some circumstances, whether in geographical proximity or through 'distant networks' (Fontes, 1966). Even a substantial portion of non-collaborating ICT entrepreneurial 'isolates' recognized this, it will be recalled. However, theirs was a somewhat parasitical proximity to a knowledgeable host community, or perhaps a quasi-monopoly.

Innovation

Naturally, innovation is a priority in the economic development process. Open innovation (Chesbrough, 2003) means more global opportunities to source and supply knowledge for innovation. Entrepreneurship is intimately interactive with the knowledge transfer process by means of which innovations occur. Institutions of a public goods nature such as universities, research institutes and centres of excellence, as well as service providers like health care, are key actors in innovation systems. Regional innovation systems are near-market carriers of support for innovation through regional development agencies in tune with the policy priorities of their national innovation system. Each region is different because they evolve regional asymmetries in the combinations of knowledge they possess or develop historically. Some are more entrepreneurial, attracting venture capital to invest in their SMEs. Some are more institutional, relying upon large organizations, firms and banking finance for development to be supported and sustained. But in all respects, the modern world of competitive markets thrives on innovation, a key source of productivity, competitiveness and growth.

Regional Innovation Systems and Platform Policies

For most regions RIS will necessarily support innovative platform policies aimed at:

- upgrading human capital
- research and design driven knowledge

- high- and medium-tech activities
- entrepreneurship
- related variety networks
- co-investment
- regional innovation systems.

In particular seeking to restructure in neighbouring sectors to the economy's key sectors makes sense. Hence even dependence on industrial age sectors such as chemicals can evolve into knowledge age 'biologics', and chemicals link to bioplastics and plasturgy from where a wide range of sectors like automotive, electronic, software, pharmaceutical, cosmeceuticals and nutraceuticals link to the agro-food and health-care industries. Less favoured regions or countries were shown in this chapter to place their dominant industry at the heart of their economic diversification plans into 'related variety' sectors and the World Bank advice to Baltic and East European economies is, similarly, to seek 'related variety'.

Finance

The means to support such restructuring by large firms and entrepreneurship by innovative suppliers finance must be found and pooled through private, public or National and Regional Co-investment Funds. These may usefully be public–private partnership funds for making grants, loans and taking business equity. The regional cases of Styria, Austria and Eastern England in the UK, where successful academic entrepreneurship has produced a new platform of related variety technologies, show the importance of there being a financial platform of different types of investment funding from early-stage seed funding to later-stage debt or equity financing. Clusters of SMEs and possibly larger firm customers are then more likely to be sustainable. But for clusters to work there needs to be appropriate innovative and entrepreneurial talent to attract investment in innovative activities. In Austria, Viennese and other biomedical and biotechnology clusters link closely with the health-care industry, a major growth sector worldwide. In UK biotechnology the same can be said, whereas in the – older – ICT industries linkage to multiple platforms is the norm for what is a pervasive general purpose technology. Venture capital has been fundamental to commercial success but publicly funded exploration research funding created the opportunities for successful knowledge exploitation.

NOTES

1. RIS1 and RIS2 from former DG 16 [Regio]; also RITTS (Regional Innovation and Technology Transfer Strategy) from the old DG 13.
2. Others were ICT, food, energy, forestry and transport.
3. This model is similar to balanced scorecard (BSC below) in having '(knowledge) vision' at the centre of a hexagon ('rye-bread' shape) with, respectively – from 2 o'clock – processes of visualization (imagination); socialization (originating); externalization (interacting); combination (cyber networks); internalization (exercising); and potentialization (futurizing). These elaborate the Socialization–Externalization–Combination–Internalization (SECI) process of knowledge-creation in Nonaka and Takeuchi (1995).

REFERENCES

Asheim, B., Cooke, P. and Martin, R. (2006), *Clusters & Regional Development: Critical Reflections & Explorations*, London: Routledge.

Boschma, R. (2005), 'Proximity and innovation: an introduction', *Regional Studies*, **39**, 41–6.

Bounds, A. (2006), 'Farmers till with a drill in the eroded fields of Flanders', *Financial Times*, 15 August, p. 11.

Breschi, S.L. and Lissoni, F. (2001), 'Knowledge spillovers and local innovation systems: a critical survey', *Industrial and Corporate Change*, **10**(4), 975–1005.

Breschi, S. and Malerba, F. (1997), 'Sectoral innovation systems: technological regimes, Schumpeterian dynamics, and spatial boundaries', in C. Edquist (ed.), *Systems of Innovation*, London: Pinter.

Carlsson, B. (ed.) (1995), *Technological Systems & Economic Performance*, Dordrecht: Kluwer.

Carlsson, B. (2007), 'Innovation systems: a survey of the literature from a Schumpeterian perspective', in H. Hanusch and A. Pyka (eds), *The Companion to Neo-Schumpeterian Economics*, Cheltenham, UK and Northampton, MA, USA: Edward Elgar.

Casper, S., Lehrer, M. and Soskice, D. (2001), 'Can high technology industries prosper in Germany? Institutional frameworks and the evolution of the German software and biotechnology industries', *Industry and Innovation*, **6**, 5–24.

Chesbrough, H. (2003), *Open Innovation*, Boston, MA: Harvard Business School Press.

Cohen, W. and Levinthal, D. (1989), 'Innovation and learning: the two faces of R&D', *The Economic Journal*, **99**, 569–96.

Cooke, P. (1992), 'Regional innovation systems: competitive regulation in the new Europe', *Geoforum*, **23**, 363–82.

Cooke, P. (1999), 'Biotechnology clusters in the UK (2001)', *Small Business Economics*, **17**, 43–59.

Cooke, P. (2001), 'Regional innovation systems, clusters and the knowledge economy', *Industrial & Corporate Change*, **10**, 945–74.

Cooke, P. (2004), 'Introduction: regional innovation systems – an evolutionary approach', in P. Cooke, M. Heidenreich and H. Braczyk (eds), *Regional Innovation Systems*, London: Routledge.

324 *Empirical findings*

Cooke, P. (2006), *Regional Innovation Systems as Public Goods*, UNIDO Policy Papers, Vienna.

Cooke, P. (2007), 'Constructing advantage in Preseli, Pembrokeshire', in J. Osmond (ed.), *The Preseli Papers*, Cardiff: Institute for Welsh Affairs.

Cooke, P. and Leydesdorff, L. (2006), 'Regional development in the knowledge-based economy: the construction of advantage', *Journal of Technology Transfer*, **31**, 5–15.

Cooke, P. and Morgan, K. (1994), 'The regional innovation system of Baden-Württemberg', *International Journal of Technology Management*, **9**, 394–420.

Cooke, P. and Morgan, K. (1998), *The Associational Economy*, Oxford: Oxford University Press.

Cooke, P. and Memedovic, O. (2003), *Strategies for Regional Innovation Systems*, Vienna: UNIDO.

Cooke, P., Boekholt, P. and Tödtling, F. (2000), *The Governance of Innovation in Europe*, London: Pinter.

Cooke, P., Heidenreich, M. and Braczyk, H. (2004), *Regional Innovation Systems*, London: Routledge.

De la Mothe, J. and Mallory, G. (2003), 'Industry–government relations in a knowledge economy: the role of constructed advantage', *PRIME Discussion Paper 02-03*, University of Ottawa, PRIME.

Department of Trade and Industry (DTI) (2006), 'Innovation in the UK: Indicators and Insights', *DTI Occasional Paper No. 6*, London: Department of Trade and Industry.

Edquist, C. (ed.) (1997), *Systems of Innovation*, London: Pinter.

Eliasson, G. (2000), 'Development in industrial technology and production competence requirements and the platform theory of on-the-job learning', in G. Eliasson (ed.), *New Emerging Industries, New Jobs and New Demands for Competence*, Stockholm: KTH; Thessaloniki: CEDEFOP.

European Commission (2006), *Constructing Regional Advantage*, Brussels: DG Research.

Florida, R. (2002), *The Rise of the Creative Class*, New York: Basic Books.

Fontes, M. (2006), 'Knowledge access at distance: strategies and practices of new biotechnology firms in emerging locations', in P. Cooke and A. Piccaluga (eds), *Regional Development in the Knowledge Economy*, London: Routledge.

Foray, D. and Freeman, C. (1993), *Technology and the Wealth of Nations: The Dynamics of Constructed Advantage*, London: Pinter.

Goozner, M. (2004), *The $800 Million Pill: The Truth Behind the Cost of New Drugs*, Berkeley, CA: University of California Press.

Harmaakorpi, V. (2006), 'Regional development platform method as a tool for regional innovation policy', *European Planning Studies*, **14**(8), 1093–112.

Harmaakorpi, V. and Melkas, H. (2005), 'Knowledge management in regional innovation networks: the case of Lahti', *European Planning Studies*, **15**(5), 641–60.

Helpman, E. (ed.) (1998), *General Purpose Technologies and Economic Growth*, Cambridge, MA: MIT Press.

Hinoul, M. (2005), *A Mutual Learning Platform for the Regions: The Leuven Experience*, Brussels: Committee of the Regions.

Lundvall, B. (ed.) (1992), *National Systems of Innovation*, London: Pinter.

McNulty, S. (2006), 'Oil majors cultivate an interest in the next generation of biofuels', *Financial Times*, 24 August, p. 6.

Nelson, R. (ed.) (1993), *National Systems of Innovation*, Oxford: Oxford University Press.

Nonaka, I. and Takeuchi, H. (1995), *The Knowledge Creating Company*, Oxford: Oxford University Press.

Parsons, W. (1995), *Public Policy*, Aldershot, UK and Brookfield, USA: Edward Elgar.

Porter, M. (1998), *On Competition*, Boston, MA: Harvard Business School Press.

Rallet, A. and Torre, A. (1998), 'On geography and technology; proximity relations in localised innovation networks', in M. Steiner (ed.), *Clusters & Regional Specialisation*, London: Pion.

Roper, S., Love, J., Cooke, P. and Clifton, N. (2005), *The Scottish Innovation System: Actors, Roles & Policies*, Edinburgh: The Scottish Executive.

Shane, S. (2004), *Academic Entrepreneurship*, Cheltenham, UK and Northampton, MA, USA: Edward Elgar.

Steiner, M. (1998), 'The discreet charm of clusters: an introduction', in M. Steiner (ed.), *Clusters & Regional Specialisation*, London: Pion.

Tiits, M., R. Kattel and Kalvet, T. (2006), *Made in Estonia*, Tartu: Institute of Baltic Studies.

Tödtling, F. and Sedlacek, S. (1997), 'The regional innovation system of Styria', *European Planning Studies*, **5**, 246–60.

Watkins, A. and Agapitova, N. (2004), 'Creating a 21st century national innovation system for a 21st century Latvian economy', *Research Working Paper 2357*, Washington, World Bank.

Index